Acoustics: Advances and Applications

Acoustics: Advances and Applications

Edited by Seth Grattan

CLANRYE
INTERNATIONAL
www.clanryeinternational.com

Clanrye International,
750 Third Avenue, 9th Floor,
New York, NY 10017, USA

ISBN: 978-1-63240-824-2

Cataloging-in-Publication Data

Acoustics : advances and applications / edited by Seth Grattan.
    p. cm.
Includes bibliographical references and index.
ISBN 978-1-63240-824-2
1. Sound. 2. Hearing. I. Grattan, Seth.
QC225.15 .A36 2019
534--dc23

For information on all Clanrye International publications
visit our website at www.clanryeinternational.com

𝓒LANRYE
𝓘NTERNATIONAL

Contents

Preface

Acoustics is a branch of physics that is concerned with the study of mechanical waves in different media such as solids, liquids and gases. The study covers the generation, propagation and reception of mechanical waves and vibrations. The subdisciplines of acoustics are archaeoacoustics, aeroacoustics, acoustic signal processing, architectural acoustics, musical acoustics, etc. An understanding of different mechanical waves such as pressure waves, surface waves, transverse and longitudinal waves is crucial for a comprehensive understanding of acoustics. The applications of acoustics are in medicine, audio and noise control, climate change monitoring, sonar for submarine detection, etc. The book aims to shed light on some of the unexplored aspects of acoustics and the recent researches in this field. The various sub-fields of acoustics along with technological progress that have future implications are glanced at in this book. It aims to equip students and experts with the advanced topics and upcoming concepts in this area of study.

This book unites the global concepts and researches in an organized manner for a comprehensive understanding of the subject. It is a ripe text for all researchers, students, scientists or anyone else who is interested in acquiring a better knowledge of this dynamic field.

I extend my sincere thanks to the contributors for such eloquent research chapters. Finally, I thank my family for being a source of support and help.

Editor

Optimization of Fixed Microphone Array in High Speed Train Noises Identification Based on Far-Field Acoustic Holography

Rujia Wang and Shaoyi Bei

School of Automotive and Transportation, Jiangsu University of Technology, Changzhou, Jiangsu, China

Correspondence should be addressed to Rujia Wang; rujia.wang@jsut.edu.cn

Academic Editor: Marc Asselineau

Acoustical holography has been widely applied for noise sources location and sound field measurement. Performance of the microphones array directly determines the sound source recognition method. Therefore, research is very important to the performance of the microphone array, its array of applications, selection, and how to design instructive. In this paper, based on acoustic holography moving sound source identification theory, the optimization method is applied in design of the microphone array, we select the main side lobe ratio and the main lobe area as the optimization objective function and then put the optimization method use in the sound source identification based on holography, and finally we designed this paper to optimize microphone array and compare the original array of equally spaced array with optimization results; by analyzing the optimization results and objectives, we get that the array can be achieved which is optimized not only to reduce the microphone but also to change objective function results, while improving the far-field acoustic holography resolving effect. Validation experiments have showed that the optimization method is suitable for high speed trains sound source identification microphone array optimization.

1. Introduction

The noise of high speed vehicles such as high speed trains is one of the severest noise pollution sources [1, 2]. Based on studies in the reconstruction of the sound field microphone array, predecessors to build a grid array cross array have inherent defects; namely, in order to ensure a small main lobe width of the sound field reconstruction to improve resolution, the need to maintain a larger size of the array, such that the spacing between adjacent array elements, is increased and causes the emergence of grating lobes, which greatly weakened the ability of the sound field reconstruction array.

Compared with beamforming, acoustic holography method can achieve a quantitative measure of moving sound source, so in recent years it has been widely studied and applied. Acoustic holography theory in the 1980s by the Williams and Maynard et al. [3–5] put forward the strict acoustic radiation based on the theory. In the 1990s, Tanaka

et al. [6] were first used the far-field acoustic holography method to measure a vehicle noise source and in 2004 by using two-dimensional arrays based on far-field acoustic holography method of analyzing tire/road noise and tire noise successfully. From 1998 to 2008, Park et al. [7, 8] established a framework for holographic mobile, using sound field space transformation method to eliminate the Doppler effect and measure and analyze a low-speed train noise. Yang et al. [9–11] first proposed the international far-field diffraction acoustic holography method and in 2010 proposed eliminating the time-domain method Doppler effect, for the first time to achieve a speed of 117 km/h, vehicle quantitative identification of noise sources outside the vehicle.

In this paper, based on formation simulate annealing optimization method to optimize the unequal a fixed spacing array, rather than a random array, for obtaining a more accurate result of the sound source identification, it can

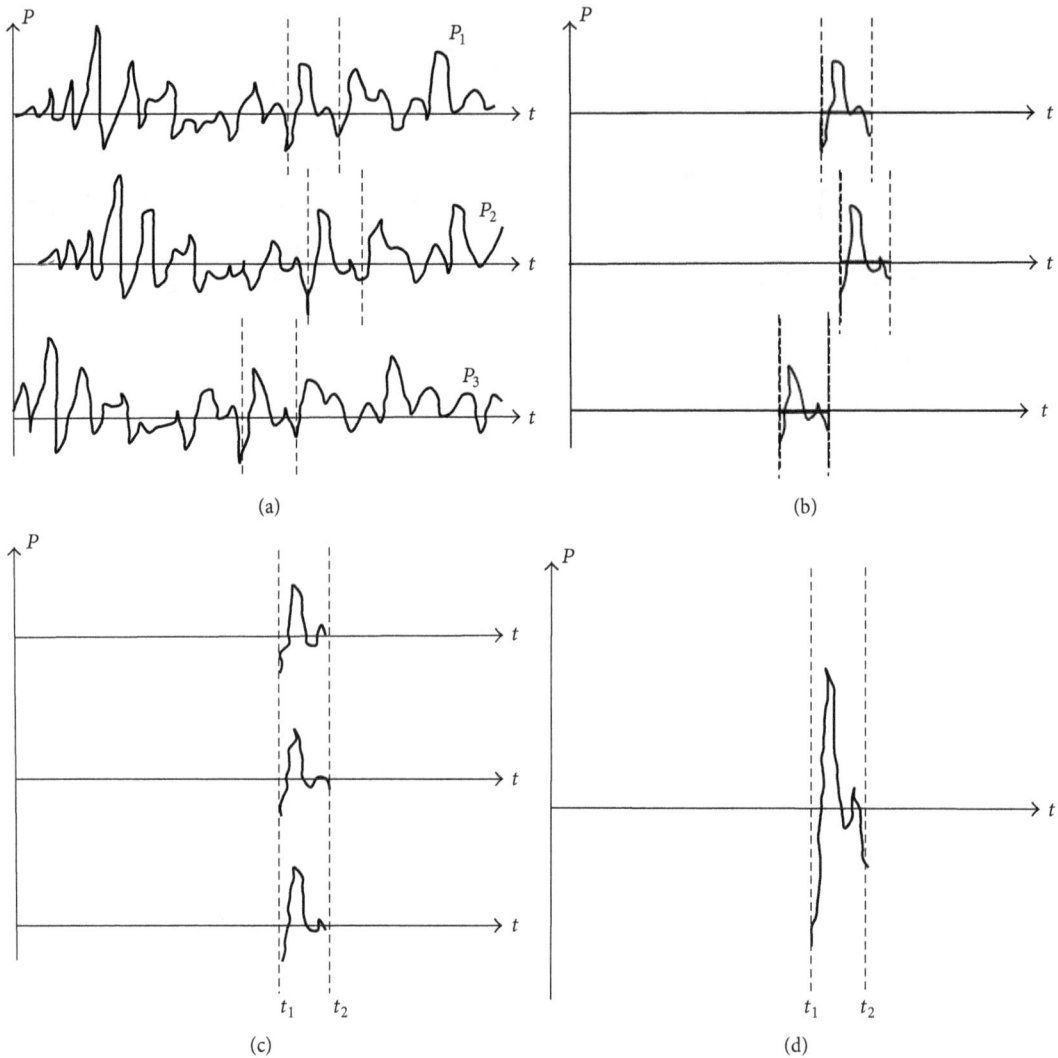

FIGURE 1: Schematic diagram of beamforming sound field reconstruction.

further improve the accuracy of identification of the sound source.

2. Acoustical Holography Far-Field Sound Source Identification Method for the Moving Sound Source

2.1. Short-Time Beamforming. Short beamforming method is based on the "delay accumulate" beamforming signal superimposed principle [12, 13]. For the reconstruction of the surface at points, the period of its reconstruction $[t_1, t_2]$ sound pressure signal within the principle is shown in Figure 1.

First, calculate the time period to be analyzed according to the respective microphone receiving the segment signal period, as shown in Figure 1(a); secondly extracted sound for each signal segments the desired pressure, as shown in Figure 1(b); then, any delay time based on acoustic wave propagation, such as the signal, is $p(t)$, provided that it

emits sound waves at time point $s(\varepsilon, \eta)$ to the microphone i, the propagation time of $r_i(t, \varepsilon, \eta)/c$, and then a signal delay processing as shown in Figure 1(c):

$$\tilde{p}_i = p_i \left(t + \frac{r_i(t, \varepsilon, \eta)}{c} \right). \tag{1}$$

Finally, the superimposed signals of each delay processing, to obtain the reconstruction result of the analysis period, are shown in Figure 1(d).

According to this principle, the sound field characteristic function reconstruction formula at any point $s(\varepsilon, \eta)$ on the reconstruction side R is shown as follows:

$$W_s(\varepsilon, \eta) = \int_{t_1}^{t_2} P^2(t, \varepsilon, \eta) \, dt,$$

$$P(t, \varepsilon, \eta) = \frac{1}{N} \sum_{i=1}^{N} p_i \left(t + \frac{r_i(t, \varepsilon, \eta)}{c} \right), \tag{2}$$

Optimization of Fixed Microphone Array in High Speed Train Noises Identification Based...

3

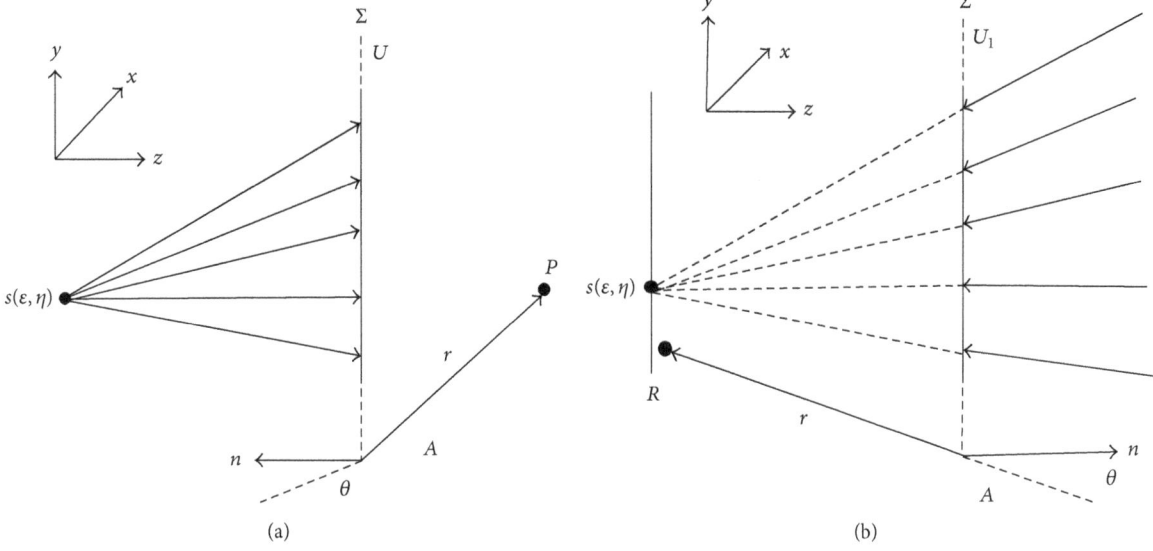

FIGURE 2: Schematic diffraction of acoustic holography.

where $W_s(\varepsilon, \eta)$ is the sound field characteristic function for any point on the surface of the sound source $s(\varepsilon, \eta)$ at time t within $t_1 - t_2$, $P(t, \varepsilon, \eta)$ is the sound source estimated characteristic function applied of beamforming method, $p_i(t)$ is the received sound source pressure signals of i_{th} microphone at time t, c is the sound velocity, N is the number of microphones, $r_i(t, \varepsilon, \eta)$ is the physical distance between the point $s(\varepsilon, \eta)$ in the sound source surface and i_{th} microphone at time t.

Based on this principle, across the entire surface of the sound source, sound field characteristic function of the distribution of the entire surface of the sound source can be obtained inside in time $[t_1, t_2]$.

2.2. Far-Field Acoustic Holography Method.
NAH (Nearfield Acoustical Holography) method is proposed by Williams et al. [14, 15] in the 1980s, and then the method and theory of NAH's applications extend to the far-field conditions. Professor Yang from Tsinghua University has proposed and established the theory of diffraction acoustic holography based on far-field acoustic holography theory. The principle is shown in Figure 2.

In Figure 2, N is the outward normal to the direction of the hologram surface, r is the radius vector, θ is the angle between the hologram surface r and outward normal direction n, Σ is the position in the measuring surface of infinite plane in space, the measurement part of the surface A is Σ, and R is assumed source point s to the plane, known as the sound source surface.

As it is showed in Figure 2(a) U can be viewed as the point sound source, which assumed that spherical wave at a spatial sound field distribution, $H(x, y, f)$, is U. In Σ plane component, in the case of U which is known, the wave propagation direction of the space of passive can be calculated

by the using of Kirchhoff diffraction integral calculation method. Any point P of the sound field formula is obtained in

$$U(P) = \frac{1}{4\pi} \iint_H \left[\left(\frac{e^{jkr}}{r} \right) \frac{\partial U}{\partial n} - U \frac{\partial}{\partial n} \left(\frac{e^{jkr}}{r} \right) \right] d\sigma. \quad (3)$$

Figure 2(b) S_1 can be viewed as a virtual sound source space. Sound field in the distribution in S_1, a sonic converge in S, spherical wave, and S sent out actually mutually conjugate wave. According to the principle of acoustic holography reconstruction as shown in (4) it is obtained based on the principle of conjugate wave convergence:

$$U(\varepsilon, \eta, f) = \frac{Ck}{j} \iint_H H * (x, y, f)$$
$$\cdot \left[1 + \left(1 - \frac{1}{jkr} \right) \frac{z_0}{r} \right] \frac{e^{-jkr}}{r} \, dx \, dy. \quad (4)$$

$k = f/c$ is the wave number, c is the sound velocity, C is a hologram constant, and r is the reconstruction of the surface point $Q(\varepsilon, \eta)$ between the hologram surface points (x, y) distance. Equation (4) is calculated according to the results of a single frequency f, continuous several frequencies. Calculating according to the energy superposition can be obtained within a frequency range of the sound pressure amplitude calculation, as shown in

$$P(\varepsilon, \eta) = \int_{f_1}^{f_2} |U(\varepsilon, \eta, f)|^2 \, df. \quad (5)$$

Reconstructing by the above method assumes that the sound pressure of the sound source Q value, as we can see in Figure 3, supposes the point traverse plane of reconstruction

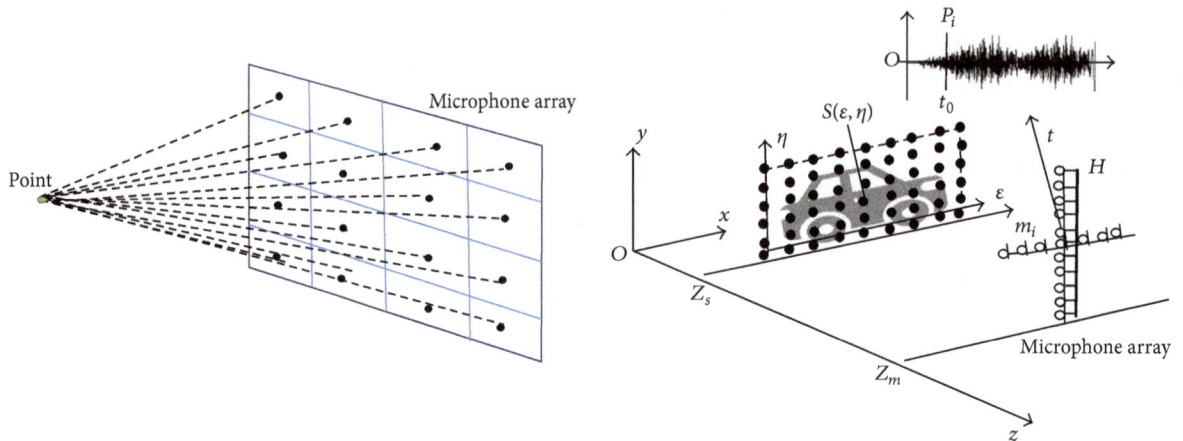

FIGURE 3: Hologram surface structure.

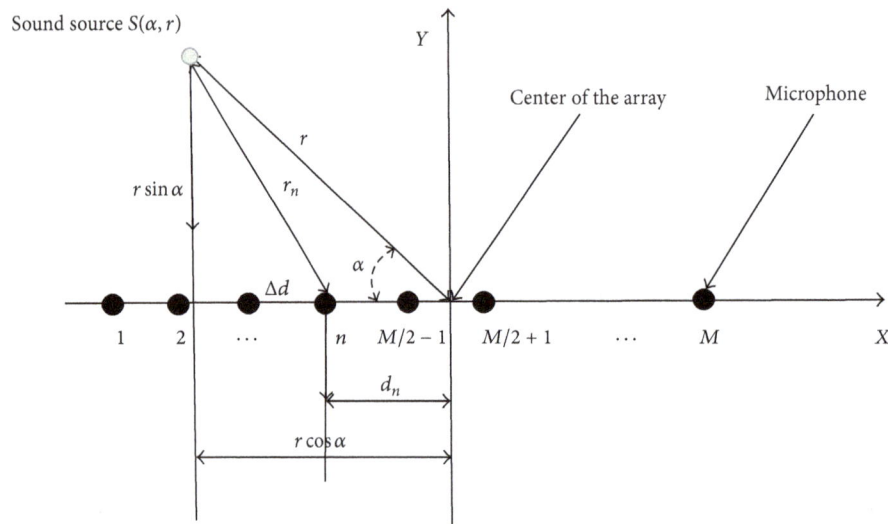

FIGURE 4: Model of uniform linear array.

and repeats this calculation; we can calculate the entire plane of the sound pressure distribution. In actual measurement, the continuous sound pressure points in hologram surface cannot be measured; therefore, (4) is discretized; as shown in Figure 3 for rules discrete arrays can be reconstructed by directly formula, as shown in

$$U\left(\varepsilon,\eta,f\right) = \frac{Ck}{j}\sum_{m=1}^{M}\sum_{n=1}^{N}H^{*}\left(m,n,f\right)$$
$$\cdot\left[1+\left(1-\frac{1}{jkr}\right)\frac{z_0}{r}\right]\frac{e^{-jkr}}{r}\,dx\,dy, \tag{6}$$

where M and N, respectively, are microphone rows and columns, $H^{*}(m,n,f)$ is the mth row and the nth column of the holographic information of the microphone, r is the reconstruction surface points $S(\varepsilon,\eta)$ mth row and the nth column from the microphone, and Δx and Δy are, respectively, microphone spacing and row distance.

3. Compared Simulation

3.1. Application in Microphones Array Sound Sources Discrimination. The microphones array's performance mainly reflected the result of spatial resolution and identification precision of source of noises; we used a microphone array of regular arrangement with equal distance as compared with an array of optimization displacement of microphones positions; in both simulations, the number of microphones is the same; but the size of the array may be different in the simulation result. In this paper, we studied the influence of parameters on the performance of the array of arrays. We used a fixed cross X-type microphone array as an example to describe the plane array performance and then analyzes the simulation and discusses the impact of microphones array displacement on the results of the identification. In this paper, we propose a word "fixed array"; it means that the microphones array is to adjust or optimize the X direction than random microphone array displacement.

TABLE 1: Simulation result of equal distance array and optimization array.

	SLR (side lobe ratio)	MLA (main lobe area)
Equal distance	0.2781	0.0635
Optimization array	0.6824	0.0310

Performance of the array is mainly reflected in the spatial resolution, such as an array of irregular random arrangement of the microphone as compared with a regular grid array having the same number of microphones, and then how to identify the source of noise at higher frequencies is very important, due to the performance of the microphone array, its array of applications, selection, and design instructive. Figure 4 shows the model of uniform linear array.

This article studies the influence of parameters on the performance of the array of arrays. A fixed cross X-type microphone array as an example describes the plane array performance analysis and process simulation and discusses the impact of microphones arranged in the form of its properties. Based simulated annealing optimization method, the MLA (main lobe area and main lobe energy/sidelobe energy) and SLR (side lobe ratio) were chosen as objective functions, whether it is feasible to get a better result by the means of reducing the numbers of microphones.

3.2. *Simulation.* Based simulated annealing optimization method, we selected two objective functions, MLA (main lobe area) and SLR (side lobe ratio), which are the parameters of measurement criteria for evaluation of identification results, and we tend to get the value of MLA, the smaller the better, and the value of SLR, the bigger the better, or to get the balance between them. The purpose of the simulation is to compare the microphones array fore-and-aft optimization and then to identify if there would be the possibility of reduction of the numbers of microphones but can achieve a better identification resolution. The simulation result can be seen in Figure 5; in the simulation, the parameters of the sound source are as follows: the speed of sound source is 120 km/h, the numbers of microphones are 29 in one fixed X array, and the distance between microphone array and the sound source is 10 m.

Calculation methods of the main lobe area (m2) and the main sidelobe ratio (main lobe energy/sidelobe energy) are shown in Table 1.

As we can get from Figure 5, the result of the simulation is listed in Table 1.

From Table 1, it can be seen that, as a result of optimization array, the value of SLR is 0.6824, and the value of MLA is 0.0310, both of the results are better than the results of equal distances microphones array, and the result is in accordance with the objectives and results of optimization we proposed at first. Then we get the 29 optimization microphones' coordinates as shown in Figure 5(b), then we

TABLE 2: Result of optimization microphones array coordinate.

Number of microphones	X	Y
1	0	0
2	1.9963	1.9963
3	−1.9963	1.9963
4	−1.9963	−1.9963
5	1.9963	−1.9963
6	2.0000	2.0000
7	−2.0000	2.0000
8	−2.0000	−2.0000
9	2.0000	−2.0000
10	0.0000	0.0000
11	0.0000	0.0000
12	0.0000	0.0000
13	0.0000	0.0000
14	0.0666	0.0666
15	−0.0666	0.0666
16	−0.0666	−0.0666
17	0.0666	−0.0666
18	1.9371	1.9371
19	−1.9371	1.9371
20	−1.9371	−1.9371
21	1.9371	−1.9371
22	1.9999	1.9999
23	−1.9999	1.9999
24	−1.9999	−1.9999
25	1.9999	−1.9999
26	0.0037	0.0037
27	−0.0037	0.0037
28	−0.0037	−0.0037
29	0.0037	−0.0037

will analyze the ways of easibility of reducing the number of microphones.

The 29 optimization microphones' coordinates are shown in Table 2.

And from Table 2, the coordinate numbers 2–5 and coordinate numbers 22–25 are very near to the coordinate [2.0000]; the error is 0.01% which can be considered as the microphones which are in the same positions; we use coordinate numbers 6–9 to replace the coordinate numbers 2–5 and coordinate numbers 22–25, and then 8 microphones can be reduced. And coordinate numbers 10–13 are the same as number 1 [0.000], which is in the same positions, and then 4 microphones can be reduced. There were only 17 microphones left. The rest of 17 microphons' simulation is carried out by using the same method, then we get the result of MLA and SLR, and there are no changes in the value of both of them.

In the above simulation, the results show the optimization method which can be achieved to optimize and improve the resolution of the objective function, while achieving the purpose of reducing the number of microphones.

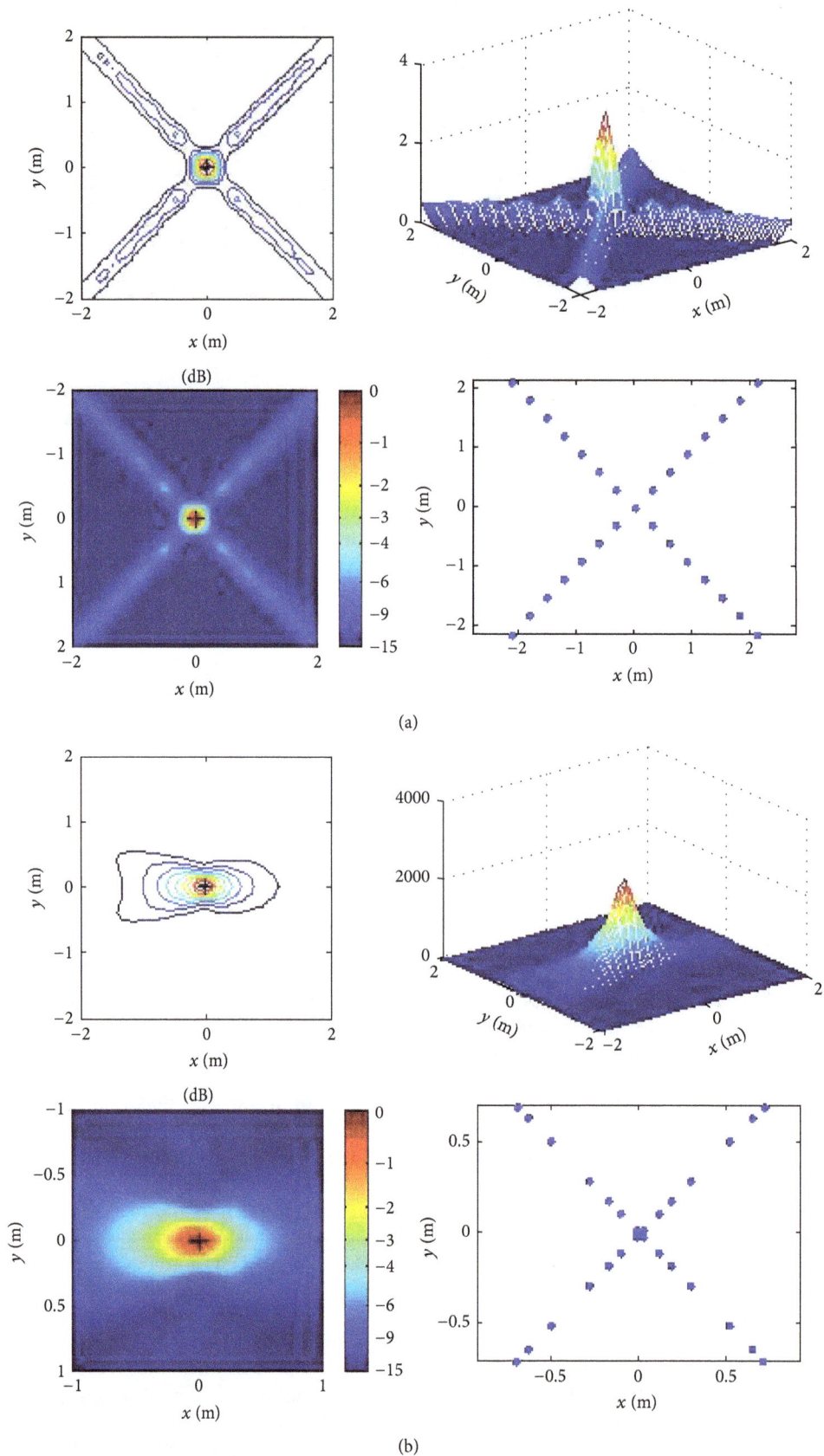

FIGURE 5: (a) Equal distance array simulation result; (b) optimization array result.

FIGURE 6: The experiment lab and microphone array.

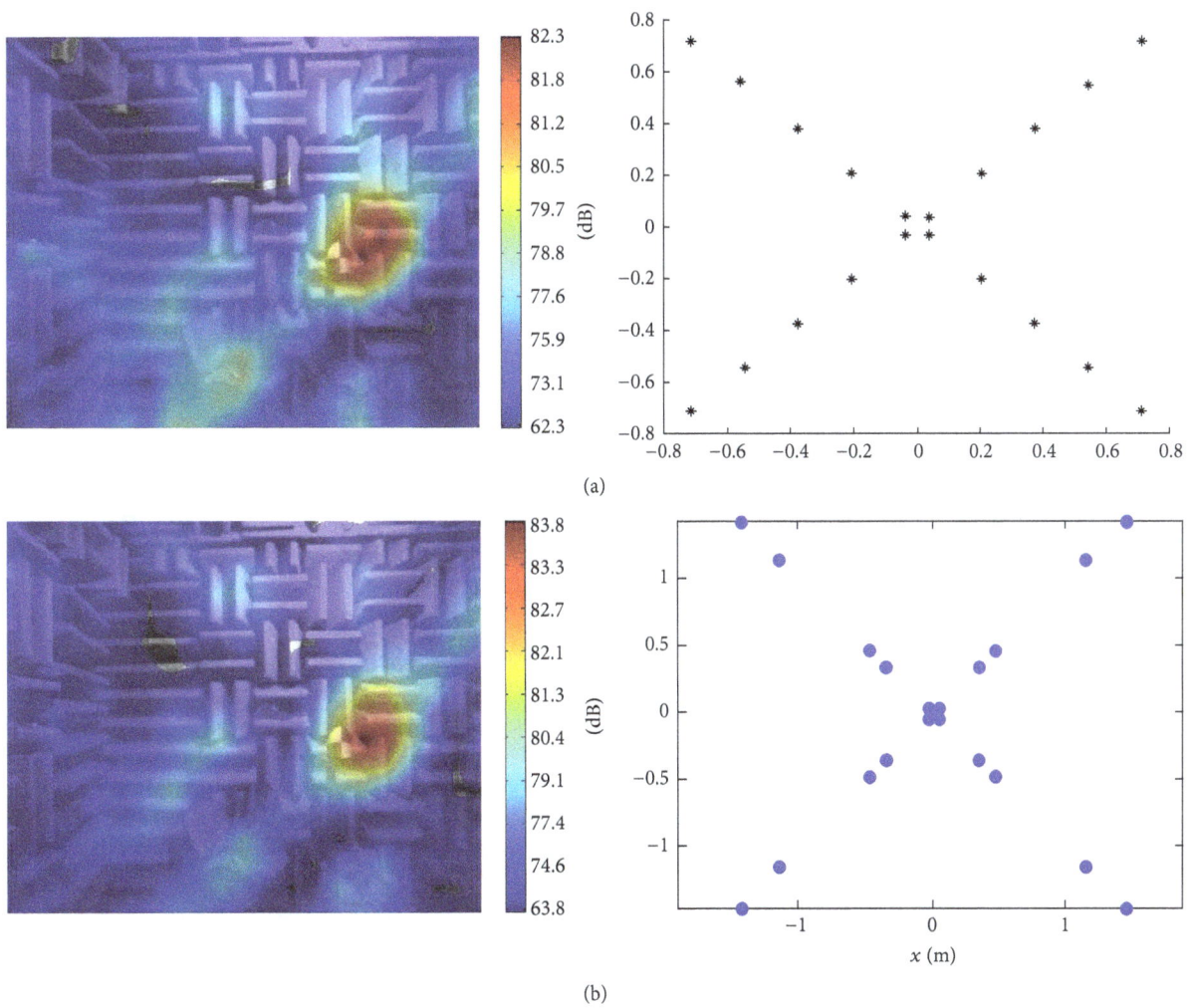

(a)

(b)

FIGURE 7: The experiment results of the static speaker in the anechoic room. (a) Result from the acoustic holography method; (b) virtual microphone array.

FIGURE 8: The measurement system and testing high speed train.

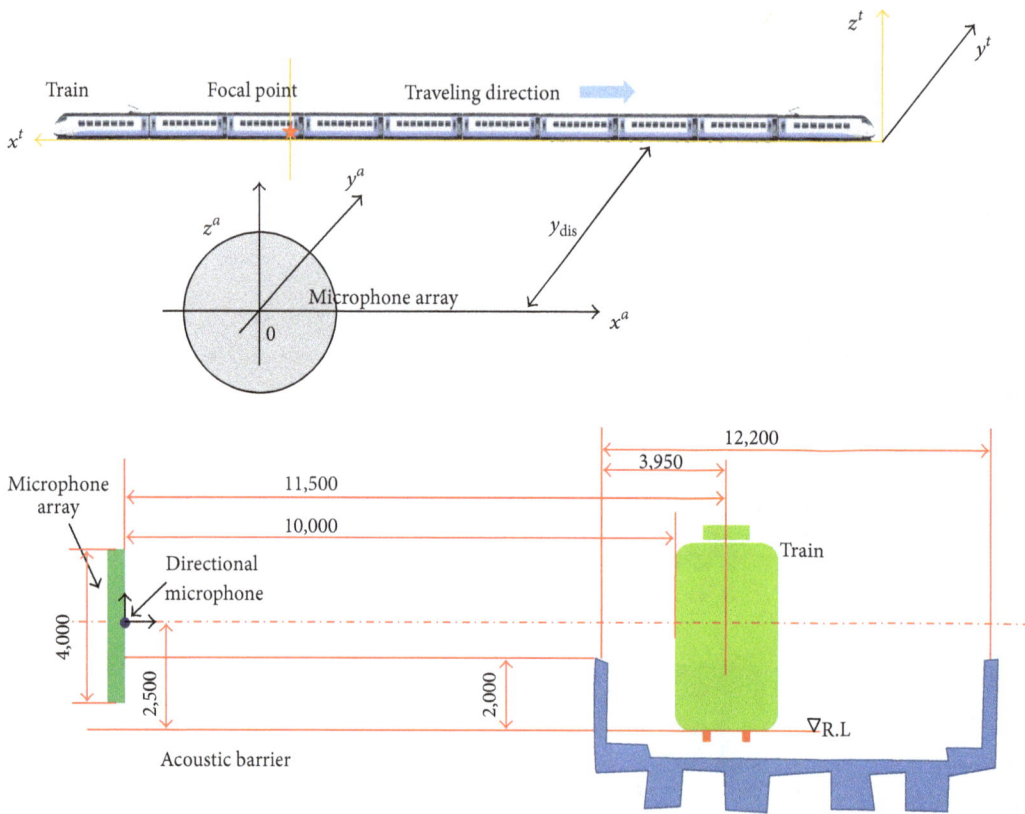

FIGURE 9: The measurement system and testing high speed train.

4. Validation Experiments

4.1. Static Sound Sources Experiment. This experiment is done in anechoic chamber. The microphone array is made of 21 microphones. The spacing between each of the microphones is 160 mm. In this experiment, the sound source is a powered loud speaker, and the noise signal is a simple harmonic sound of 1 kHz. The experiment lab and experiment device are shown in Figure 6.

As can be seen from Figure 7 and the simulation result in Table 3, comparing both results, there is hardly a change

in the value of MLA and SLR (also the result of optimization of MLA (0.0045) is better than equal distance microphones array) and no change in the relative position; to some extent, the result is nearly the same; it is concluded that the 19 microphones nearly have the same solution with 21 microphones.

4.2. High Speed Moving Sound Sources Experiment. The validation experiment is done in circle test ground for high speed train; we choose the high speed train which was

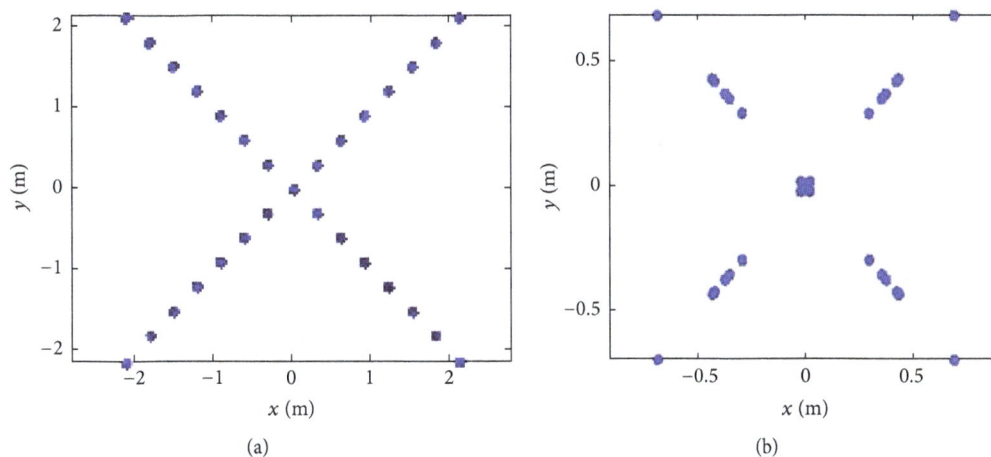

FIGURE 10: Microphones array. (a) The equal distance; (b) the optimization array.

TABLE 3: Comparison of the results.

	Equal distance of 21 microphones	
	SLR	MLA
(a)	0.5216	0.0068
	Optimization of 19 microphones	
	SLR	MLA
(b)	0.4531	0.0045

TABLE 4: Comparison of the results.

	Equal distance of 29 microphones	
	SLR	MLA
(a)	0.6535	0.0035
	Optimization of 17 microphones	
	SLR	MLA
(b)	0.6442	0.0033

processing test to testify the simulation results. We set the microphones array which is made of 29 microphones and then use the 17 microphones optimization array to process the sound data. The parameters in this experiment are as follows: the sound source is radiated by one high speed train, and the noise signal is a simple harmonic sound of 500 Hz–1 kHz. The experiment system is shown in Figures 8–10. The main validation experiment device is as follows: conventional signal measurement and analysis equipment, B&K company's set of high-precision multimicrophone set, 32 sets of high performance acoustic sensors, and an array of developed measurement systems, X-type array.

The experimental conditions are shown in Figure 9. In this experiment, the sound source data analysis frequency is 500–700 Hz, the speed of testing train is 125 km/h, the compare results are shown in Figure 11, one is the result of identification high speed train sound sources with 29 equal distance microphones' array, and the other is the result of identification high speed train sound sources with optimization microphones array.

The results of experiment are shown in Figure 11 and Table 4.

As can be seen from the result, both Figure 11 and Table 4, compared with the results, there is hardly change in the value of MLA and SLR (also the result of optimization of MLA (0.0033) is better than equal distance microphones array) and no change in the relative position; to some extent, the result

is nearly the same; it is concluded that the 17 microphones nearly have the same solution with 29 microphones.

5. Conclusion

In this paper, based on the fixed X microphones array, select the main SLR (side lobe ratio) and the MLA (main lobe area) as the optimization objective function, by using optimization method to optimize the microphone array. Compare simulation results of equal distance microphones array with optimization array found that reduced the number of optimization microphones; there were no changes in the purpose function value. Also, validation experiment has done to testify the simulation result; it is drawn from the results that the experiment result can be by reducing the number of microphones but does not have changes in the purpose values. So the experiments with limited conditions can be done, as to use fewer microphones to get the similar result. In the next study, it will consider the angle between surface reconstruction of sound pressure and measurement plane by further optimization method.

Competing Interests

The authors declare that they have no competing interests.

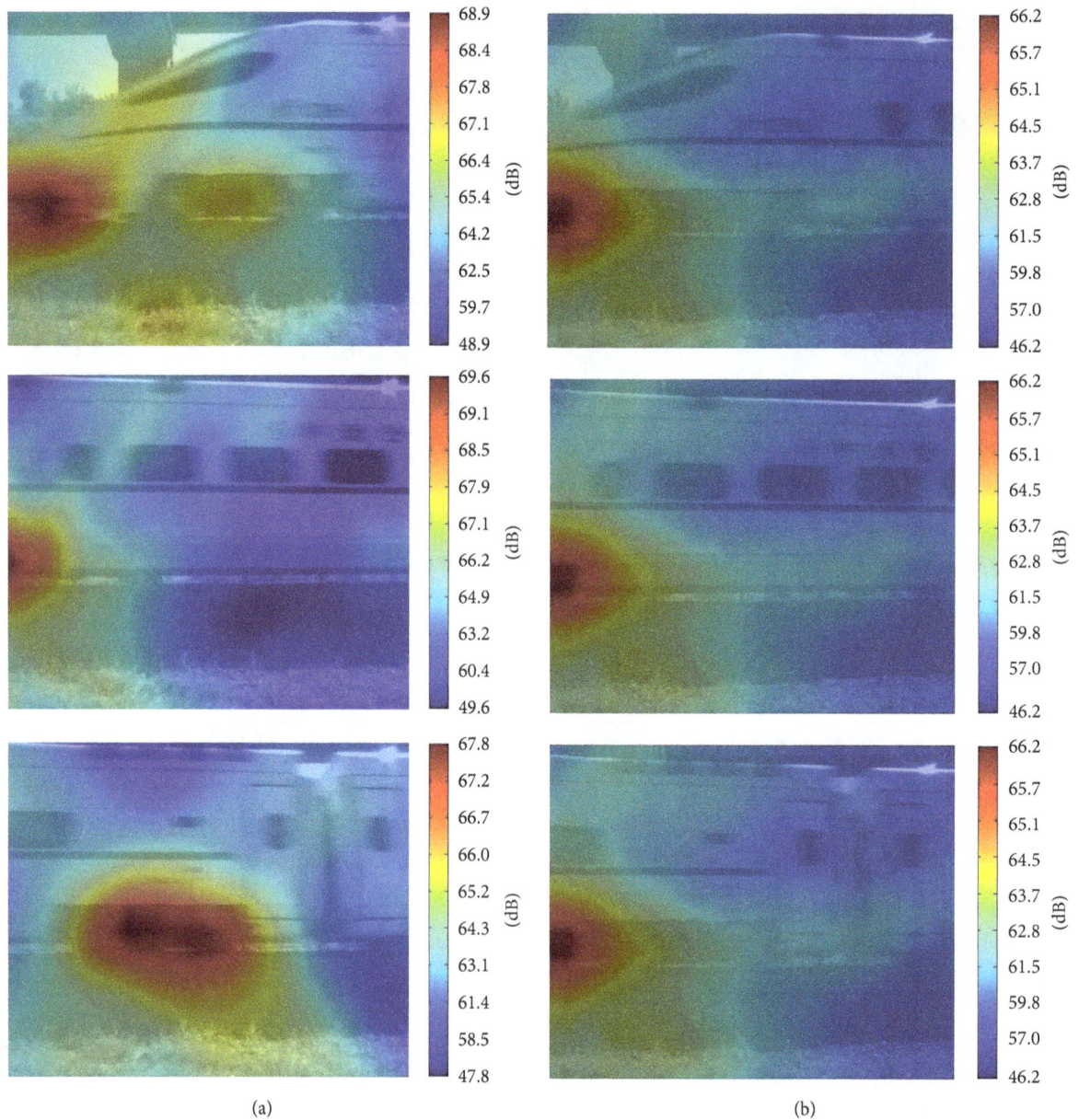

FIGURE 11: The comparison of validation experiment results of the high speed train with equal distance microphones array (a) and optimization array (b).

References

[1] H. Pu and J. Weikang, "Application of sound intensity technique in identify the noise sources of maglev train," *Journal of Railway Engineering Society*, vol. 5, pp. 5–8, 2005.

[2] W. F. King III and D. Bechert, "On the sources of wayside noise generated by high-speed trains," *Journal of Sound and Vibration*, vol. 66, no. 3, pp. 311–332, 1979.

[3] Y. Takano, K. Terada, F. Aizawa, A. Iida, and H. Fujita, "Development of a 2-dimensional microphone array measurement system for noise sources of fast moving vehicles," in *Proceedings of the International Congress on Noise Control Engineering (InterNoise '92)*, pp. 1175–1178, Toronto, Canada, July 1992.

[4] H. Kook, G. B. Moebs, P. Davies, and J. S. Bolton, "Efficient procedure for visualizing the sound field radiated by vehicles during standardized passby tests," *Journal of Sound and Vibration*, vol. 233, no. 1, pp. 137–156, 2000.

[5] Y. Takano, "Development of visualization system for high-speed noise sources with a microphone array and a visual sensor," in *Proceedings of the International Congress on Noise Control Engineering (InterNoise '03)*, Seogwipo, Republic of Korea, August 2003.

[6] M. Genescà, J. Romeu, T. Pàmies, and A. Sánchez, "Real time aircraft fly-over noise discrimination," *Journal of Sound and Vibration*, vol. 323, no. 1-2, pp. 112–129, 2009.

[7] S. Gade, J. Hald, and B. Ginn, "Refined beamforming with increased spatial resolution," in *Proceedings of the 41st International Congress and Exposition on Noise Control Engineering (INTER-NOISE '12)*, pp. 3571–3578, New York, NY, USA, August 2012.

[8] K. Saijyou and H. Uchida, "Data extrapolation method for boundary element method-based near-field acoustical holography," *Journal of the Acoustical Society of America*, vol. 115, no. 2, pp. 785–796, 2004.

[9] Z. Wang and S. F. Wu, "Helmholtz equation-least-squares method for reconstructing the acoustic pressure field," *Journal of the Acoustical Society of America*, vol. 102, no. 4, pp. 2020–2032, 1997.

[10] R. Steiner and A. N. Kaelin, "Sound field reconstruction of moving noise sources by means of acoustical holography," in *Proceedings of the InterNoise 1998*, Christchurch, New Zealand, November 1998.

[11] S.-H. Park and Y.-H. Kim, "Visualization of pass-by noise by means of moving frame acoustic holography," *Journal of the Acoustical Society of America*, vol. 110, no. 5, pp. 2326–2339, 2001.

[12] C.-S. Park and Y.-H. Kim, "Time domain visualization using acoustic holography implemented by temporal and spatial complex envelope," *Journal of the Acoustical Society of America*, vol. 126, no. 4, pp. 1659–1662, 2009.

[13] D. G. Yang, S. F. Zheng, Y. K. Li, X. M. Lian, and X. Y. Jiang, "Research on acoustic holography method for the identification of sound source," *Acta Acustica*, vol. 26, no. 2, pp. 156–160, 2001.

[14] Y. Diange, Z. Sifa, L. Bing, L. Keqiang, and L. Xiaomin, "Video visualization for moving sound sources based on binoculars stereo and acoustic holography," *Chinese Journal of Acoustics*, vol. 30, no. 2, pp. 203–213, 2011.

[15] D. Yang, Z. Wang, B. Li, Y. Luo, and X. Lian, "Quantitative measurement of pass-by noise radiated by vehicles running at high speeds," *Journal of Sound and Vibration*, vol. 330, no. 7, pp. 1352–1364, 2011.

2

Electricity Generation Characteristics of Energy-Harvesting System with Piezoelectric Element Using Mechanical-Acoustic Coupling

Hirotarou Tsuchiya, Hiroyuki Moriyama, and Satoru Iwamori

Course of Science and Technology, Graduate School of Tokai University, 4-1-1 Kitakaname, Hiratsuka, Kanagawa, Japan

Correspondence should be addressed to Hiroyuki Moriyama; moriyama@keyaki.cc.u-tokai.ac.jp

Academic Editor: Mohammad Tawfik

This paper describes the electricity generation characteristics of a new energy-harvesting system with piezoelectric elements. The proposed system is composed of a rigid cylinder and thin plates at both ends. The piezoelectric elements are installed at the centers of both plates, and one side of each plate is subjected to a harmonic point force. In this system, vibration energy is converted into electrical energy via electromechanical coupling between the plate vibration and piezoelectric effect. In addition, the plate vibration excited by the point force induces a self-sustained vibration at the other plate via mechanical-acoustic coupling between the plate vibrations and an internal sound field into the cylindrical enclosure. Therefore, the electricity generation characteristics should be considered as an electromechanical-acoustic coupling problem. The characteristics are estimated theoretically and experimentally from the electric power in the electricity generation, the mechanical power supplied to the plate, and the electricity generation efficiency that is derived from the ratio of both power. In particular, the electricity generation efficiency is one of the most appropriate factors to evaluate a performance of electricity generation systems. Thus, the effect of mechanical-acoustic coupling is principally evaluated by examining the electricity generation efficiency.

1. Introduction

To deal with depletion of fossil fuels and to materialize a low-carbon society, not only the improvement of energy saving technologies but also the creation of new energy sources has been attempted in a lot of studies. Scavenging untapped vibration energy and converting it into usable electric energy via piezoelectric materials has attracted considerable attention and has been regarded as one of new energy sources [1]. Typical energy harvesters adopt a simple cantilever configuration to generate electric energy via piezoelectric materials, which are attached to or embedded in host structures, and the behavior is governed by electromechanical coupling phenomena.

In general, the flexural stiffness of a beam structure is considerably intensified in comparison with that of a piezoelectric element, so that most of the strain energy is stored in the beam structure. To enhance the conversion efficiency, two methods have been adopted in many cases: the optimization of piezoelectric element placement and the use of a large element or many elements. To further improve the conversion efficiency, mechanical impedance matching method, which was derived from using spacers between the piezoelectric element and beam structure and tuning for the size of the piezoelectric element, was proposed [2]. These structural vibrations are caused by vibrators and various power sources. For instance, a self-sustained oscillation caused by placing a plate into a flow whose critical velocity was overpassed (so-called fluttering) is a well-known phenomenon. To utilize such a fluttering phenomenon for energy-harvesting, the plate on which the piezoelectric elements were arranged was used, and the effect of their arrangement along the flow axis was considered. Then an optimization of the arrangement was performed among some positions and dimensions of piezoelectric elements [3].

The authors have also been interested in a mechanical-acoustic coupling problem. The representative example was

TABLE 1: Mechanical and electrical properties of plate and piezoelectric element.

	Density [kg/m³]	Young's modulus [GPa]	Poisson's ratio
Aluminum plate	2680	70.6	0.33
Electrode plate	6900	100	0.35
Piezoelectric material	8400	132	0.30
	$\gamma^{\varepsilon} = 2.213 \times 10^{-9}$ [F/m], $d_{31} = -3.7 \times 10^{-12}$ [m/V]		

investigated as an architectural acoustic problem via a coupled panel-cavity system consisting of a rectangular box with slightly absorbing walls and a simply supported panel. The effect of the panel characteristics on the decay behavior of the sound field in the cavity was considered both theoretically and experimentally [4, 5]. In an attempt to control noise in an airplane, an analytical model for investigating the coupling between the sound field in an aircraft cabin and the vibrations of the rear pressure bulkhead was proposed [6, 7]. A cylindrical structure adopted as the analytical model, in which the rear pressure bulkhead at one end of the cylinder was assumed to be a circular plate, was examined under various conditions. The plate was supported at its edges by springs whose stiffness could be adjusted to simulate the various support conditions. These investigations clarified the influence of the support conditions on the sound pressure of an internal sound field coupled with the vibration of the end plate. We used the above-mentioned analytical model [6, 7] of a cylindrical structure with plates at both ends to investigate the effect of the excitation frequency, at which the coupling system becomes nonperiodic owing to the application of excitation forces of different frequencies to the respective end plates, on the mechanical-acoustic coupling [8]. Then the effect of the excitation position with respect to the nodal lines on the appearance of vibration modes on the plates was investigated [9]. Finally, because a phase difference between both plate vibrations affects strongly mechanical-acoustic coupling, this coupling was considered in the cases, in which one end plate and both end plates were excited by a harmonic point force, respectively [10, 11]. In particular, [10] made reference to the electricity generation with piezoelectric elements using coupling, as well as this paper. However, the effect of coupling was not clarified and has remained unclear, because the theoretical consideration was not carried out and the effect was not compared between the electricity generations using and not using coupling.

On the other hand, to suppress the above vibration and acoustic energy that were amplified by mechanical-acoustic coupling, an analytical model that included the installation of passive devices on the vibration system was proposed as the electromechanical-acoustic system. The effect was fully validated in the numerical approach owing to the damping technique using an RL shunt circuit, in which the piezoelectric device was incorporated, to the frequency characteristics of the coupling system [12]. Furthermore, this work was developed by adopting a more complex coupling system that comprised the plate equipped with shunted piezoelectric elements and interior and exterior acoustic fields coupled with plate vibration. This system was analyzed by means of a coupled finite element/boundary element method to control

the noise radiation and sound transmission of the vibrating plate by passive piezoelectric techniques [13].

To develop a new electricity generation system, we adopt an analytical model similar to the above-mentioned cylindrical structure with plates at both ends, because the vibration area of the model on which piezoelectric elements can be installed is twice as large as that in the case of a single plate. The cylinder length is varied over a wide range while changing the plate thickness, while the harmonic point force is applied to one end plate and its frequency is selected to cause the plate to vibrate in the fundamental mode. The plate vibration induces electricity generation via electromechanical coupling with the piezoelectric effect of the surface-mounted piezoelectric element, while the plate vibration of the excitation side oscillates the other plate via mechanical-acoustic coupling. Consequently, the electromechanical-acoustic coupling problem must be considered and is estimated theoretically and experimentally from the electric power caused by the electricity generation, the mechanical power supplied to the plate, and the electricity generation efficiency that is derived from the ratio of both powers. In particular, by focusing on the electricity generation efficiency as the most significant characteristic, we verify that the performance of the proposed system is improved by using mechanical-acoustic coupling in comparison with using only the plate vibration without coupling. These results can be applied to power generation floors of roads, bridges, and so on, which can ensure a space to cause coupling. Recently, the monitoring system to grasp road surface information and tire conditions is developed and the vibration power generation is adopted as its power source. Since a tire has an inside space, we expect that these results are also useful at such a system design.

2. Analytical Method

2.1. Analytical Model. The analytical model consists of a cavity with two circular end plates, as shown in Figure 1. The plates are supported by translational and rotational springs distributed at constant intervals and the support conditions are determined by their respective spring stiffness T and stiffness R. The plates of radius r_c and thickness h_c have Young's modulus E_c and Poisson's ratio v_c. On the surfaces of both plates, piezoelectric elements are installed at the centers of the plates and have radius r_p, thickness h_p, Young's modulus E_p, and Poisson's ratio v_p. Then an electrode plate is sandwiched between the above plate and piezoelectric element and has radius r_b, thickness h_b, Young's modulus E_b, and Poisson's ratio v_b. The suffixes c, p, and b herein indicate the circular plate, piezoelectric element, and

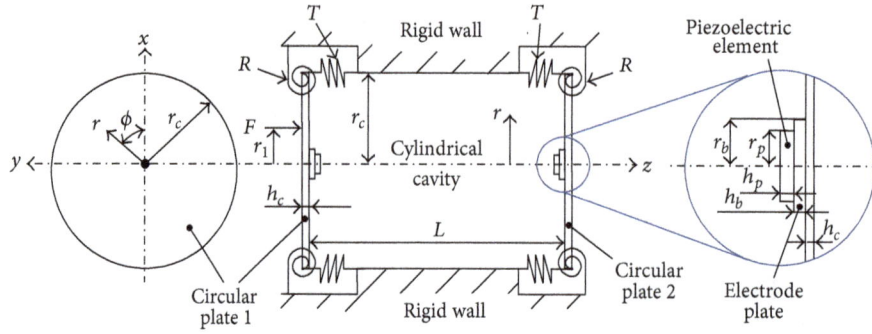

FIGURE 1: Configuration of analytical model.

electrode plate. On the other hand, the sound field, which is assumed to be cylindrical, has the same radius as that of the plates and varying length L because the resonance frequency depends on the length. The boundary conditions are considered structurally and acoustically rigid at the lateral wall between the structure and sound field. The coordinates used are radius r, angle ϕ between the planes of the plates and the cross-sectional plane of the cavity, and distance z along the cylinder axis. The periodic point force F is applied to plate 1 at distance r_1 and angle ϕ_1. The natural frequency of the plates is employed as the excitation frequency.

To formulate the plate motion, Hamilton's principle is applied to the analytical model:

$$\delta H$$

$$= \delta \int_{t_0}^{t_1} \left(T_c + T_p + T_b - U_c - U_p - U_b - U_s + W \right) dt \quad (1)$$

$$= 0,$$

where H is the Hamiltonian; T_c, T_p, and T_b are the kinetic energy of the circular plates, piezoelectric elements and electrode plates, respectively; and U_c, U_p, and U_b are their respective potential energies. U_s is the elastic energy stored in the springs, and W is the total work done on plate 1 by the point force and on both plates by the sound pressure inside the cavity. Finally, t_0 and t_1 are two arbitrary times.

w_{c1} and w_{c2} are the flexural displacements of plates 1 and 2, and w_{p1} and w_{p2} are those of the piezoelectric elements installed on the plates. Then suffixes 1 and 2 indicate plates 1 and 2, respectively. They are found by substituting X_{cnm}^s of (3) for the plate modes into (2) as suitable trial functions. The flexural displacements of the piezoelectric elements are identical to those of the plates, respectively, because it is assumed that the piezoelectric elements adhere completely to each circular plate through the electrode plate.

$$w_{c1} = w_{p1} = \sum_{s=0}^{1} \sum_{n=0}^{\infty} \sum_{m=0}^{\infty} X_{nm}^s A_{1nm}^s e^{j(\omega t + \alpha_1)},$$

$$w_{c2} = w_{p2} = \sum_{s=0}^{1} \sum_{n=0}^{\infty} \sum_{m=0}^{\infty} X_{nm}^s A_{2nm}^s e^{j(\omega t + \alpha_2)}, \quad (2)$$

$$X_{nm}^s = \sin\left(n\phi + \frac{s\pi}{2}\right)\left(\frac{r}{r_c}\right)^m, \quad (3)$$

where n, m, and s are, respectively, the circumferential order, radial order, and symmetry index with respect to the plate vibration. A_{1nm}^s and A_{2nm}^s are coefficients to be determined, ω is the angular frequency of the harmonic point force acting on the plate, and t is the elapsed time. α_1 and α_2 are the phases of the respective plate vibrations. In this analysis, α_1 is set to $0°$, and α_2 ranges from $0°$ to $180°$.

2.2. Modelling of Piezoelectric Part. Only the piezoelectric part of plate 1 is used to explain its modelling in this section. The relationships of stress σ_{p1}, strain ε_{p1}, electric displacement D_1, and electric field E_1 are as follows:

$$\begin{Bmatrix} \sigma_{p1} \\ D_1 \end{Bmatrix} = \begin{bmatrix} E_p^E & -e^T \\ e & \gamma^\varepsilon \end{bmatrix} \begin{Bmatrix} \varepsilon_{p1} \\ E_1 \end{Bmatrix}. \quad (4)$$

E_p^E signifies Young's modulus that was measured at a constant electric field, and γ^ε indicates the dielectric constant that was measured at a constant strain. The above equation expresses relationships between electrical and mechanical characteristics of a piezoelectric element, and the stress is concretely related to the electric field by the piezoelectric coupling coefficient e. The piezoelectric coupling coefficient is expressed as

$$e = d_{31} E_p^E, \quad (5)$$

where d_{31} is the piezoelectric strain constant, in which the electric field occurs in the normal direction of the in-plane strain.

If the first variation is carried out in terms of the piezoelectric part on plate 1, we can obtain

$$\delta \int_{t_0}^{t_1} \left(T_p - U_p \right) dt = \frac{1}{2} \int_{t_0}^{t_1} \int_{V_{p1}} \{ \rho_p \dot{w}_{p1} \delta \dot{w}_{p1}$$

$$- \left(E_p^E \varepsilon_{p1} - e E_1 \right) \delta \varepsilon_{p1}$$

$$+ \left(e \varepsilon_{p1} + \gamma^\varepsilon E_1 \right) \delta E_1 \} dV_{p1} dt, \quad (6)$$

where V_{p1} is the volume of the piezoelectric element on plate 1. Here, the strain and electric charge are assumed as follows:

$$\varepsilon_{p1} = -z \left\{ \frac{\partial^2 w_{p1}}{\partial r^2} + v \left(\frac{1}{r} \frac{\partial w_{p1}}{\partial r} + \frac{1}{r^2} \frac{\partial^2 w_{p1}}{\partial \phi^2} \right) \right\}$$

$$= -z X_{nm}^{s\prime\prime} A_{1nm}^{s} e^{j(\omega t + \alpha_1)}, \tag{7}$$

$$q_1 = B_{1nm}^{s} e^{j(\omega t + \alpha_1)}. \tag{8}$$

B_{1nm}^{s} is coefficient to be determined in this analysis as well as A_{1nm}^{s}. Then the electric field E_1 is expressed by using its electric charge:

$$E_1 = Y_{nm} v_1 = -R_p \dot{q}_1, \tag{9}$$

$$Y_{nm} = \begin{cases} -\dfrac{1}{h_p} & \dfrac{h_c}{2} < z < \dfrac{h_c}{2} + h_b + h_p, \\ 0 & -\dfrac{h_c}{2} < z < \dfrac{h_c}{2} + h_b. \end{cases} \tag{10}$$

In this case, v_1 is the voltage that occurs in the electric field. We assume that the electric potential across the piezoelectric element is constant, since it is in the field that is not applied to the plate. Thus, Y_{nm} is defined as described above. R_p is the overall resistance value in an electricity generation circuit.

In order to easily express the electromechanical equation that is induced from (6), the elements $M_{p1nmm'}^{s}$ and $K_{p1nmm'}^{s}$ of the mass and stiffness matrices can be denoted as

$$M_{p1nmm'}^{s} = \int_{V_{p1}} \rho_p X_{nm}^{s} X_{nm'}^{s} dV_{p1},$$

$$K_{p1nmm'}^{s} = \int_{V_{p1}} z^2 X_{nm}^{s\prime\prime} E_p^E X_{nm'}^{s\prime\prime} dV_{p1}. \tag{11}$$

The index m' is also of a radial order and has a transposed relation to m. The elements θ_1 and C_{p1} of the electromechanical coupling and capacitance matrices are defined as

$$\theta_1 = -\int_{V_{p1}} z \rho_p X_{nm}^{s} e Y_{nm'} dV_{p1},$$

$$C_{p1} = \int_{V_{p1}} Y_{nm} \gamma^{\varepsilon} Y_{nm'} dV_{p1}. \tag{12}$$

2.3. Governing Equations of Electromechanical Coupling.
Here, the governing equation of electromechanical coupling with respect to plate 1 is determined. This equation defines the mechanical motion and electrical characteristics of this coupling system. The sound field inside the cavity or the vibration of plate 2 is not taken into consideration. The equation is obtained by applying the flexural displacement w_{c1} of (2) to the Hamiltonian H of (1) in the same manner as (6) of the piezoelectric part that was derived in the previous section. Then the motion of plate 1 having a piezoelectric part

and the electricity generation behavior of its piezoelectric element are governed by the following, respectively:

$$\sum_{m'=0}^{\infty} \left[\left\{ K_{c1nmm'}^{s} \left(1 + j\eta_c \right) + K_{p1nmm'}^{s} \left(1 + j\eta_p \right) \right. \right.$$

$$+ K_{b1nmm'}^{s} \left(1 + j\eta_b \right)$$

$$- \omega^2 \left(M_{c1nmm'}^{s} + M_{p1nmm'}^{s} + M_{b1nmm'}^{s} \right) \right\}$$

$$\left. + r_c F_{sn} \left\{ T_1 + \left(\frac{m}{r_c} \right) \left(\frac{m'}{r_c} \right) R_1 \right\} \right] A_{1nm'}^{s} e^{j(\omega t + \alpha_1)}$$

$$- \sum_{m'=0}^{\infty} \theta_1 v B_{1nm'}^{s} e^{j(\omega t + \alpha_1)} = \mathbf{F}_{nm}^{s} e^{j(\omega t + \alpha_1)}, \tag{13}$$

$$\sum_{m'=0}^{\infty} C_{p1}^{-1} \theta_1 A_{1nm'}^{s} e^{j(\omega t + \alpha_1)} = \sum_{m'=0}^{\infty} \left(j\omega R_p + C_{p1}^{-1} \right)$$

$$\cdot B_{1nm'}^{s} e^{j(\omega t + \alpha_1)}. \tag{14}$$

In (13), $K_{c1nmm'}^{s}$, $K_{b1nmm'}^{s}$ and $M_{c1nmm'}^{s}$, $M_{b1nmm'}^{s}$ are elements of the symmetrical stiffness and mass matrices with respect to the circular and electrode plates, respectively, because the index m' has a transposed relation to m, as well as $M_{p1nmm'}^{s}$ and $K_{p1nmm'}^{s}$. η_c, η_p, and η_b are the structural damping factors of the circular plate, piezoelectric element, and electrode plate, respectively. Moreover, F_{sn} is a coefficient that is determined by the indices n and s, and $\mathbf{F}_{nm}^{s}$ is a load vector that expresses the point force. The details of F_{sn} and the element F_{nm}^{s} of the load vector are as follows:

$$F_{sn} = \begin{cases} \pi, & \text{at } n \neq 0, \\ 0, & \text{at } n = 0, \ s = 0, \\ 2\pi, & \text{at } n = 0, \ s = 1, \end{cases} \tag{15}$$

$$F_{nm}^{s} = \int_{A_1} F \delta \left(r - r_1 \right) \delta \left(\phi - \phi_1 \right) X_{nm}^{s} dA_1.$$

Here, δ is the delta function associated with the point force on plate 1, whose area is denoted by A_1.

Since A_{1nm}^{s} and B_{1nm}^{s} can be obtained simultaneously from (13) and (14), the behavior of the plate vibration and the electricity generation under electromechanical coupling can be determined from (2) and (8). However, only the vibration of plate 1 is an analysis object and the behavior is assumed to be harmonic, so that $\exp\{j(\omega t + \alpha_1)\}$ can be eliminated in (13) and (14).

2.4. Governing Equations of Electromechanical-Acoustic Coupling.
In this section, electromechanical-acoustic coupling is determined by adding a sound field inside the cavity to the electromechanical coupling introduced in the previous section. Naturally, the vibration of plate 2 is also added on the assumption that this coupling is phenomena based on the analytical model in Figure 1. The variation of (1) is carried out with respect to both plates. Consequently, the extremum

of the Hamiltonian yields Euler's equations, which are the equations of motion of the respective plates.

$$
\begin{aligned}
\sum_{m'=0}^{\infty} & \Big[\big\{ K^s_{c1nmm'} (1+j\eta_c) + K^s_{p1nmm'} (1+j\eta_p) \\
& + K^s_{b1nmm'} (1+j\eta_b) \\
& - \omega^2 \big(M^s_{c1nmm'} + M^s_{p1nmm'} + M^s_{b1nmm'} \big) \big\} \\
& + r_c F_{sn} \Big\{ T_1 + \Big(\frac{m}{r_c}\Big)\Big(\frac{m'}{r_c}\Big) R_1 \Big\} \Big] A^s_{1nm'} e^{j\alpha_1} \\
& - \sum_{m'=0}^{\infty} \theta_1 v B^s_{1nm'} e^{j\alpha_1} = \mathbf{F}^s_{nm} e^{j\alpha_1} - \mathbf{P}^s_{1nm},
\end{aligned}
\tag{16}
$$

$$
\begin{aligned}
\sum_{m'=0}^{\infty} & \Big[\big\{ K^s_{c2nmm'} (1+j\eta_c) + K^s_{p2nmm'} (1+j\eta_p) \\
& + K^s_{b2nmm'} (1+j\eta_b) \\
& - \omega^2 \big(M^s_{c2nmm'} + M^s_{p2nmm'} + M^s_{b2nmm'} \big) \big\} \\
& + r_c F_{sn} \Big\{ T_2 + \Big(\frac{m}{r_c}\Big)\Big(\frac{m'}{r_c}\Big) R_2 \Big\} \Big] A^s_{2nm'} e^{j\alpha_2} \\
& - \sum_{m'=0}^{\infty} \theta_2 v B^s_{2nm'} e^{j\alpha_2} = \mathbf{P}^s_{2nm}.
\end{aligned}
\tag{17}
$$

Equations (16) and (17) are obviously the motion equations of plates 1 and 2, respectively, from the subscripts of each element constructing both equations, since the elements correspond to them in (13). However, $\exp(j\alpha_1)$ and $\exp(j\alpha_2)$ remained in (16) and (17), respectively, since both vibrations have an effect on each other via mechanical-acoustic coupling. Equation (16) has the point force and acoustic excitation terms on the right-hand side, because plate 1 is simultaneously subjected to both excitations. However, the point force is not applied to plate 2, with the result that (17) consists of only the acoustic term, which also functions as the coupling term between each plate vibration and the sound field. The elements of the acoustic excitation vectors are expressed as

$$
\begin{aligned}
P^s_{1nm} &= \int_{A_1} P_s X^s_{nm} \, dA_1, \\
P^s_{2nm} &= \int_{A_2} P_s X^s_{nm} \, dA_2.
\end{aligned}
\tag{18}
$$

Here, P_s is the sound pressure at an arbitrary point on the boundary surface of the plates, and A_2 signifies the area of plate 2.

This procedure also derives the equations representing electrical characteristics in this coupling system. They are as follows:

$$
\begin{aligned}
\sum_{m'=0}^{\infty} C^{-1}_{p1} \theta_1 A^s_{1nm'} &= \sum_{m'=0}^{\infty} \big(j\omega R_p + C^{-1}_{p1} \big) B^s_{1nm'}, \\
\sum_{m'=0}^{\infty} C^{-1}_{p2} \theta_2 A^s_{2nm'} &= \sum_{m'=0}^{\infty} \big(j\omega R_p + C^{-1}_{p2} \big) B^s_{2nm'}.
\end{aligned}
\tag{19}
$$

The above elements constructing the respective equations correspond to those of (14), as well as (16) and (17).

In this investigation, the acoustic modal shape Z^s_{npq} and angular resonance frequency ω_{npq} in the cavity (where the indices n, p, and q indicate the circumferential, radial, and longitudinal orders, resp.) are defined as

$$
\begin{aligned}
Z^s_{npq} &= \sin\Big(n\phi + \frac{s\pi}{2} \Big) J_n\big(\lambda_{np} r \big) \cos\Big\{ \Big(\frac{q\pi}{L}\Big) z \Big\}, \\
\omega_{npq} &= c \Big\{ \lambda_{np}{}^2 + \Big(\frac{q\pi}{L}\Big)^2 \Big\}^{1/2},
\end{aligned}
\tag{20}
$$

where J_n is the nth-order Bessel function, c is the cavity speed of sound, and λ_{np} is the pth solution of an eigenvalue problem for a circular sound field having modes (n, p) divided by the plate radius r_c. The boundary conditions between the plate vibrations and sound field on the respective plate surfaces are found by assuming continuity of the velocities on the plates:

$$
\begin{aligned}
\Big(\frac{\partial P_s}{\partial \mathbf{u}} \Big)_{z=0} &= \rho_s \omega^2 w_{c1}, \\
\Big(\frac{\partial P_s}{\partial \mathbf{u}} \Big)_{z=L} &= -\rho_s \omega^2 w_{c2},
\end{aligned}
\tag{21}
$$

where P_s is the sound pressure within and on the surface bounding the medium and $\mathbf{u}$ is the unit normal to the boundary surface (positive towards the outside). However, $\partial P_s / \partial \mathbf{u}$ is 0 on the lateral wall of the cylinder since the wall remains rigid.

Because the analytical model has two boundary surfaces, we can apply Green's function G to (21), so that the sound pressure P_s inside the cavity becomes

$$
P_s = -\int_{A_1} G \rho_s \omega^2 w_{c1} \, dA_1 + \int_{A_2} G \rho_s \omega^2 w_{c2} \, dA_2.
\tag{22}
$$

In addition, P_s can also be expressed as follows [6, 7]:

$$
P_s = \rho_s c^2 \sum_{p=1}^{\infty} \sum_{q=0}^{\infty} \frac{P^s_{npq} Z^s_{npq}}{M^s_{npq}},
\tag{23}
$$

where ρ_s is the fluid density in the cavity, P^s_{npq} is the pressure coefficient to be determined, and M^s_{npq} is the mean value of $Z^s_{npq}{}^2$ averaged over the sound field.

The equation relating (22) and (23) is obtained by applying Green's function [14] to an arbitrary acoustic mode (n, p, q) as

$$
\begin{aligned}
\big(& \omega^2_{npq} - \omega^2 \big) P^s_{npq} \\
&= -\frac{\omega^2}{V_s} \Big(\int_{A_1} Z^s_{npq} w_{c1} \, dA_1 + \int_{A_2} Z^s_{npq} w_{c2} \, dA_2 \Big).
\end{aligned}
\tag{24}
$$

Substituting (2) for w_{c1} and w_{c2} and considering a modal damping factor η_s, (24) can be rewritten as

$$\left(\omega_{npq}^2 + j\eta_s\omega_{npq}\omega - \omega^2\right) P_{npq}^s$$
$$= \frac{A\omega^2}{V_s}\left(-\sum_{m=0}^{\infty} I_1 A_{1nm}^s e^{j\alpha_1} + \sum_{m=0}^{\infty} I_2 A_{2nm}^s e^{j\alpha_2}\right), \tag{25}$$

$$I_1 = \frac{1}{A}\int_{A_1} X_{nm}^s Z_{npq}^s \mathrm{d}A_1,$$
$$I_2 = \frac{1}{A}\int_{A_2} X_{nm}^s Z_{npq}^s \mathrm{d}A_2, \tag{26}$$

where A is the total surface area of the plates, V_s is the cavity volume, and then I_1 and I_2 are the spatial coupling coefficients. Moreover, substituting (23) for P_s and applying I_1 and I_2 to the integrals in (18), the acoustic excitation terms P_{1nm}^s and P_{2nm}^s can be expressed with respect to an arbitrary vibration mode (n,m) as

$$P_{1nm}^s = \rho_s c^2 A \sum_{p=1}^{\infty}\sum_{q=0}^{\infty} \frac{I_1 P_{npq}^s}{M_{npq}^s},$$
$$P_{2nm}^s = \rho_s c^2 A \sum_{p=1}^{\infty}\sum_{q=0}^{\infty} \frac{I_2 P_{npq}^s}{M_{npq}^s}. \tag{27}$$

Finally, replacing P_{npq}^s in (27) with those in (25) and then inserting them in (16) and (17), we can complete the electromechanical-acoustic coupling equations in conjunction with (19). The details on the respective right-hand sides of (16) and (17) are

$$\mathbf{F}_{nm}^s e^{j\alpha_1} - \mathbf{P}_{1nm}^s = \mathbf{F}_{nm}^s e^{j\alpha_1} + \frac{\rho_s c^2\omega^2 A^2}{V_s}$$
$$\cdot\sum_{m'=0}^{\infty}\sum_{p=1}^{\infty}\sum_{q=0}^{\infty} \frac{I_1\left(I_1 A_{1nm'}^s e^{j\alpha_1} - I_2 A_{2nm'}^s e^{j\alpha_2}\right)}{M_{npq}^s\left(\omega_{npq}^2 + j\eta_s\omega_{npq}\omega - \omega^2\right)},$$
$$\mathbf{P}_{2nm}^s = \frac{\rho_s c^2\omega^2 A^2}{V_s}$$
$$\cdot\sum_{m'=0}^{\infty}\sum_{p=1}^{\infty}\sum_{q=0}^{\infty} \frac{I_2\left(I_1 A_{1nm'}^s e^{j\alpha_1} - I_2 A_{2nm'}^s e^{j\alpha_2}\right)}{M_{npq}^s\left(\omega_{npq}^2 + j\eta_s\omega_{npq}\omega - \omega^2\right)}. \tag{28}$$

On the right-hand sides, $\mathbf{P}_{1nm}^s$ and $\mathbf{P}_{2nm}^s$ show the acoustic excitation for plates 1 and 2, respectively. The acoustic excitation terms have both I_1 and I_2 since the acoustic mode of the sound field is coupled with the vibration modes of the respective plates. In actual calculation, the relationships between A_{1nm}^s and B_{1nm}^s and between A_{2nm}^s and B_{2nm}^s are obtained from (19) and are applied to (16) and (17), respectively, with the result that A_{1nm}^s, A_{2nm}^s, B_{1nm}^s, and B_{2nm}^s can be derived by solving the above simultaneous equations (16) and (17).

The flexural displacements w_{c1} and w_{c2} of plates 1 and 2 are obtained from (2) by employing A_{1nm}^s and A_{2nm}^s determined above. The voltage v_1 based on the vibration of plate 1 is derived from (8) and (9) by employing the above B_{1nm}^s. Then the voltage v_2 based on the vibration of plate 2 is also obtained in the same manner by employing the above B_{2nm}^s. However, the procedure to calculate P_s involves substituting P_{npq}^s in (25) for that in (23). Solved from (29), into which P_s is inserted, over the entire volume of the cavity, the sound pressure P_v is estimated by the logarithmic value L_{pv} relative to $P_0 = 4\times10^{-10}\,\mathrm{N^2/m^4}$, as shown in (30) as follows:

$$P_v = \frac{1}{2V_s}\int_{V_s} P_s P_s^* \mathrm{d}V_s, \tag{29}$$

$$L_{pv} = 10\log\frac{P_v}{P_0}, \tag{30}$$

where P_c^* is the conjugate component.

3. Experimental Apparatus and Method

Figure 2 shows the configuration of the experimental apparatus used in this study. The structure consists of a steel cylinder with circular aluminum end plates whose thickness h_c varies at 2.0, 2.5, 3.0, and 4.0 mm. The cylinder has an inner radius of 153 mm, and the length L can range from 500 to 2000 mm to emulate the analytical model. Plate 1 is subjected to the point force, whose frequency makes the plate excite in the $(0,0)$ mode; and amplitude F is controlled to be 1 N, excited by a small vibrator. The position of the point force r_1 is normalized by radius r_c and is set to $r_1/r_c = 0.4$.

In the excitation experiment, to estimate the mechanical power P_m supplied to plate 1 by the small vibrator, an acceleration sensor is installed near the position of the point force on plate 1, and P_m is predicted from the point force and acceleration a_1. The phase difference between the plate vibrations is also measured owing to the installation of the acceleration sensor at the same position on plate 2, resulting in significant effects on the mechanical-acoustic coupling. To estimate the internal acoustic characteristics, the sound pressure level in the cavity is measured using condenser microphones with a probe tube. The tips of the probe tubes are located near the plates and the cylinder wall, which are the approximate locations of the maximum sound pressure level when the sound field becomes resonant.

To perform the electricity generation experiment, the piezoelectric element is used. It is comprised of a piezoelectric part constructed of ceramics and an electrode part constructed of brass, which have radiuses r_p and r_b of 12.5 and 17.5 mm and thicknesses h_p and h_b of 0.23 and 0.30 mm, respectively. The piezoelectric elements are installed at the centers of both plates. The electric powers P_{e1} and P_{e2} generated by the expansion and contraction of the piezoelectric elements on plates 1 and 2 are discharged through the resistance circuit, which consists of three resistors having resistances R_v, R_i, and R_c, as shown in Figure 2(b). R_v and R_i are the resistances of the voltmeter and ammeter, which are built into the wattmeter, and are 2 MΩ and 2 mΩ, respectively, while R_c is the resistance of the resistor connected outside the wattmeter and is 97.5 kΩ. To grasp the effect of mechanical-acoustic coupling on energy-harvesting, the electric power

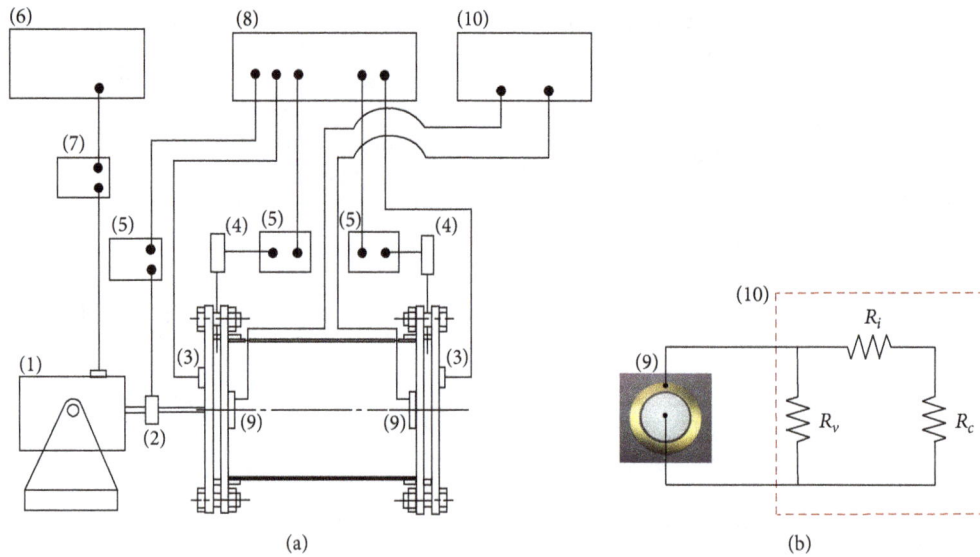

FIGURE 2: Configuration of experimental apparatus: (a) measurement system and (b) electrical circuit of energy-harvesting device. (1) Vibration generator, (2) load cell, (3) acceleration sensor, (4) condenser microphone, (5) amplifier, (6) multifunction generator, (7) power supply, (8) FFT analyzer, (9) piezoelectric element, and (10) power meter.

and other data are also measured without the cylinder (i.e., in the electricity generation under the vibration of only plate 1) and are estimated in comparison with those with cylinder. In such an estimation, electricity generation efficiency is used and is derived from the electric power normalized by the mechanical power P_m supplied to plate 1 by the vibrator. However, the electricity generation efficiencies, which are obtained from P_{e1}, P_{e2}, and their total electric power P_e, are denoted by P_{em1}, P_{em2}, and P_{em}, respectively, when electromechanical coupling is taken into consideration.

4. Results and Discussion

4.1. Electricity Generation Characteristics Based on Plate Vibration. The electricity generation caused by the piezoelectric element, which adheres to the center of plate 1 excited by the vibrator, is considered, so that the cavity and plate 2 are not assumed in the analytical model of Figure 1 and are actually eliminated in the experimental apparatus of Figure 2. The plate and piezoelectric element are the same as the experimental apparatus in the respective dimensions and have mechanical and electrical properties as shown in Table 1. However, only h_c = 3 mm is employed as the thickness of plate 1 in this section. The support condition of plate 1, which has flexural rigidity D [= $E_c h_c^3/\{12(1-\nu_c^2)\}$], is expressed by the nondimensional stiffness parameters T_n (= $T_c r_c^3/D$) and R_n (= $R_c r_c/D$). If R_n ranges from 0 to 10^8 when T_n is 10^8, the support condition can be assumed, from a simple support to a clamped support.

Figure 3 shows the theoretical and experimental accelerations a_1 at the excitation point as functions of the excitation frequency f. Before obtaining these results, we carried out experimental modal analysis and made sure of the natural frequency of plate 1. Then the actual condition adopts $T_n = 10^8$

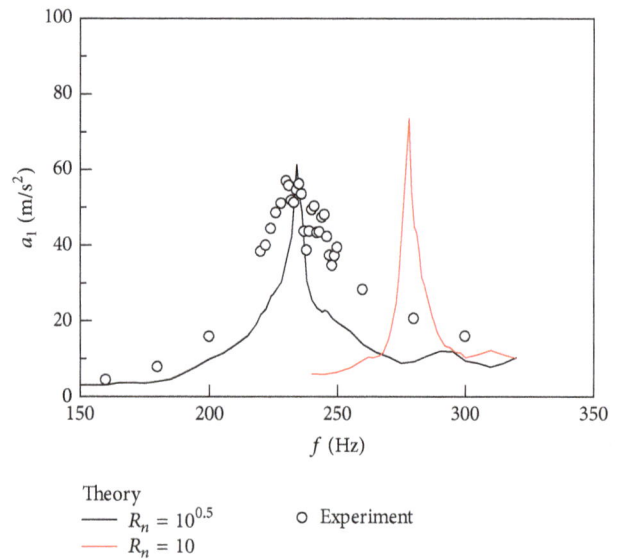

FIGURE 3: Acceleration at excitation point as function of excitation frequency.

and R_n = 10^1 to get closer to the experimental support condition. Because this support condition brings the natural frequency to 280 Hz, the corresponding a_1 reaches the peak at the natural frequency. However, the experimental a_1 peaks around f = 230 Hz and the actual natural frequency is shifted to a lower frequency region than the above natural frequency. Under a constraint of the experimental apparatus, the load cell cannot be directly installed on plate 1, so that the point force is applied to plate 1 via a stick from the vibration generator. Therefore, an additional mass derived from the stick contributes to shift the natural frequency to the lower frequency region, not avoided in this experiment.

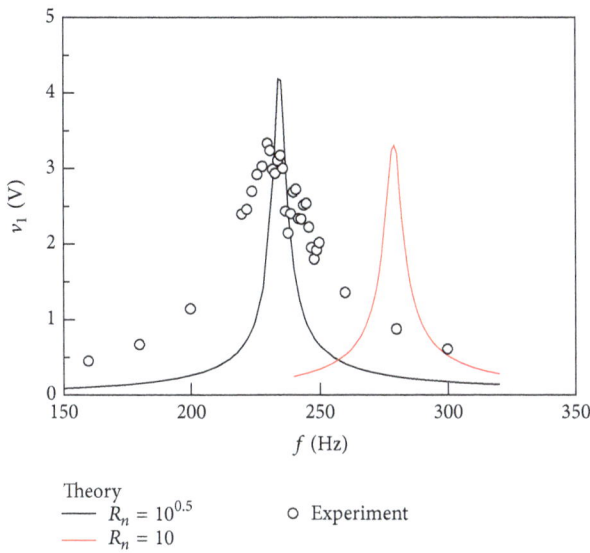

FIGURE 4: Voltage with electricity generation as function of excitation frequency.

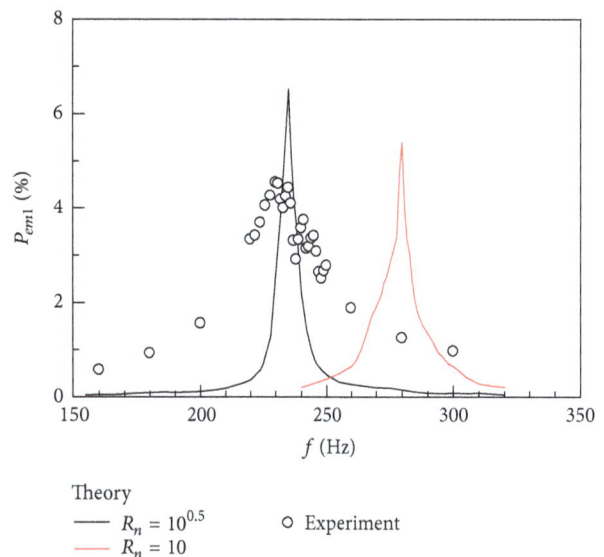

FIGURE 5: Relationship between electric power caused by electricity generation and mechanical power supplied from vibrator.

Moreover, because the shift in the natural frequency cannot be explained only by such an additional mass, characteristics of the vibration generator are also considered as other factors. The flexural deflection of plate 1 excited by the vibration generator is suppressed by the generator characteristics and the excitation position in comparison with the free vibration. In general, a vibration system receiving damping force becomes lower in the natural frequency than a free vibration system. Because of those, we suppose that such a shift in the natural frequency is caused by some factors.

Since it is difficult to theoretically express the actual situation at the moment, a theoretical consideration is attempted by shifting the natural frequency of plate 1 to the experimental peak frequency owing to substituting $10^{0.5}$ for R_n. By dealing with such an attempt, its behavior is similar to the experimental a_1. Figure 4 shows the voltage v_1 with electricity generation as functions of f. The theoretical v_1 has peaks at $f = 234$ and 278 Hz that are derived from $R_n = 10^{0.5}$ and 10^1, respectively, and the experimental v_1 is maximized around $f = 230$ Hz. These correspond to a_1 very well.

In electricity generation by means of the plate vibration with piezoelectric elements, coupling between the structural vibration and electric field (i.e., electromechanical coupling) should be considered to relate the in-plane stress to the applied electric field. The electricity generation characteristics are considered by the electric power P_{e1} via the piezoelectric element and the mechanical power P_m supplied to plate 1 by the vibrator. Figure 5 shows the relationship between P_{e1} and P_m at $f = 234$ and 280 Hz of $R_b = 10^{0.5}$. The experimental results corresponding to those are also shown. In this figure, both axes are expressed logarithmically, because P_{e1} at $f = 280$ Hz is considerably smaller than that of $f = 234$ Hz even if we can make sure of it from Figure 4 and is hardly exhibited together with that at $f = 234$ Hz on the ordinary axes. With respect to the relationship between P_{e1} and P_m, linear changes mean that P_{e1} is directly proportional to P_m,

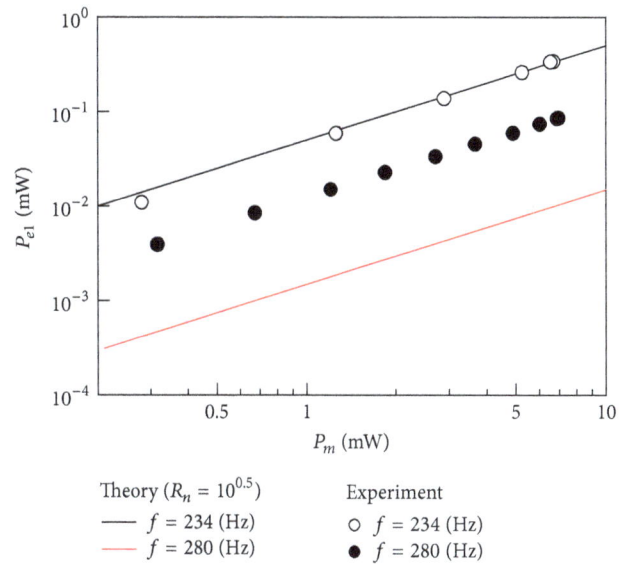

FIGURE 6: Variations in electricity generation efficiency with excitation frequency.

and the difference of P_{e1} in the changes indicates that of the rate of changes in P_{e1} for P_m; that is, the difference is identical to that of inclinations of changes in P_{e1} for P_m on the ordinary axes.

The theoretical and experimental P_{e1}, which is derived from the peak voltages in Figure 4, increases similarly to P_m, while the theoretical P_{e1} at $f = 280$ Hz is suppressed in comparison with that at $f = 234$ Hz, and the difference from the experimental P_{e1} at $f = 280$ Hz is expanded. The additional mass of the stick affects not only the natural frequency but also the mechanical power supplied to the plate 1 because a force derived from the additional mass is added to the point force, so that P_{e1} also increases in response to P_m.

Since the effect is remarkable relatively at f = 280 Hz that is far away from the peak frequency, we consider that these experimental results are somewhat overestimated.

Figure 6 shows the electricity generation efficiency P_{em1} as functions of f and also the comparison between theoretical and experimental results in Figures 3 and 4. The theoretical P_{em1} of R_n = $10^{0.5}$ and 10^1 peaks at f = 234 and 280 Hz, respectively, and decreases as the natural frequency increases. The experimental P_{em1} varies similarly to the above P_{em1} of R_n = $10^{0.5}$, with the result that it demonstrates the theoretical consideration and that such an electricity generation system does not function effectively in any frequency range except the natural frequency region.

4.2. Electricity Generation Characteristics Using Mechanical-Acoustic Coupling.
In this section, the cylinder and plate 2 are added to the above theoretical and experimental models which were introduced in the previous section; that is, they are models as shown in Figures 1 and 2(a). In the analytical model, the dimensions and so on of plate 2 are identical to those of plate 1, the support condition adopts T_n = 10^8 and R_n = 10^1 in both plates, and the point force F is set to 1 N. The cylinder has the same radius as that of plates 1 and 2, and its length L ranges from 100 to 2000 mm.

Figure 7 shows the sound pressure level L_{pv}, which is averaged over the entire volume of the cavity and is maximized at each L when h_c and f are set to 3 mm and 280 Hz, respectively, and the phase α_2 ranges from 0 to 180° as functions of L. The theoretical level L_{pv} peaks at 610, 1230, and 1840 mm. The peaks are caused by the promotion of mechanical-acoustic coupling between the plate vibration and acoustic modes. Then the acoustic modes are the $(0, 0, 1)$, $(0, 0, 2)$, and $(0, 0, 3)$ modes whose plane modal shape is similar to that of plate vibration mode $(0, 0)$. To validate these theoretical results, the sound pressure levels L_{p1} and L_{p2}, which are measured near plates 1 and 2, are also indicated. The experimental peaks occur around the lengths where L_{pv} peaks, whereas L_{p1} decreases remarkably around L = 950 and 1600 mm in the process of shifting acoustic modes because of changing L.

To consider the effect of mechanical-acoustic coupling on electricity generation characteristics, Figure 8 shows the experimental accelerations a_1 and a_2 of plates 1 and 2 as functions of L under the above coupling. Since plate 1 is excited at f = 280 Hz via the stick connected with the vibrator, a_1 remains low and almost constant over the entire range of L, having small increases at L = 620, 1250, and 1880 mm. These results can be also predicted from those in Figure 3. On the other hand, the cylindrical sound field connecting with plate 1 is formed via mechanical-acoustic coupling, and plate 2 is excited by the sound field without the point force. As a result, a_2 of plate 2 has specific peaks at L = 620, 1250, and 1880 mm, suppressed in the other ranges of L, because the natural frequency of plate 2 becomes 280 Hz. Figure 9 shows the voltages v_1 and v_2 based on the electricity generations of plates 1 and 2 as functions of L. v_1 and v_2 are directly proportional to a_1 and a_2, respectively, depending on the in-plane strain of the plate that is determined by the

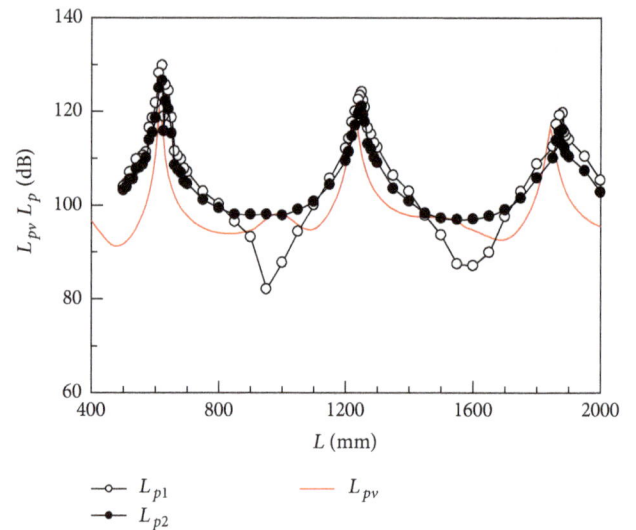

FIGURE 7: Sound pressure level inside cavity as function of cylinder length.

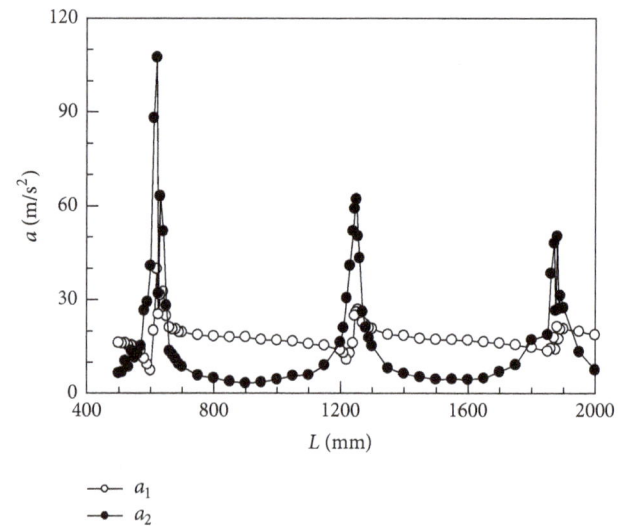

FIGURE 8: Acceleration at both plates as function of cylinder length.

out-of-plane deflection of the plate. Therefore, v_1 and v_2 have a tendency that is similar to that of a_1 and a_2; in particular, v_2 is much larger than v_1 at L = 620 mm, at which coupling with the $(0, 0, 1)$ mode is promoted, as well as the behavior of a_1 and a_2.

The electricity generation efficiencies P_{em1}, P_{em2}, and P_{em} are calculated from the relationship between electric power via the piezoelectric element and mechanical power supplied to plate 1. They are shown with changing L in Figure 10. They are derived from the above results so that P_{em1} of plate 1 remains almost constant over the entire range of L, and P_{em2} of plate 2 increases greatly at L = 620, 1250, and 1880 mm. It is natural that P_{em} peaks at those lengths and remains almost constant in the other ranges of L, calculated from both electric power values of plates 1 and 2. Because

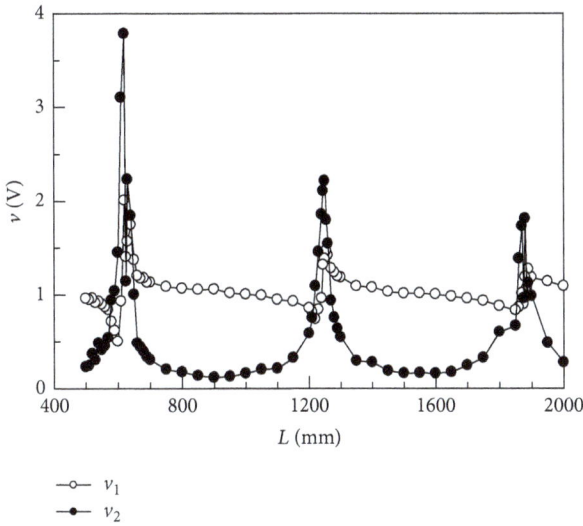

FIGURE 9: Voltages with respective electricity generations at both plates as function of cylinder length.

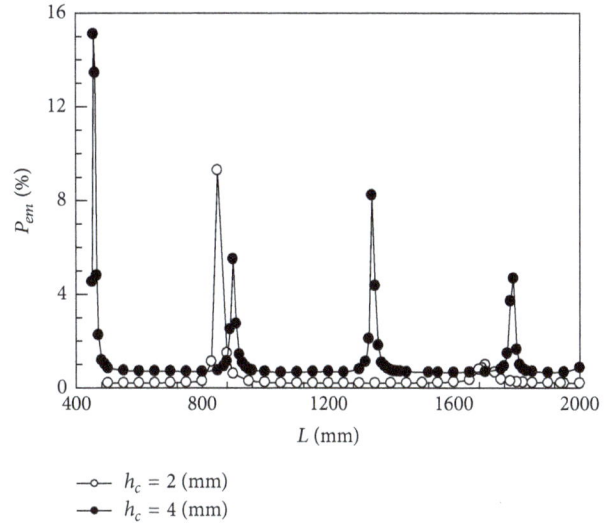

FIGURE 11: Comparison between electricity generation efficiencies when plate thicknesses are 2 and 4 mm.

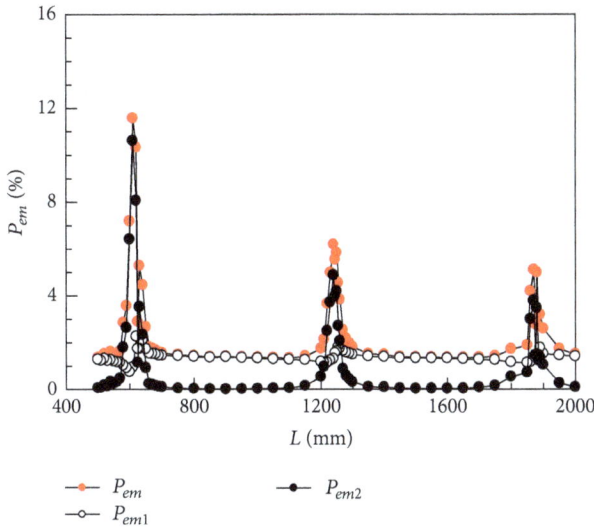

FIGURE 10: Variations in electricity generation efficiency with cylinder length.

the above characteristics of the electricity generation efficiency are affected by mechanical-acoustic coupling, whose characteristics are influenced by the relationship between the natural frequencies of the plate vibrations and internal sound field, the other thicknesses are also employed in this study. Figure 11 shows P_{em} of $h_c = 2$ and 4 mm, whose respective natural frequencies are 190 and 390 Hz, as functions of L. P_{em} reaches the respective maximum values caused by coupling with the $(0, 0, 1)$ mode at $L = 850$ and 455 mm when $h_c = 2$ and 4 mm. The promotion of mechanical-acoustic coupling has been confirmed by theoretical and experimental acoustic characteristics.

4.3. Variation in Electricity Generation Efficiency with Plate Thickness. As described in the previous section, the electricity generation characteristics are strongly affected by the shift

in the natural frequency of plates 1 and 2 as their thicknesses change. Before discussing the effect of mechanical-acoustic coupling on the electricity generation efficiency, we must make sure of the efficiency P_{em1} under only the plate vibration without coupling.

Figure 12 shows, theoretically and experimentally, such an efficiency as functions of the thickness h_c. In the theoretical results, $T_n = 10^8$ and $R_n = 10^1$ are adopted as the support condition. Since the electric power caused by the electricity generation and the mechanical power supplied to the plate (i.e., the input and output power) depend on those of the flexural displacement and vibration velocity, decreases in the flexural rigidity from thinning the plate and increases in the natural frequency from thickening the plate are factors to increase both powers. In general, the flexural displacement is inversely proportional to h_c^3, and the natural frequency is directly proportional to h_c, so that P_{em1} is inversely proportional to h_c^2 if the point force is constant in the entire range of h_c.

The same figure also indicates the total P_{em} with mechanical-acoustic coupling. P_{em} increases greatly with h_c up to approximately $h_c = 4$ mm, so that the efficiency exceeds that in the electricity generation without coupling and the difference between both efficiencies is expanded with increasing h_c. On the other hand, the theoretical P_{em} decreases with increasing h_c in subsequent variations. The maximum value reaches approximately 14%, which appears at approximately $h_c = 4$ mm. These results are derived from specific conditions in this study such as the dimensions, mechanical, and electric properties, and so on of the plate and the piezoelectric element and its electrode, so that we cannot conclude that those results are generalized. However, it is remarkable that the experimental results demonstrate the validity of the theoretical analysis and verify the usefulness of this electricity generation system.

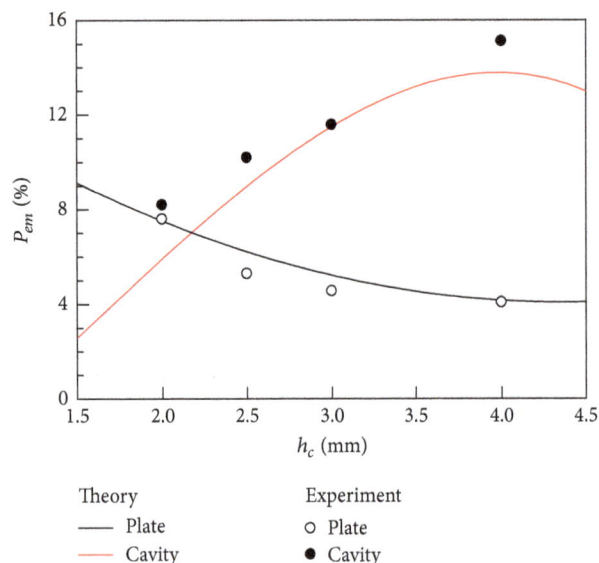

FIGURE 12: Variations in electricity generation efficiency with plate thickness.

5. Conclusion

We improved the electricity generation characteristics derived from the vibration of a plate with a piezoelectric element installed at its center. The cylinder had such plates at both ends. One side of each plate was subjected to a harmonic point force. We used a mechanical-acoustic coupling between the plate vibrations and internal sound field. In this study, the effect of mechanical-acoustic coupling on energy-harvesting was estimated theoretically and experimentally from the electricity generation efficiency. In particular, electromechanical-acoustic coupling was taken into consideration in the theoretical study.

For electricity generation caused only by the plate vibration without mechanical-acoustic coupling, the electric power and electricity generation efficiency are maximized at the natural frequency of the generation system and significantly exceed those in the frequency range except for the natural frequency region. However, the natural frequency of the system, in which the stick to apply the mechanical power to the plate is contained, is somewhat lower than that of the plate.

With respect to the electricity generation system using mechanical-acoustic coupling, it is confirmed experimentally that the electric power of the nonexcitation side increases significantly in comparison with that of the excitation side, in which the plate is excited by the point force of its natural frequency. The theoretical and experimental studies reveal that the electricity generation efficiency is improved by using mechanical-acoustic coupling in comparison with that of the electricity generation caused only by the plate vibration without coupling.

Competing Interests

The authors declare that they have no competing interests.

Acknowledgments

This work was partially supported by JSPS KAKENHI Grant no. 15K05874.

References

[1] S. R. Anton and H. A. Sodano, "A review of power harvesting using piezoelectric materials (2003–2006)," *Smart Materials and Structures*, vol. 16, no. 3, pp. R1–R21, 2007.

[2] K. Yamada, H. Matsuhisa, and H. Utsuno, "Improvement of efficiency of piezoelectric element attached to beam based on mechanical impedance matching," *Journal of Sound and Vibration*, vol. 333, no. 1, pp. 52–79, 2014.

[3] M. Piñeirua, O. Doaré, and S. Michelin, "Influence and optimization of the electrodes position in a piezoelectric energy harvesting flag," *Journal of Sound and Vibration*, vol. 346, no. 1, pp. 200–215, 2015.

[4] J. Pan and D. A. Bies, "The effect of fluid-structural coupling on sound waves in an enclosure—theoretical part," *Journal of the Acoustical Society of America*, vol. 87, no. 2, pp. 691–707, 1990.

[5] J. Pan and D. A. Bies, "The effect of fluid-structural coupling on sound waves in an enclosure—experimental part," *Journal of the Acoustical Society of America*, vol. 87, no. 2, pp. 708–717, 1990.

[6] L. Cheng and J. Nicolas, "Radiation of sound into a cylindrical enclosure from a point-driven end plate with general boundary conditions," *Journal of the Acoustical Society of America*, vol. 91, no. 3, pp. 1504–1513, 1992.

[7] L. Cheng, "Fluid-structural coupling of a plate-ended cylindrical shell: vibration and internal sound field," *Journal of Sound and Vibration*, vol. 174, no. 5, pp. 641–654, 1994.

[8] H. Moriyama and Y. Tabei, "Acoustic characteristics inside cylindrical structure with end plates excited at different frequencies," *Journal of Visualization*, vol. 7, no. 1, pp. 93–101, 2004.

[9] H. Moriyama, Y. Tabei, and N. Masuda, "Acoustic characteristics of a sound field inside a cylindrical structure with excited end plates: influence of excitation position on vibro-acoustic coupling," *Acoustical Science and Technology*, vol. 26, no. 6, pp. 477–485, 2005.

[10] H. Moriyama, H. Tsuchiya, and Y. Oshinoya, "Energy harvesting with piezoelectric element using vibroacoustic coupling phenomenon," *Advances in Acoustics and Vibration*, vol. 2013, Article ID 126035, 11 pages, 2013.

[11] A. Kojima, H. Moriyama, and Y. Oshinoya, "Vibroacoustic coupling of cylindrical enclosure with excited end plates," *Acoustical Science and Technology*, vol. 33, no. 3, pp. 180–189, 2012.

[12] W. Larbi, J.-F. Deü, and R. Ohayon, "Finite element formulation of smart piezoelectric composite plates coupled with acoustic fluid," *Composite Structures*, vol. 94, no. 2, pp. 501–509, 2012.

[13] W. Larbi, J.-F. Deü, R. Ohayon, and R. Sampaio, "Coupled FEM/BEM for control of noise radiation and sound transmission using piezoelectric shunt damping," *Applied Acoustics*, vol. 86, pp. 146–153, 2014.

[14] P. M. Morse and K. U. Ingard, *Theoretical Acoustics*, McGraw-Hill, New York, NY, USA, 1968.

3

Corn Husk Fiber-Polyester Composites as Sound Absorber: Nonacoustical and Acoustical Properties

Nasmi Herlina Sari,[1] **I. N. G Wardana,**[2] **Yudy Surya Irawan,**[2] **and Eko Siswanto**[2]

[1]*Department of Mechanical Engineering, Faculty of Engineering, Mataram University, Nusa Tenggara Barat, Indonesia*
[2]*Department of Mechanical Engineering, Faculty of Engineering, Brawijaya University, East Java, Indonesia*

Correspondence should be addressed to Nasmi Herlina Sari; n.herlinasari@unram.ac.id

Academic Editor: Marc Thomas

This study investigates the acoustical and nonacoustical properties of composites using corn husk fiber (CHF) and unsaturated polyester as the sound-absorbing materials. The influence of the volume fraction of CHF on acoustic performance was experimentally investigated. In addition, the nonacoustical properties, such as air-flow resistivity, porosity, and mechanical properties of composites have been analyzed. The results show that the sound absorptions at low frequencies are determined by the number of lumens in fiber, particularly the absorption coefficient, which increases the amount of fiber. For high-frequency sound, the absorption coefficient is determined by the arrangement of fibers in the composite. An absorption coefficient is close to zero when the fibers are arranged in a conventional pattern; however, when they are arranged in a random pattern, a high absorption coefficient can be obtained. The bond interface between the fiber and resin enhances its mechanical properties, which increases the longevity of the composite panel.

1. Introduction

Noise pollution and waste management are two problems that need to be solved in modern societies. The use of newly developed alternative materials to absorb the noise considerably minimizes these problems. Hence, the inexpensive, easily created, thin, and lightweight composite materials that can absorb sound waves in broader frequency fields are highly desirable.

The fibrous sound-absorbing materials have been extensively investigated [1–5]. Biot studies [1, 2] provide an approach for the propagation of elastic waves in the fluid medium-saturated porous material at high and low frequencies, where factors such as pore geometry, fluid, and medium having comparable densities are required to be considered. Delany and Bazley [3] developed a simple model for estimating the sound of absorption coefficients and characteristics of impedance of different types of fibrous absorbent materials. Lee and Chen [4] developed Acoustic Transmission Analysis (ATA) model to estimate the acoustic absorption of a multilayer absorbers. Attenborough [5] developed a model for

estimating the acoustical characteristics of fibrous absorbents soils and sands using flow resistivity formulae.

The polymer has been widely utilized as a matrix in fiber composites because it is easily formed from a material that has physical and acoustical properties [6–13]. Veerakumar and Selvakumar [6] studied acoustic properties for composite made from kapok fiber with polypropylene fiber, which were found to demonstrate good sound absorption behavior in the frequency range 250–2000 Hz. Jailani et al. [7] studied on panels made from coconut coir fibers which have been conducted to analyze compression effect of the panel on the acoustic performance. The coir fiber panel is a good sound absorber at 1.5–5 kHz. Zulkifli et al. [8] investigated the effect of the porous layer backing and a perforated panel on the sound absorption coefficient of coconut coir fiber. They indicated that increasing the thickness material of the panel will improve the sound absorption ability, especially in the low-frequency range at 600–2400 Hz. Chen et al. [9] studied the sound-absorbing properties of ramie fiber-reinforced polylactic acid materials. Putra et al. [10] studied the potential of waste fibers from paddy mixed with polyurethane as acoustic material and found that the absorption coefficient is greater than 0.5 from 1 kHz and can reach the average value of 0.8 above 2.5 kHz. Bastos et al. [11] developed vegetable fibers: coconut, palm, sisal, and acaı as sound-absorbing panels. Measurement scale reverberation chamber exposed promising results from acoustic performance for all panels. Flammability, odor, fungal growth, and aging tests have been performed on samples to identify their practical ability. Koizumi et al. [12] developed bamboo fiber as sound-absorbing material. They reported that the bamboo fiber material has equivalent acoustics properties with glass wool. Jayamani and Hamdan [13] studied sound absorption coefficient of urea-formaldehyde and polypropylene mixed with kenaf fiber. They reported that the kenaf fiber reinforced with polypropylene demonstrates higher sound absorption coefficients than kenaf fiber reinforced with urea-formaldehyde. These previous studies represented that a better understanding of the microstructure and physical parameters of a material could help in developing high-performance acoustic materials.

This study primarily investigates the effect of adding corn husk fibers (CHFs) on acoustical and nonacoustics properties of polyester composites. In addition, the effects of fiber content on the tensile properties and microstructures via SEM have been analyzed. The results of this study could contribute to engineering applications, especially as sound absorbers.

2. Materials and Methods

2.1. Materials and Sample Preparation. CHF is the main raw material used in this study. The fiber contains 46.15% cellulose, 33.79% hemicellulose, and 3.92% lignin. It has been treated with 5% sodium hydroxide (NaOH) for 2 h. The scheme of reaction is given as follows:

$$CHF - OH + NaOH \longrightarrow CHF - O\text{-}Na^+ + H_2O \quad (1)$$

Chemical reactions have been removing impurities on the fiber surface. The CHF was rinsed five times with mineral water in other to remove NaOH sticking from the fiber surfaces. They were dried in natural sunlight to remove any residual moisture and were then preserved in a dry box with a humidity of 40%. The chemical contents of treated CHF are 54.37% cellulose, 22.37% hemicellulose, and 5.64% lignin. The average of diameter of a single CHF is 0.133 ± 0.03 mm, measured by a Mitutoyo digital micrometer.

The unsaturated polyester resin 2250 BW-EX has a viscosity of 6–8 poise (25°C), the tensile strength of 8.8 Kg/mm², a tensile modulus of 500 Kg/mm², the flexural strength of 2.5 Kg/mm², and elongation of 2.3%.

The weight of polyester resin and CHFs were measured before processing so as to determine the volume fraction of CHFs and polyester in the resulting composite. The composition of different sound absorbers is summarized in Table 1. The mixtures were hot pressed at 100°C and 0.3 MPa for 4 min into a round mold with a diameter of 32 mm, followed by cooling to room temperature at 5 MPa to obtain a round shape to fit in the impedance tube during the sound

TABLE 1: The composition of the composite (mean values in volume fraction).

Sample	CHF (%)	Polyester resin (%)
PF-E	20	80
PF-G	40	60
PF-H	50	50
PF-I	60	40
PF-K	70	30
PF-M	80	20

absorption test. All the absorber materials were obtained with a diameter of 29 mm and thickness of 20 mm. Six different samples were used for acoustical and porosity tests.

2.2. Porosity. The connected porosity of composite sample was nonacoustically measured using the method of water saturation used by Vašina et al. [14]. All the samples were dried at 105°C for 24 h. Subsequently, they were weighed before being left in a vacuum vessel to saturate under water; the density of water is $\rho_w = 1000$ Kg·m^{-3}. After 24 h, they were carefully removed and weighed again. The porosity was computed using $\varepsilon = V_a/V_s$, where V_a is the volume of the composite occupied by the water and V_s is the total volume of the composite. The volume of water can be computed using $V_w = (m_2 - m_1)/\rho_w$, where m_2 and m_1 are the wet and the dry masses of the composite (Kg), respectively.

2.3. Air-Flow Resistivity. There are several empirical and semiempirical equations in the literature that can be used to estimate the flow resistivity of absorber materials based upon fiber radius and material porosity or the bulk density of the materials [14–16]. The air-flow resistivity of the samples used in this study is expressed in [16]

$$R = \frac{6.8\mu (1 - \varepsilon)^{1.296}}{a^2 \varepsilon^3}, \quad (2)$$

where μ is the viscosity of air (1.84×10^{-5} Pa·s), ε is the porosity, and a is the radius of the fiber.

2.4. Tortuosity. The following empirical formula was used to calculate tortuosity (σ) in terms of porosity. The tortuosity is expressed in [5]

$$\sigma = 1 + \frac{(1 - \varepsilon)}{(2\varepsilon)}. \quad (3)$$

2.5. Sound Absorption Measurement. The acoustic properties of the composite sample were measured using a two-microphone transfer-function method, according to ASTM E-1050-98/ISO 10534-2 standards. The testing apparatus was part of complete acoustic material testing system Brüel & Kjaer (type 4206, Brüel & Kjær), as it is seen in Figure 1. A small tube setup was employed to measure different acoustical parameters in the frequency range of 100 Hz–6.4 kHz. At one end of the tube, a loudspeaker was situated

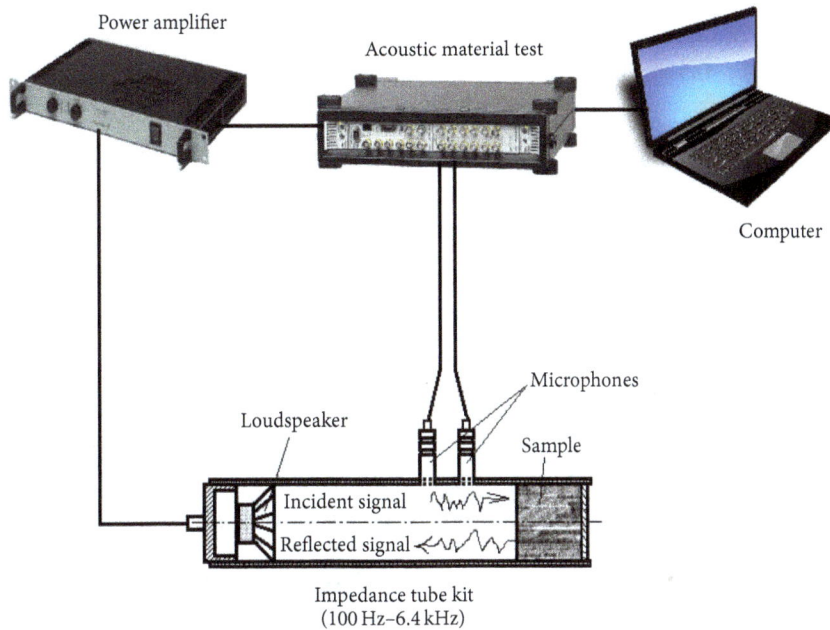

FIGURE 1: Impedance tube kit (type 4206, Brüel & Kjær).

to act as a sound source and the test material was placed at the other end to measure sound absorption properties. Two acoustic microphones (type 4187, Brüel & Kjær) were located in front of the sample to record the incident sound from the loudspeaker and the reflected sound from the material. The recorded signals in the analyzer in terms of the transfer function between the microphones were processed using Brüel & Kjær material testing software to obtain the absorption coefficient of the sample under test. Each set of the experiment was repeated three times in order to have average measurements.

2.6. Mechanical Properties. The tensile and Young's modulus were determined using a Tensilon RTG-1310 universal testing machine with a load cell of 10 kN. All the samples of composites were tested after conditioning for 24 h in a standard testing atmosphere of 70% relative humidity and 28°C. According to the ASTM D 3039 standard, a gauge length of 150 mm and a crosshead speed of 5 mm/min were used for tensile testing. The sample size was 250 mm × 25.4 mm × 6 mm. In total 21 samples were tested for each sample condition and the average and standard deviation values were reported.

2.7. Scanning Electron Microscope. The surface morphologies of composites were observed using an Inspect-S50 scanning electron microscope with field emission gun. An accelerating voltage of 10 kV was used to collect SEM images on the surface of the sample. The morphologies of the composites were observed and analyzed via SEM at room temperature. Before testing, the samples were sliced and mounted onto SEM stubs using double-sided adhesive tape. They were gold sputtered for 5 min to a thickness of approximately 10 nm under pressure of 0.1 torr and 18 mA current to make the sample conductive. SEM micrographs were recorded at different magnifications to ensure clear images.

3. Results and Discussion

3.1. Nonacoustic Composites Properties. Large differences were observed in nonacoustical properties of the composite samples, because of their different microstructures as a result of the addition of the CHF in the polyester. This diversity is very interesting because it provides different porous microstructures and consequently different acoustic properties. Porosity, tortuosity, and flow resistivity values are listed in Table 2.

Increasing the amount of fiber volume fraction in the polyester resin increases the porosity and decreases both tortuosity and air-flow resistivity in the absorbent material (seen Table 2). The porosity value of the sample PF-M was 0.8267, which was higher than those of the other samples used in this study. The presence of lumen in the fiber indicates that the porosity of the sample increases when the number of fibers increases (Figure 2). In other words, the lower value of porosity and higher value of the flow resistivity of the sample can be attributed to the higher volume fraction of polyester resin.

All the composite samples demonstrate an open pore structure wherein the pores are interconnected. This is one of the most important factors for noise absorption because such a structure decreases air-flow resistivity and thus the dissipation of the wave energy in the pores. In these samples, the multiscale fiber structure with the *lumen* inside fiber bundle has a pore shape, and the pore size can differ by several orders of magnitude (Figures 2(a) and 2(b)).

Table 2: Nonacoustical properties of samples.

Sample	Thickness (mm)	Density (Kg·m^3)	Porosity ε	Air-flow resistivity, R (Pa·s·m^{-2})	Tortuosity, σ
PF-E	20	640.5	0.6474	44,980	1.272
PF-G	20	383.4	0.7053	29,353	1.208
PF-H	20	304.1	0.7247	25,424	1.190
PF-I	20	244.1	0.7457	21,576	1.171
PF-K	20	198.0	0.7582	19,568	1.160
PF-M	20	158.3	0.7954	14,435	1.128

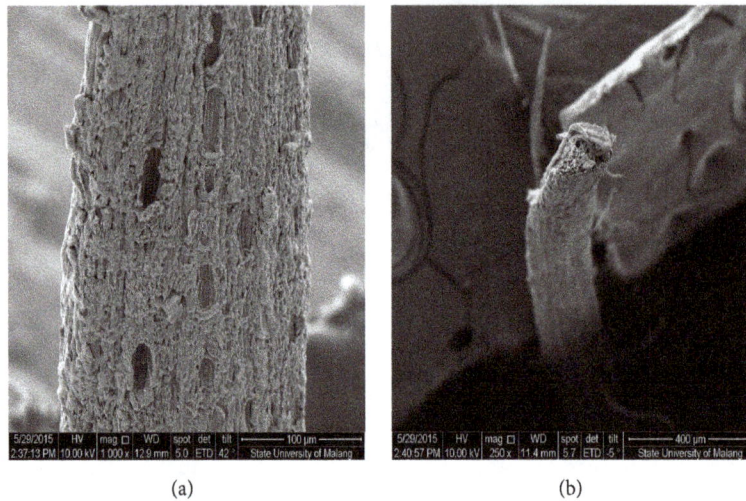

(a)   (b)

Figure 2: SEM photomicrographs of corn husk fibers 5% NaOH treated: (a) surface and (b) cross-sectional features.

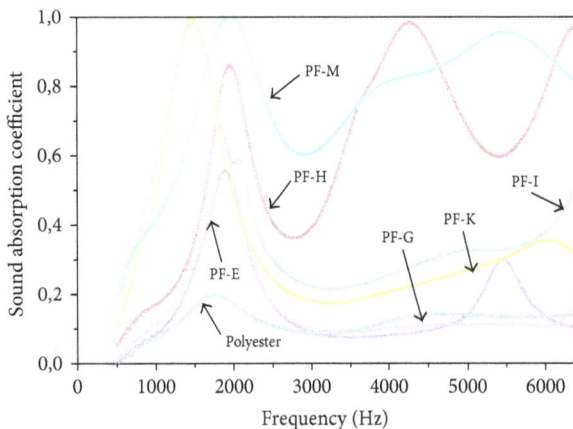

Figure 3: The sound absorption coefficients of composite samples.

3.2. *Acoustical Properties.* The normal sound absorption properties for all samples of CHF-polyester composites are graphically illustrated in Figure 3. The zero value in the y-axis indicates perfect sound reflection, and the value of one implies complete sound absorption. These results show that all composite samples demonstrated an increase in sound absorption coefficient in the range of 1 kHz–2.5 kHz. This is because lumen inside the fiber bundle increases the amount

of fiber, which results in high absorption coefficient. The additional thermal energy is dissipated more rapidly due to the increased frictional surface area. The sound absorption coefficient of the PF-M sample is therefore correspondingly higher than those of the other samples. The sound waves propagate vibration energy through the air spaces in the individual lumina inside the fiber. A part of this sound energy is converted into heat in the lumina, which is then absorbed by the surrounding walls. The larger the air cavities and lumina inside the fiber, the longer the wavelength of the sound that is absorbed, so more dominant at low frequencies. SEM micrograph analysis (Figures 6(a), 6(c), 6(e), 6(g), 6(i), and 6(k)) illustrates that there are many *lumens* inside the fiber and continuous channels in the porous structure of polyester composites.

At frequencies above 2 kHz, the sound absorption capability of PF-E, PF-G, PF-I, and PF-K samples decreases. The decrease caused by the interface of the fiber/resin and orderly fiber arrangement that cause the higher value of the flow resistivity of the sample makes movements of the sound difficult to pass through the samples. An absorption coefficient is close to zero when the fibers are arranged in a conventional pattern. SEM micrographs (Figure 6) illustrate that there is a distinct interface between fibers and resin in all samples. Interface surfaces between fibers and resin of PF-E, PF-G, PF-I, and PF-K samples (Figures 6(b), 6(d), 6(h),

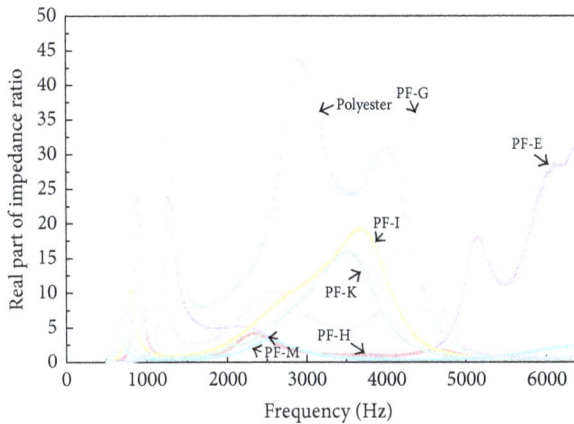

FIGURE 4: The real part of the impedance ratio of different samples.

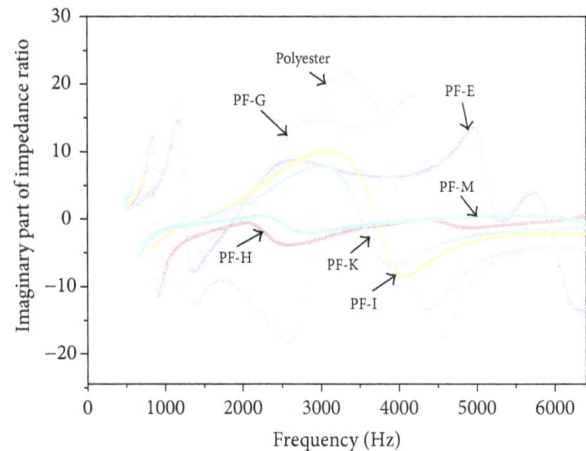

FIGURE 5: The imaginary part of the impedance ratio of samples.

and 6(j), resp.) are quite dense and contain orderly fiber bundles arrangement. Interface strength not only influences composite mechanical property but also influences sound absorption quality. When sufficient amount of resin is used, the interface area between the fibers and the resin is smooth and strong (Figures 6(a) and 6(b)). When the incident sound waves are continuously transmitted onto a composite interface, the sound waves will be reflected and refracted and the acoustic damping phenomena will consume a small amount of energy, reducing heat losses and thus obtaining a lower absorption coefficient at high frequencies. This would also explain why the sound absorption of PF-E is lower than those of sample patterns of composites which are similar.

Sample pure polyester resin (PE) had the absorption coefficient under 0.2. Although polyester may be a valuable option in noise absorption applications, these results discourage its use as a sound-absorbing material.

Figure 3 also shows that the PF-H and PF-M samples demonstrated the ability to absorb sound at high frequencies above 4 kHz. This is due to the random distribution of the fiber. The random distribution of the fibers in the fibrous absorbent materials allows the sound waves to hit the lumen of the fiber bundle and strengthen the sound absorption effect; a high absorption coefficient can be obtained. SEM micrographs (Figures 6(e), 6(f), 6(k), and 6(l)) illustrate the random distribution of the fibers in PF-H and PF-M samples.

Figures 4 and 5 show that the real part is the resistance associated with energy losses and the imaginary part is the reactance associated with phase changes, respectively. In this case, we can observe a better performance of PF-H and PF-M samples than other materials studied. Increasing the amount of fiber reduces the number of impedance values and material stiffness. The reduced impedance values increase the fraction of wave energy that can be transmitted through the sample.

Furthermore, sound absorption at lower frequencies (over 1.0–2 kHz) is desirable for automotive applications because of this frequency range according to noise from the wind, engine running, tires, road, and conversation, thereby making CHF-polyester composites a promising candidate for automotive interior sound absorption.

3.3. Mechanical Properties. Theoretically, there should be an interaction between hydrophobic polyester and hydrophilic cellulose. The disappearance of the noncellulose material on the surface of the fiber enables surface interaction with the polyester matrix. The void fraction is mainly formed because the composites have not been consolidated (not sufficiently pressed to form a contiguous solid structure) in order to manufacture composites.

Figures 7 and 8 show that the increase in the fiber volume fraction leads to increase in the tensile strength value and Young's modulus of the composite from 18.81 ± 8.5 to 25.73 ± 3.19 MPa. The increase in mechanical strength can be attributed to the bond interface between the fibers and resin. The mechanical properties of the PF-M sample (or composite sample with 20% resin and 80% CHF) are therefore correspondingly higher than those of the other samples.

For PF-H sample, there was a 12.53% decrease in the tensile strength values, with a strength value of 20.40 ± 1.1 MPa. The probability of the overlapping of multiple CHF in composite samples thereby causes the weaker transference of load between fiber and matrix occurring due to the poor interfacial adhesion causing lowering in the mechanical properties. However, the value of the modulus of elasticity of the sample PF-H is higher than that of the material used in this study, contributing to the sound absorption of the sample.

The tensile strength value of the PF-E sample is the lower compared to other samples. This is due to the fiber volume fraction less than the other samples. The tensile strength of the fiber of 237.43 MPa is higher than the tensile strength of the resin.

4. Conclusions

In this study, a CHF-polyester sound absorber was proposed and the sound absorption capability of the material was significantly enhanced through the simple method. The presence of a number of lumen structures in the fiber bundle facilitates sound absorption at low frequencies in the range

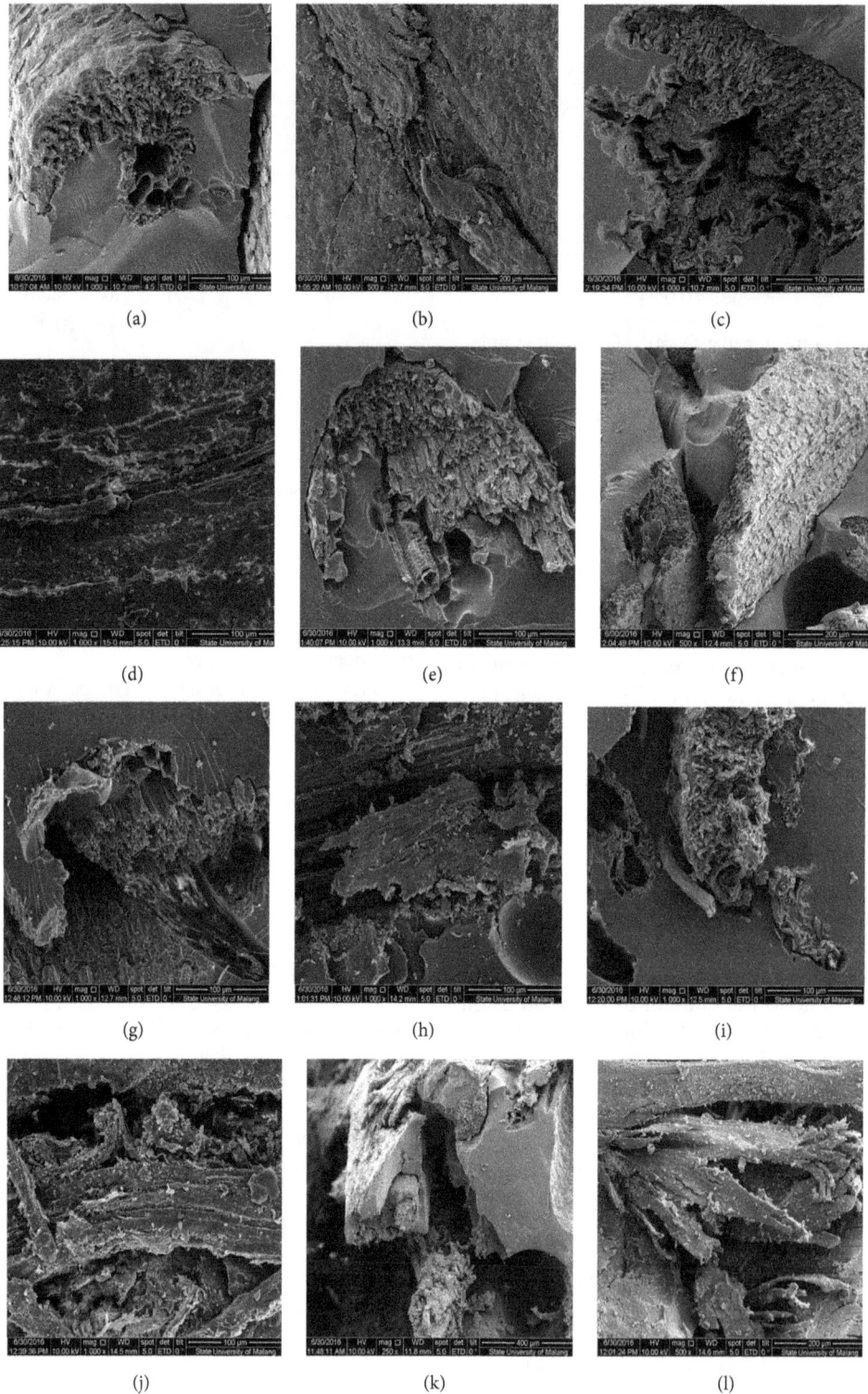

FIGURE 6: Scanning electron microscope (SEM) images of surfaces of the samples and cross-sectional features of composite samples. (a, b) PF-E, (c, d) PF-G, (e, f) PF-H, (g, h) PF-I, (i, j) PF-K, and (k, l) PF-M.

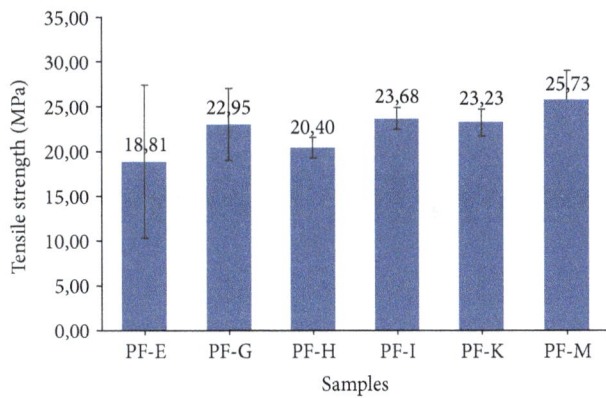

FIGURE 7: Tensile strength of each sample.

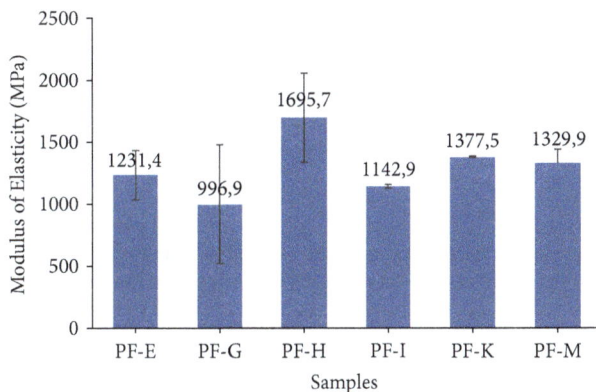

FIGURE 8: Modulus of elasticity of each sample.

of 1 kHz–2 kHz. The interface between the surface of the fiber/resin and orderly arrangement of fibers within the resin of PF-E, PF-G, PF-I, and PF-K samples caused a decrease in the sound absorption properties at frequencies above 2 kHz. High frequencies above 4 kHz (PF-H and P F-M samples) are obtained due to the random distribution of the fiber.

Increased resin lowers friction between the fibers, reducing heat losses and subsequently its sound absorption coefficient.

All samples used in this study have the potential to be used as sound-absorbing materials. These results indicate that alternative high-performance sound-absorbing materials could be obtained using CHF, which can solve environmental problems and reduce noise pollution.

Competing Interests

The authors declare that there is no conflict of interests regarding the publication of this paper.

References

[1] M. A. Biot, "Theory of propagation of elastic waves in a fluid-saturated porous solid. I. Low-frequency range," *The Journal of the Acoustical Society of America*, vol. 28, no. 2, pp. 168–178, 1956.

[2] M. A. Biot, "Theory of propagation of elastic waves in a fluid-saturated porous solid. II. Higher frequency range," *The Journal of the Acoustical Society of America*, vol. 28, no. 2, pp. 179–191, 1956.

[3] M. E. Delany and E. N. Bazley, "Acoustical properties of fibrous absorbent materials," *Applied Acoustics*, vol. 3, no. 2, pp. 105–116, 1969.

[4] F.-C. Lee and W.-H. Chen, "Acoustic transmission analysis of multi-layer absorbers," *Journal of Sound and Vibration*, vol. 248, no. 4, pp. 621–634, 2001.

[5] K. Attenborough, "Acoustical characteristics of rigid fibrous absorbents and granular materials," *The Journal of the Acoustical Society of America*, vol. 73, no. 3, pp. 785–799, 1983.

[6] A. Veerakumar and N. Selvakumar, "A preliminary investigation on kapok/polypropylene nonwoven composite for sound absorption," *Indian Journal of Fibre and Textile Research*, vol. 37, no. 4, pp. 385–388, 2012.

[7] M. Jailani, M. Nor, and R. Zulkifli, "Effect of compression on the acoustic absorption of coir fiber," *America Journal of Applied Sciences*, vol. 7, no. 9, pp. 1285–1290, 2010.

[8] R. Zulkifli, Zulkarnain, and M. J. M. Nor, "Noise control using coconut coir fiber sound absorber with porous layer backing and perforated panel," *American Journal of Applied Sciences*, vol. 7, no. 2, pp. 260–264, 2010.

[9] D. Chen, J. Li, and J. Ren, "Study on sound absorption property of ramie fiber reinforced poly(L-lactic acid) composites: morphology and properties," *Composites Part A: Applied Science and Manufacturing*, vol. 41, no. 8, pp. 1012–1018, 2010.

[10] A. Putra, Y. Abdullah, H. Efendy, W. M. F. W. Mohamad, and N. L. Salleh, "Biomass from paddy waste fibers as sustainable acoustic material," *Advances in Acoustics and Vibration*, vol. 2013, Article ID 605932, 7 pages, 2013.

[11] L. P. Bastos, G. D. S. V. De Melo, and N. S. Soeiro, "Panels manufactured from vegetable fibers: an alternative approach for controlling noises in indoor environments," *Advances in Acoustics and Vibration*, vol. 2012, Article ID 698737, 9 pages, 2012.

[12] T. Koizumi, N. Tsujiuchi, and A. Adachi, "The development of sound absorbing materials using natural bamboo fibers," *High Performance Structures and Materials*, vol. 4, pp. 157–166, 2002.

[13] E. Jayamani and S. Hamdan, "Sound absorption coefficients natural fibre reinforced composites," *Advanced Materials Research*, vol. 701, pp. 53–58, 2013.

[14] M. Vašina, D. C. Hughes, K. V. Horoshenkov, and L. Lapčík Jr., "The acoustical properties of consolidated expanded clay granulates," *Applied Acoustics*, vol. 67, no. 8, pp. 787–796, 2006.

[15] F. P. Mechel, *Formulas of Acoustics*, Springer, Berlin, Germany, 2nd edition, 2008.

[16] R. Maderuelo-Sanz, A. V. Nadal-Gisbert, J. E. Crespo-Amorós, and F. Parres-García, "A novel sound absorber with recycled fibers coming from end of life tires (ELTs)," *Applied Acoustics*, vol. 73, no. 4, pp. 402–408, 2012.

4

CAA of an Air-Cooling System for Electronic Devices

Sven Münsterjohann,[1] **Jens Grabinger,**[2] **Stefan Becker,**[1] **and Manfred Kaltenbacher**[3]

[1]*Friedrich-Alexander-Universität Erlangen-Nürnberg, Cauerstr. 4, 91058 Erlangen, Germany*
[2]*SIMetris GmbH, Am Weichselgarten 7, 91058 Erlangen, Germany*
[3]*Vienna University of Technology, Getreidemarkt 9, 1060 Wien, Austria*

Correspondence should be addressed to Sven Münsterjohann; sven.muensterjohann@fau.de

Academic Editor: Marc Thomas

This paper presents the workflow and the results of fluid dynamics and aeroacoustic simulations for an air-cooling system as used in electronic devices. The setup represents a generic electronic device with several electronic assemblies with forced convection cooling by two axial fans. The aeroacoustic performance is computed using a hybrid method. In a first step, two unsteady CFD simulations using the Unsteady Reynolds-Averaged Navier-Stokes simulation with Shear Stress Transport (URANS-SST) turbulence model and the Scale Adaptive Simulation with Shear Stress Transport (SAS-SST) models were performed. Based on the unsteady flow results, the acoustic source terms were calculated using Lighthill's acoustic analogy. Propagation of the flow-induced sound was computed using the Finite Element Method. Finally, the results of the acoustic simulation are compared with measurements and show good agreement.

1. Introduction

In today's world, the dimensions of electronic components are continuously decreasing. While many electronic devices are consuming less energy than a few years ago, the density of the chips assembled on a printed circuit board is increasing. This higher density leads to less heat dissipation by natural convection, which demands forced convection in order to prevent overheating. Convection can be forced by water or air cooling, with the latter using radial or axial fans to generate an air flow through the devices. As many of these air-cooled devices, measurement equipment or video game consoles, are operated by humans, their noise emission is an important attribute. On the one hand, the noise emission has to be low to please the users; on the other hand, these devices have to meet strict industrial regulations with a maximum allowable noise emission in the working environment. To meet these requirements, the prediction of the noise emission is an essential tool to avoid claims and restrictions. In addition to the experimental investigation of prototypes, the numerical estimation of the noise emission is a useful tool even before a first prototype is built. Acoustic measurements mostly provide an overall sound pressure level of the device without

the possibility of closer detection of the acoustic sources. An exception is the microphone array measurement, but the spatial resolution depends strongly on the frequency under investigation. Furthermore, nearly all electronic devices have housing that makes the localization of acoustic sources on the inside virtually impossible. Numerical simulations are not subject to these restrictions and thus provide a strong tool for acoustic optimizations.

In this paper, the numerical calculation of the noise emission of a generic electronic 19-inch slide-in device with two 120 mm axial fans is presented. The computation facilitated an aeroacoustic hybrid approach that uses the Finite Volume (FV) solver ANSYS CFX for the flow simulation and the in-house Finite Element (FE) solver CFS++ [1] for the acoustic simulation. The whole process starting from generation of the CAD model, CFD mesh generation, CFD setup, and solving to acoustic mesh generation and simulation of acoustic wave propagation is described. Two different turbulence models, URANS-SST and SAS-SST, will be compared with a focus on flow field quantities, that is, pressure and velocity, and sound pressure. While the URANS-SST model is only capable of resolving large scaled, triggered phenomena like vortex shedding behind a cylinder or rotor-stator interactions, the SAS-SST model can resolve the turbulence, given that temporal and spatial discretization are fine enough. This will especially improve the acoustic results in frequency ranges where broadband noise caused by turbulence is generated that cannot be resolved by the URANS model. Hence, the SAS-SST model is a good choice for capturing system-triggered as well as turbulence generated noise.

2. Aeroacoustics in Cooling Systems

Most work performed on the acoustics of cooling systems in electronic devices, that is, computers or notebooks, is done experimentally. Baugh [2] investigated the changes in hydrodynamic and acoustic behavior due to the inlet restrictions in notebooks. By scaling the hydrodynamic performance of fans to isoacoustic fan curves under the condition of inlet restrictions, Baugh found a way to compare the in-system performance and acoustic behavior of different fans if the pressure drop in the system is known. Nantais et al. [3] performed acoustic measurements on the cooling solutions for different graphics processing units and found that the noise to be emitted was dependent not only on the fan size and the cooling requirements, but also on the circumferential speed of the fan and the design of the cooling system. Huang's group carried out research on computer cooling fan noise. Huang [4] used analytic and empirical models to study the influence of rotor-stator interaction by decomposing this source into axial thrust, circumferential speed, and radial force using point force formulation. He investigated the influence of the number of blades and motor struts and the distribution of the source on the noise emission. Later, Huang and Wang [5] combined experimental investigations and an analysis of the rotor-stator interaction with the point source formulation. They found three main sources within the investigated four-strutted seven-bladed fan: the interaction between the blades and the struts, the additional size of one strut holding the electric wiring for the motor, and the incomplete bell mouth at the intake of the fan.

So far, most work has been performed by applying experimental and/or analytic approaches. Defoe and Novak [6] gave a brief review of methods for computational aeroacoustic (CAA) in electronic devices. They discussed different CAA methods: direct CAA, acoustic analogies, boundary element methods, and broadband methods. Comparing the results obtained by other authors with these different methods, the conclusion was drawn that the use of acoustic analogy is the best trade-off between computational effort and a detailed description of the acoustic sound field.

Using numerical methods our intention is to provide further insight into the sound generation in air-cooling systems of electronic devices. With a combination of CFD and CAA and modal analysis of the housing, a wide range of effects in the final acoustic spectra can be explained.

3. Theory

Various approaches have been proposed for both flow simulation and acoustic simulation, and the theories behind these approaches are explained in Sections 3.1 and 3.2.

3.1. Flow Simulation. The turbulence models applied in the flow simulation are the Shear Stress Transport (SST) turbulence model and the Scale Adaptive Simulation (SAS) developed by Menter [7]. As the acoustic source terms are directly derived from the flow field, sufficient resolution of the flow phenomena producing the acoustic sound emission is essential. The unsteady flow field can be simulated using different turbulence modeling approaches. The Unsteady Reynolds-Averaged Navier-Stokes equations in combination with the k-ϵ, the k-ω, or the SST model are the common way with the least demands regarding spatial and temporal resolution. More information is gained by using the SAS model, which adapts the turbulence modeling with respect to the spatial and temporal resolution. Further details of the flow field are simulated by a detached eddy simulation (DES) or a large eddy simulation (LES). In contrast to the SAS model, which falls back to a URANS solution when the discretization (temporal or spatial) is too coarse, the DES or LES model has special requirements regarding discretization. The best, but also the most expensive, solution is retrieved by a direct numerical simulation (DNS) of the flow. In this case, the discretization must be able to resolve the majority of the turbulent scales in time and space.

The CFD simulations in this paper were performed using the URANS-SST model and the SAS-SST model.

3.1.1. Shear Stress Transport (SST) Turbulence Model. In the SST k-ω model Menter [7] combined the advantages of the k-ω model [8] and k-ϵ model [9]. Although the k-ϵ model has its benefits in free stream regions, the k-ω model leads to a more physical resolution of the flow in near-wall regions. With that knowledge, Menter derived the SST k-ω model: two functions, F_1 and F_2, dynamically blend between the original k-ω model and a transformed k-ϵ model. The blending is dependent on the wall distance y and the thickness of the boundary layer at that location. The two-equation set of the k-ω model is multiplied by blending function F_1. Accordingly, the transformed k-ϵ model, which has a transport equation for ϵ depending on ω, is multiplied by $(1 - F_1)$. The corresponding transport equations of the models are added, resulting in the equation set of the SST k-ω model [7].

3.1.2. Scale Adaptive Simulation (SAS) Approach. Three types of two-equation turbulence models have been developed in the past: the k-ω model by Kolmogorov [9], the k-ϵ model by Launder and Spalding [8], and the k-kL model by Rotta [10]. Compared with the k-ω and k-ϵ transport equations, Rotta found a formulation of the transport equations that hold a natural length scale. The advantage of this length scale is the possibility of dynamically reacting to resolved structures within the flow. In turn, the first and third spatial derivatives of the velocity are inherent in the transport equations.

Based on Rotta's formulation, Menter and Egorov [11, 12] showed that turbulent transport equations include the second instead of the third derivative of the velocity (as used in Rotta's approach). The turbulent transport quantities form a $\sqrt{k}L$ model that is known as the SAS model (Scale

Adaptive Simulation). The first formulation of the model [12] has undergone some modifications, resulting in the current version of the SAS model [13], which is also implemented in ANSYS CFX 14.0+. Assuming that the turbulent structure is resolved within the flow field and not modeled, the turbulent frequency ω is increased and finally less modeling of the turbulence occurs due to a reduced eddy viscosity. The advantage of the model is obvious: the SAS-SST model dynamically reduces the modeling provided that turbulent structures can be resolved within the flow field (except for a limiter function to keep a minimum value for eddy viscosity [14]).

3.2. Acoustic Simulation. The process used here to compute the flow-induced sound is a hybrid method; that is, the acoustic simulation is performed as a second step after the results of the flow simulation have become available. The propagation of flow-induced sound is governed by Lighthill's inhomogeneous wave equation [15]:

$$\frac{1}{c_0^2}\frac{\partial^2 p'}{\partial t^2} - \frac{\partial^2 p'}{\partial x_i^2} = \frac{\partial^2 T_{ij}}{\partial x_i \partial x_j}, \tag{1}$$

where c_0 is the average speed of sound in air and p' is a fluctuating pressure, which approaches the acoustic pressure outside the flow region (for details, see [1]). The wave equation is loaded on the right-hand side with the second-order spatial derivative of the Lighthill tensor:

$$\mathbf{T}_{ij} = \rho u_i u_j + \left((p - p_0) - c_0^2 (\rho - \rho_0) \right) \delta_{ij} - \tau_{ij}, \tag{2}$$

which depends on data for the turbulent flow precomputed by the flow simulation, such as the velocity u_i, the aerodynamic pressure p, the density ρ, and the viscous stress tensor τ_{ij}. For isentropic flow at low Mach number, this can be approximated by $\mathbf{T}_{ij} \approx \rho_0 u_i u_j$, where ρ_0 is the average density of air.

The inhomogeneous wave equation is solved using the Finite Element Method (FEM). Before the spatial and temporal discretization can be applied, the variational formulation of (1) must be derived [16]. This is done by multiplying with a test function w and integrating the equation over the simulation domain Ω:

$$\int_\Omega w \left(\frac{1}{c_0^2}\frac{\partial^2 p'}{\partial t^2} - \frac{\partial^2 p'}{\partial x_i^2} - \frac{\partial^2 T_{ij}}{\partial x_i \partial x_j} \right) d\Omega = 0. \tag{3}$$

Green's integral theorem is then applied in order to reduce the second-order spatial derivatives to first order:

$$\int_\Omega \frac{1}{c_0^2} w \frac{\partial^2 p'}{\partial t^2} d\Omega + \int_\Omega \frac{\partial w}{\partial x_i} \frac{\partial p'}{\partial x_i} d\Omega - \int_\Gamma w \frac{\partial p'}{\partial \mathbf{n}} d\Gamma$$
$$= -\int_\Omega \frac{\partial w}{\partial x_i} \frac{\partial T_{ij}}{\partial x_j} d\Omega, \tag{4}$$

where $\Gamma = \partial\Omega$ is the boundary of the simulation domain and $\mathbf{n}$ is the surface normal vector on Γ. If we set the

boundary integral to zero, an acoustically hard wall (i.e., ideally reflecting) is assumed at the boundary. In order to achieve free-field radiation where the boundary is purely artificial (denoted by Γ_a), we use an absorbing boundary condition expressed by

$$\int_\Gamma w \frac{\partial p'}{\partial \mathbf{n}} d\Gamma = -\int_{\Gamma_a} \frac{1}{c_0} w \frac{\partial p'}{\partial t} d\Gamma. \tag{5}$$

Subsequently, the computational domain is discretized using Lagrangian finite elements of first order. This allows us to set up a linear system of equations that needs to be solved in each time step in order to obtain the transient sound field.

In order to achieve good accuracy of the numerical solution, the following procedure has proven to provide the best results. First, the source term on the right-hand side of (4) is computed using the FE approach on the fine CFD grid. Then the source terms are transferred to another grid, which is better suited for the wave equation, using an energy-conserving interpolation technique. The sound emission from these sources is then computed using FEM on the acoustic grid.

4. Experimental Setup

The air-cooling of a generic electronic device was subjected to the investigations and simulations performed. As the interaction of the flow field with the obstacles, that is, electronic chips, pins, and heat sinks, can be a source of noise due to the production of, for example, vortex shedding or turbulent shear stresses, these geometric details are present in the experimental and numerical setup.

4.1. Generic Electronic Device. The model of the generic electronic device can be divided into four regions: the inflow area in front of the fans, the fans producing the air flow used for cooling, the settling chamber to generate an even flow through each card slot, and the card slots where the electronic components are located. In the real system, which served as a model for the generic device under investigation, the settling chamber is used to even the flow through the different card slots, especially if not all slots are in use.

The overall dimensions of the electronic device (Figure 1) are $520 \times 153 \times 338$ mm. It holds eleven card slots, each with an electronic component installed inside. The flow is generated by two axial, five-bladed 119 mm fans with an outer rotor diameter of 112.9 mm. The design of the electronic components is based on real components for personal computers (i.e., network, sound, and graphic cards).

4.2. Measurements. The acoustic measurements were performed in the anechoic chamber of the University of Erlangen-Nuremberg. The generic electronic device was configured in the same way as for the simulations (Figure 1(b)): two installed fans with a rotational speed of 5000 rpm and a settling chamber depth of 70 mm.

Two Brüel & Kjær Type 4189 microphones were placed at a distance of 1 m from the generic electronic device. The microphone facing the inlet area is aligned in the center and

(a) Schematic drawing

(b) Real setup

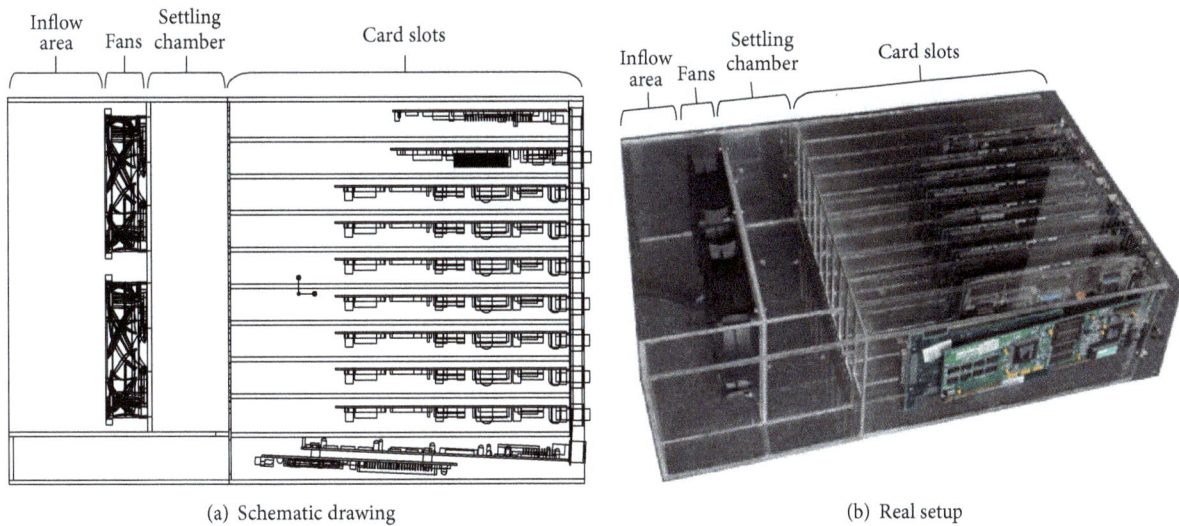

FIGURE 1: Generic electric device.

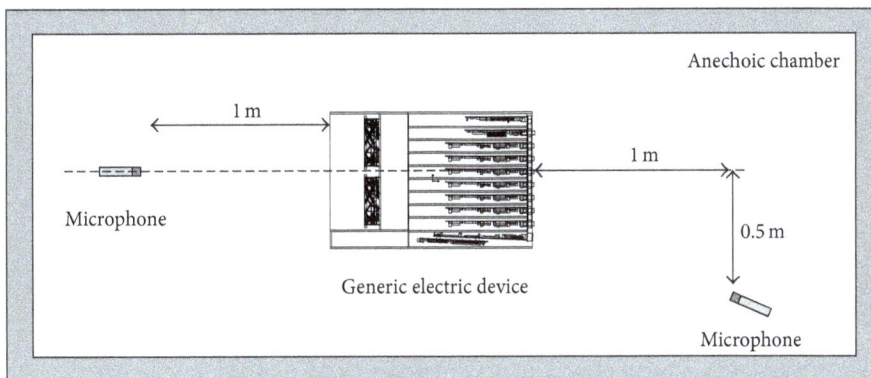

FIGURE 2: Setup for acoustic measurements.

the microphone at the outlet is installed approximately 50 cm eccentric to be out of the free stream generated by the air flow through the device (Figure 2). In the vertical direction, both microphones are placed at half of the device's height.

The results of the acoustic measurements were compared with the acoustic data received from the simulations. The measurements were performed with a sampling rate of 50 kHz and a measuring time of 30 s.

5. Simulation Process of a Generic Electronic Device

The first step of the simulation process was the generation of the CAD model, which served for both the CFD and acoustic mesh generation. During the CAD build process of the model, attention was paid to accommodating major electronic structures on the printed circuit boards, that is, through-hole devices and comparatively large surface-mounted devices (see Figure 3(b)). The whole generic electronic device was modeled in CAD including an additional inlet and outlet plenum. The manufacturer's original CAD data for the fan were simplified at the casing (removal of wires

and reinforcing ribs) and at the hub (removal of ventilation slots for cooling of motor) to ensure proper mesh generation (Figure 3(a)). Two different meshes were generated, one for the flow simulation and the other for the acoustic simulation. For the acoustic simulation, the mesh can be coarser than the CFD mesh, because the acoustic wavelength is usually greater than the turbulent length scales of the flow.

5.1. Flow Simulation. The mesh was generated using the ANSYS Workbench Mesher. The unstructured mesh consists of tetrahedra and prism elements holding a total element number of 39.6 million and 13.4 million nodes, respectively; 9.2 million elements and 3.3 million nodes correspond to each fan as shown in Figure 4(a). The whole flow region was divided into several subregions with each one having been meshed separately and being connected to its adjacent neighbor regions using interfaces. The boundary layer within the fans has a thickness of about 1×10^{-5} m and is resolved with 12 prism layers. The regions of the electronic cards consist, overall, of 16.1 million elements and 5.3 million nodes. Figure 4(b) shows exemplarily the mesh for one electronic card. The boundary layer on the walls of the card slots is

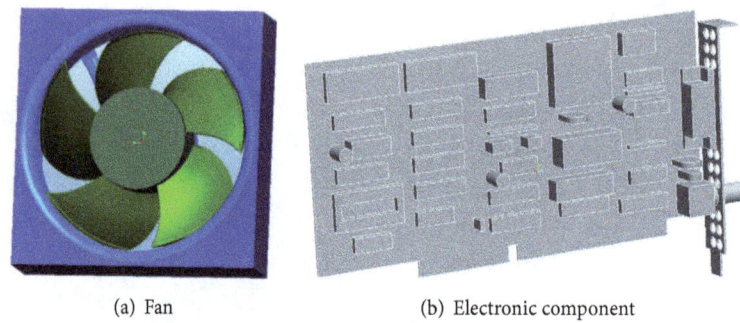

(a) Fan   (b) Electronic component

FIGURE 3: CAD models of the components.

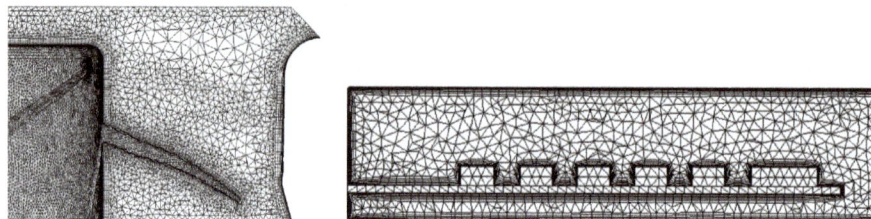

(a) Fan domain   (b) Region of an electronic card

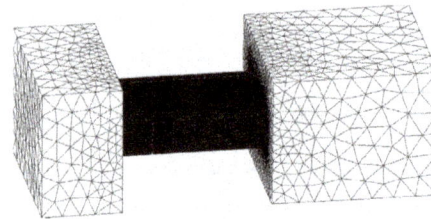

(c) Whole simulation domain including inlet and outlet plenum

FIGURE 4: CFD mesh.

resolved with ten layers (transition ratio: 0.5; growth rate: 1.3) and the electronic components on the cards are encased in seven prism layers with a total height of 2×10^{-3} m and a growth rate of 1.3. The remaining 5.1 million elements and 1.5 million nodes belong to the inlet and outlet regions and also the settling chamber incorporating 10–12 prism layers (see Figure 4(c)).

The flow simulations were performed with two different turbulence models: the Unsteady RANS-SST model and the SAS-SST model. Both CFD simulations involve the same setup except for the turbulence models. The fans are rotating with a frequency of 5000 rpm and are linked to the nonrotating domains, that is, the inflow region and the settling chamber (Figure 1), with a transient rotor-stator interface. The fluid maintains the characteristics of air at 25°C with a constant density. Owing to the low Mach number of only 0.01 at the blade tips, incompressibility of the fluid can be assumed. The CFD simulations are performed in an isothermal manner. Our preliminary experiments showed that the influence of temperature on the sound emission, at least in the temperature range that electronic devices can withstand, that is, up to 80°C, is negligible. A mass flow condition with a value of $0.15 \, \text{kg s}^{-1}$ is applied to the inlet

boundary. The outlet is represented by an opening boundary condition with a relative pressure of 0 Pa. The time step size is $\Delta t = 10^{-4}$ s and the number of inner iterations is set to eight.

5.2. Acoustic Simulation. The acoustic simulations were performed using the in-house FE solver CFS++ (Coupled Field Simulation in C++), which is being developed jointly at the Technical University of Vienna and the University of Erlangen-Nuremberg. First, the flow-induced sound sources were computed according to (4). This operation was done on the CFD mesh in order to preserve the full spatial resolution of the flow computation. Good numerical accuracy is critical in this step; otherwise the spatial derivatives required to compute sound sources from the flow data would produce too much numerical noise. Subsequently, the sound sources were interpolated using an energy-conserving scheme onto the mesh used to compute sound propagation.

The need for a separate acoustic mesh is due to requirements in spatial resolution that are completely different compared with the flow computation. The CFD mesh needs a small cell size in boundary layers and in regions where the flow is turbulent. On the other hand, the CFD mesh can be rather coarse where the flow is laminar. In contrast, the

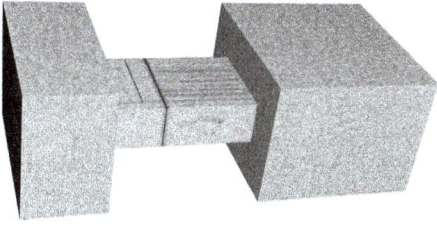

FIGURE 5: Acoustic mesh.

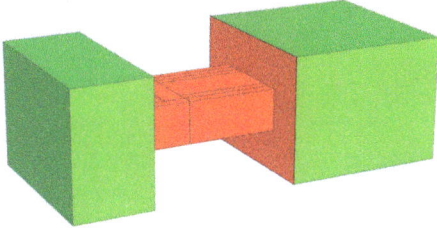

FIGURE 6: Acoustic boundary conditions: acoustically hard walls (red) and absorbing boundary (green).

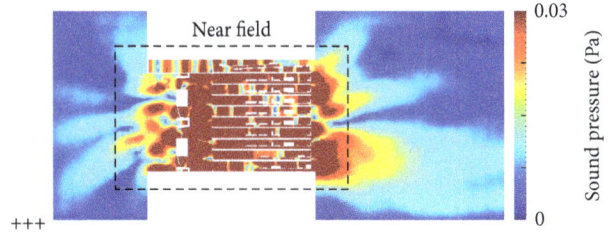

FIGURE 7: Time-harmonic magnitude of sound pressure generated by a monofrequent sound source at 4.0 kHz; shown is the cross-section through the axes of both fans.

acoustic mesh does not need to resolve boundary layers, but it must properly resolve the smallest wavelength in the whole domain. That is why we used an unstructured tetrahedral mesh without prism layers at rigid boundaries (Figure 5), but with an almost uniform element size. The maximum edge length of the tetrahedra was set to 8 mm. The element size was refined only slightly at the surface of the fans and electronic components in order to account for the actual shapes of the rigid boundaries. The time step size was the same as in the flow simulation. With the Nyquist-Shannon sampling theorem and a temporal resolution of $\Delta t = 1\times10^{-4}$ s (see Section 5.1), the maximum resolved acoustic frequency is f_{max} = 5000 Hz, resulting in a minimal acoustic wavelength of λ_{min} = c_0/f_{max} = 0.069 m that is resolved by at least nine mesh nodes. This spatial resolution leads to a mesh size of three million elements with 621 000 nodes. Approximately 93 000 elements with 24 000 nodes are located in each fan.

The acoustic simulation was performed with the same fluid properties as the flow simulation, that is, air at 25°C. To avoid nonphysical reflections of the acoustic waves, absorbing boundary conditions were applied to the inlet and outlet plenum on the same surfaces as the inlet and outlet boundary conditions of the flow simulation. The surfaces of the electronic device's housing and the particular surfaces of the inlet and outlet plenum contacting the housing were acoustically hard walls (Figure 6).

A frozen rotor model was used for the axial fans in the acoustic simulation. Furthermore, the simulation domain does not extend to either of the microphone positions, so the sound field had to be extrapolated to the microphone positions. This is allowed provided that the near field around the device is fully contained in the simulation domain. This means that the absorbing boundaries are in the far field and the sound pressure on these boundaries can be extrapolated to arbitrary points outside the simulation domain using

the inverse distance rule. As a rough approximation for the stretch of the device's near field we used the equation for the near-field length of a baffled piston, namely, $N = D^2/4\lambda_{min} = D^2 f_{max}/4c_0$. Thereby, D = 33 cm is the length of the longer edge of the device's rectangular inlet opening and serves as the characteristic size in the baffled-piston model. Strictly speaking, this formula is only valid for circular pistons. However, we can obtain a worst-case approximation of the near-field length by using the longer edge for the characteristic length D and the maximum frequency. With the size of the inlet plenum as the maximum near-field length N_{max} = 28 cm, we determined the maximum frequency that allows for extrapolation to be f_{max} = $4c_0 N_{max} D^{-2}$ ≈ 3.6 kHz. This approximation was confirmed by a preliminary simulation with an artificial sound source inside the device radiating sine waves at 4.0 kHz (cf. Figure 7).

During the first 30 ms of the acoustic simulation, the sound sources were tapered with a quadratic sine function in order to achieve a fade-in of the sources. The reason for this is that sound sources were computed from the flow only after the flow through the device had fully developed its turbulent structure. Instantaneously applying the full-scale sources to the initially quiescent acoustic medium would result in numerical acoustic waves that can have an adverse effect on the real physical solution [17]. This effect can be avoided using a fade-in of the sound sources. For time stepping a second order Newmark scheme was used.

A comparison of two different boundary conditions, namely, an ABC and a PML, was performed. An artificial point source was placed within the device and the sound pressure with an absorbing boundary condition (ABC) and a perfectly matching layer (PML) was calculated with a harmonic analysis in the range from 10 to 2000 Hz in 10 Hz steps. Figure 8 shows the difference of simulated sound pressure for ABC and PML relative to the source strength s:

$$\text{err}\,(f) = \frac{|p_{ABC}\,(f) - p_{PML}\,(f)|}{s}. \tag{6}$$

The frequency range from 200 to 350 Hz shows the largest differences between ABC and PML boundary condition. For frequencies larger than 500 Hz almost no difference can be found. Below 500 Hz we observe that the ABC cannot fully damp out acoustic modes that are purely artificial and arise from using a finite-size computational domain to model an unbounded domain. However, we can filter this spurious

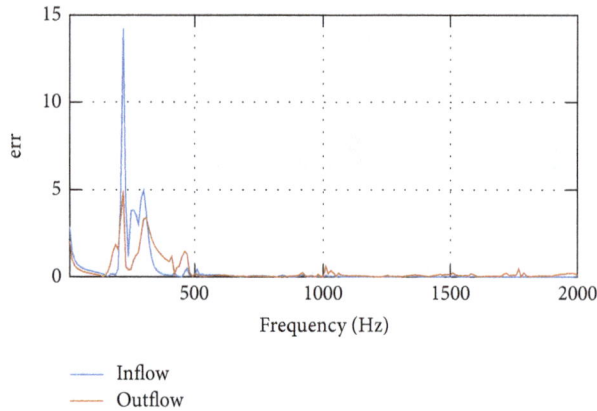

FIGURE 8: Comparison of ABC and PML boundary condition.

effect from the results of the simulation using ABC, because it occurs at discrete frequencies only. Despite the differences between the ABC and PML boundary conditions an ABC was chosen due to much smaller computational efforts of the ABC over the PML.

In addition to the sound propagation simulations based on the aeroacoustic sources, the acoustic eigenmodes were calculated using the same acoustic setup. These results will be used to identify peaks in the acoustic spectra that are not triggered by obvious effects like blade passing.

5.3. Structural Simulation. As the results of the acoustic simulations will only include aeroacoustic sources but are compared to measurements, which also include structure-borne sound, an additional structural modal analysis has been carried out. The goal of this finite element simulation was to identify structural eigenmodes contributing to the noise emission which are present in the measurements but not in the CAA simulations.

6. Results

6.1. Structural Eigenmodes. Only the first two eigenmodes were identified as their eigenfrequencies are already higher than the acoustic frequency range of the simulations. Figure 9 depicts both eigenmodes. The first eigenmode shows a vibration at the inlet section of the electronic device with an eigenfrequency of 2320 Hz. The second eigenmode causes a deflection of the mainboard area at the outlet with an eigenfrequency of 2484 Hz. Consequently, no comparison errors of acoustic simulation and measurement are to be expected due to structural eigenmodes in a range up to 2 kHz.

6.2. Flow Simulation. The simulation time of the flow field was 0.21 s with 2100 time steps, in the case of the URANS-SST, and 0.64 s with 6400 steps in case of the SAS-SST. The flow field was first initialized with a steady-state simulation; then the setup was changed to the unsteady case. Before exporting the flow data, the flow field was allowed to adapt to the unsteady conditions for approximately 0.2 s. The exported quantities are the hydrodynamic pressure and the

three velocity components. The URANS simulation took 115,200 CPU hours to complete (including initialization) and the SAS simulation 157,000 hours. The calculations were performed at the HPC center of the University of Erlangen-Nuremberg.

The time step size of 1×10^{-4} s in combination with the fine spatial resolution of the fan domains results in an average CFL number of 3.5 for the fan domains. While the highest values are to be found in the blade tip gap, the values in the boundary layer of the blades vary between 3 and 8 with a decrease in direction to lower perimeters. A lower CFL number can be achieved by further decreasing the time step size. Figure 10 holds a distribution of the CFL number in a cut plane through both fan axes and in a cut plane perpendicular to one fan axis of the URANS simulation.

Figures 11 and 12 show the velocity and the pressure field on a cut plane through the axis of the fans for the URANS-SST and SAS-SST simulations. A comparison of these two figures demonstrates the advantages of the SAS model: the resolved scales are finer and represent the flow field in more detail. Accordingly, the fluctuations incorporate higher amplitudes at higher frequencies in the SAS-SST simulation compared with the URANS-SST simulation. This is important as the frequency range of the calculated acoustic source terms is based on the fluctuations existing in the flow field. The fact that the SAS model is less modeling but more resolving with respect to the turbulent structures is underlined by Figure 13, which highlights the decrease in the eddy viscosity.

Optimizations of the air-cooling can be derived from the flow field plots. The card slots that are placed in line with the hub receive lower flow rates and, correspondingly, less cooling. Depending on the inflow angle into the card slots, separation areas form due to detachment of the flow, as depicted in Figures 11(a) and 12(a). The implementation of baffle plates would lead to more homogeneous inflow conditions with smaller inflow angles and hence less separation.

The Q-criterion is a common method for visualizing vortex regions. It is defined as the second invariant of the velocity gradient tensor. Dominant vortex regions (Figure 14) can be found behind fans, in the inflow region of the card slots and at the electronic parts soldered onto the printed circuit boards.

The relative velocity and pressure distribution in Figure 15 around the fan blade depicts a flow separation area close to the leading edge on the suction side. Due to the high rotational speeds of 5000 rpm in combination with large angle of attack the flow detaches from the blade. This causes vortex shedding in the blade wake, demonstrated by some eddies located close to the trailing edge in the pressure plot. Additionally, a stagnation point at the leading edge can be identified which corresponds with a pressure maximum.

A comparison of the hydraulic pressure fluctuations for six different positions on the blade surface is shown in Figure 16. The SAS model introduces less damping than the URANS model and thus leads to a slower decay of the fluctuations. Both models produce dominant peaks at the blade passing frequency and all harmonics. But the SAS model is also capable of capturing fluctuations in between

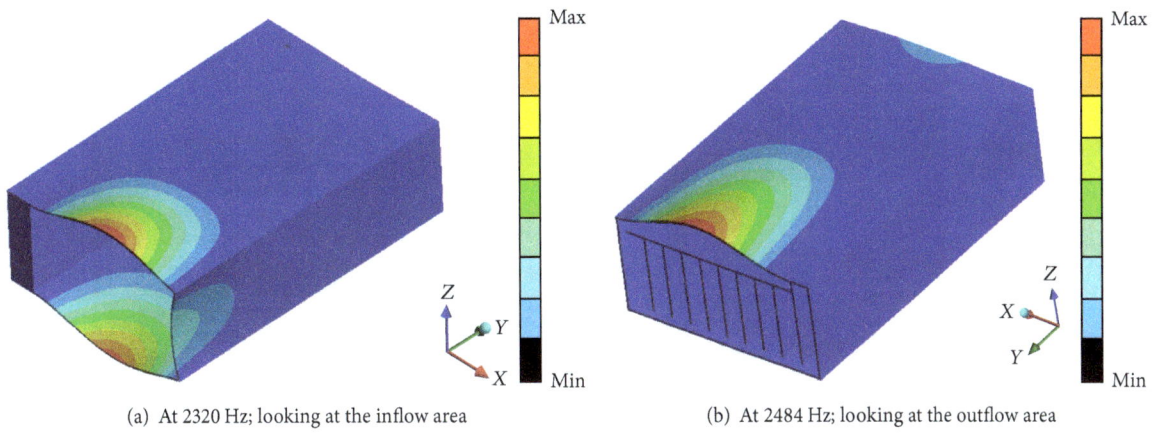

(a) At 2320 Hz; looking at the inflow area

(b) At 2484 Hz; looking at the outflow area

FIGURE 9: Structural eigenmodes (displacement shown).

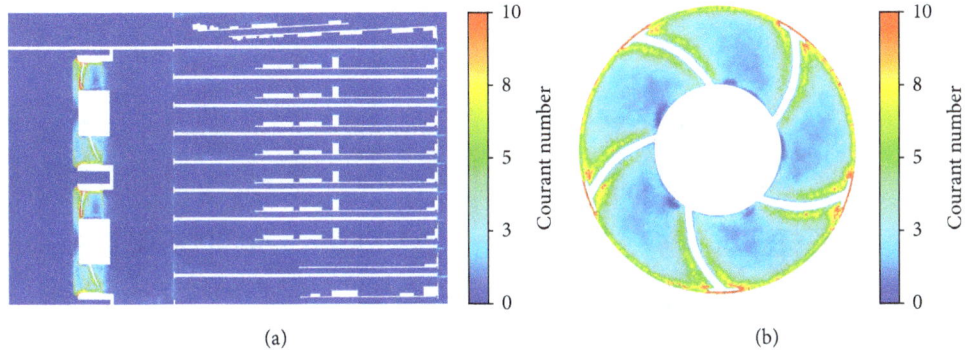

(a)

(b)

FIGURE 10: CFL number on a cut plane through both fan axes (a) and perpendicular to a fan axis (b).

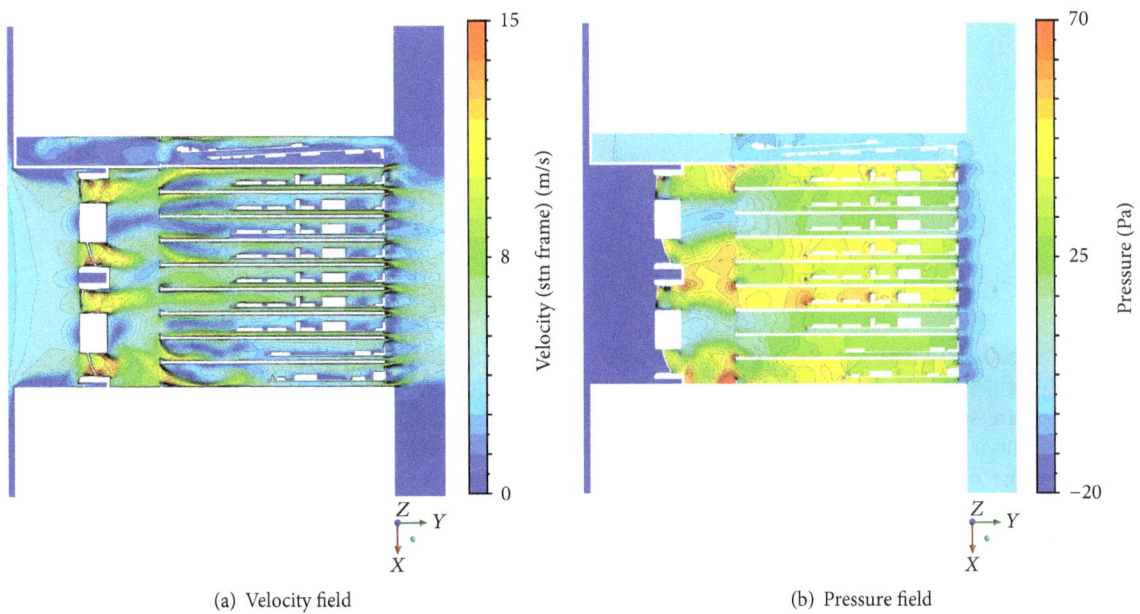

(a) Velocity field

(b) Pressure field

FIGURE 11: URANS simulation.

(a) Velocity field

(b) Pressure field

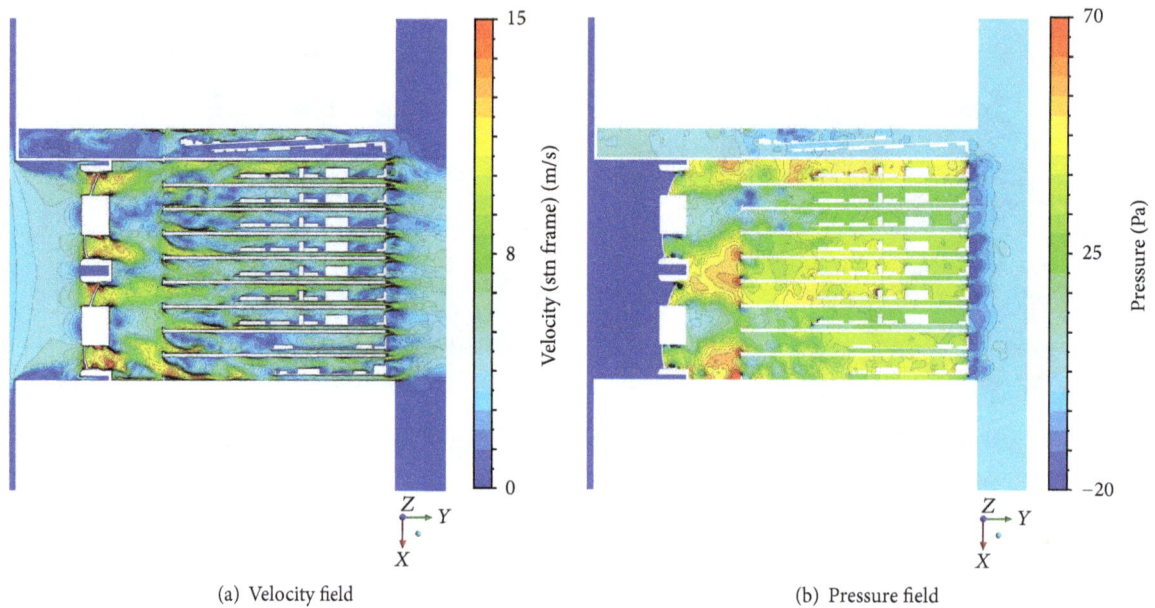

FIGURE 12: SAS simulation.

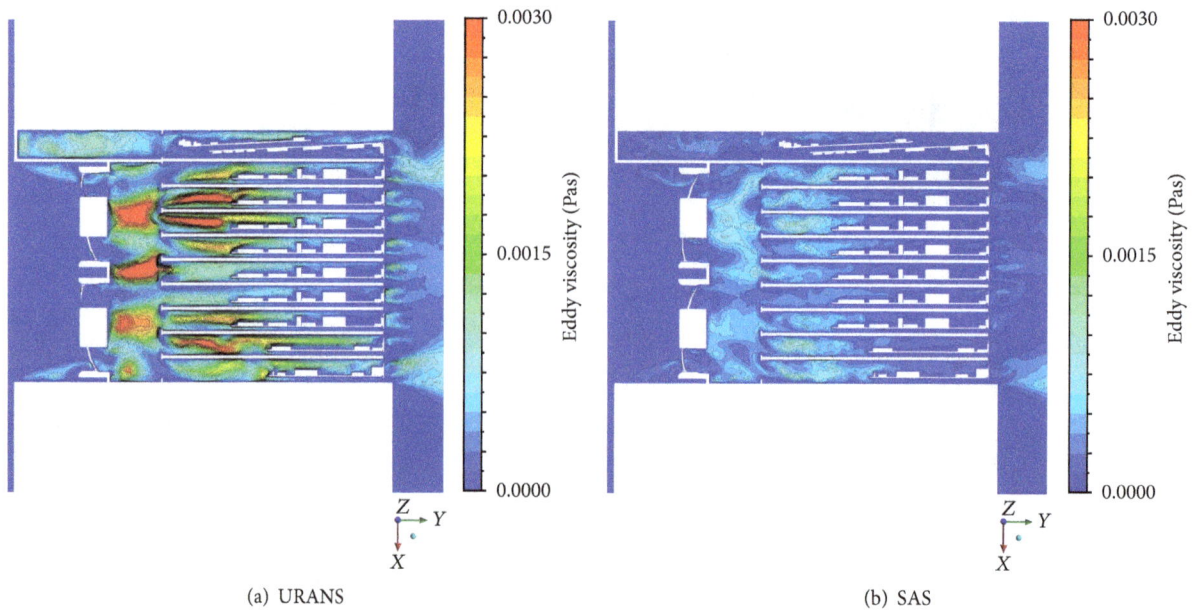

(a) URANS

(b) SAS

FIGURE 13: Eddy viscosity.

these dominant effects, resulting in higher pressure amplitudes. Also some of the effects explained in Table 1 can be found in the pressure spectrum of the trailing edge point. Particularly rotor-stator and blade-stator interaction results in an increase of the pressure amplitudes. In general, the leading edge is exposed to the highest fluctuations as it first cuts the incoming eddies.

The mean velocity downstream of the fans is plotted in Figure 17. Both turbulence models lead to very similar trends. The SAS model generates a slightly more detailed velocity profile in the downstream blade wake in regions of high velocity gradients. The velocity profiles smoothen with an

increasing distance to the fans due to the momentum transfer in the fluid.

Likewise, the spatial pressure distribution is shown in Figure 18. Here, SAS and URANS produce very similar spatial distributions and lead to the same rise in pressure with an increasing distance to the fans as the dynamic pressure component is converted into the static one.

While the time-averaged pressure results in no significant differences between the URANS and SAS models, the advantages of the SAS model become obvious for the wall pressure spectra given in Figure 19 being in accordance with Figure 16. The SAS model is much better resolving pressure

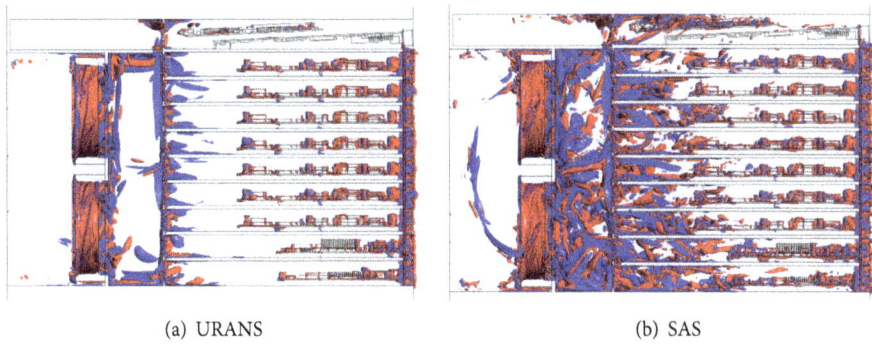

(a) URANS

(b) SAS

FIGURE 14: Q-criterion on isosurfaces of $\pm 3 \times 10^5\ \mathrm{s}^{-2}$ (blue: $-3 \times 10^5\ \mathrm{s}^{-2}$; red: $+3 \times 10^5\ \mathrm{s}^{-2}$).

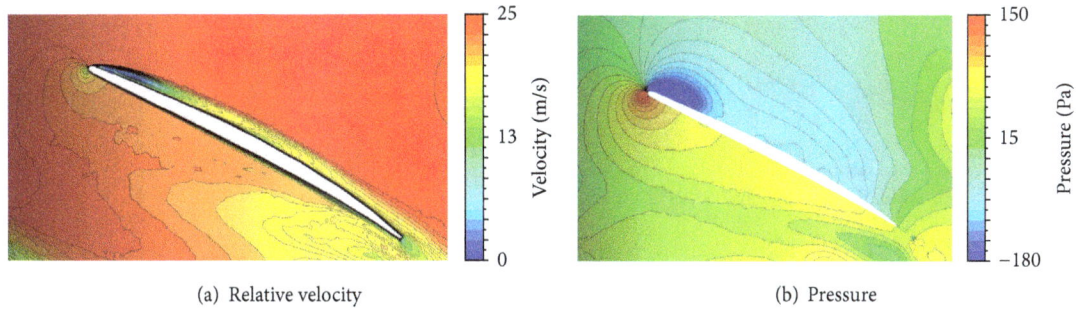

(a) Relative velocity

(b) Pressure

FIGURE 15: Relative velocity and pressure distribution on a cylindrical plane with $D_{\mathrm{Plane}} = 0.84D$ for SAS simulation.

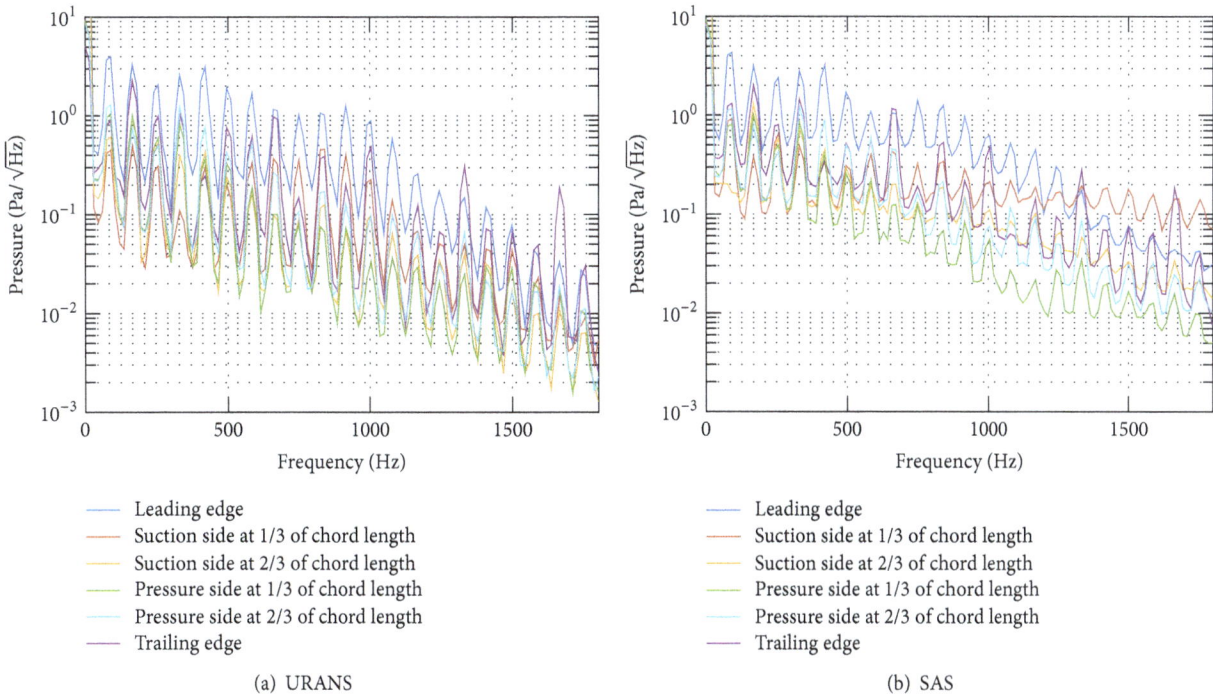

Leading edge
Suction side at 1/3 of chord length
Suction side at 2/3 of chord length
Pressure side at 1/3 of chord length
Pressure side at 2/3 of chord length
Trailing edge

Leading edge
Suction side at 1/3 of chord length
Suction side at 2/3 of chord length
Pressure side at 1/3 of chord length
Pressure side at 2/3 of chord length
Trailing edge

(a) URANS

(b) SAS

FIGURE 16: Fourier transform of hydraulic pressure at six different positions on the blade surface at a diameter of $0.84D$: leading edge, suction side at 1/3 of chord length, suction side at 2/3 of chord length, pressure side at 1/3 of chord length, pressure side at 2/3 of chord length, and trailing edge.

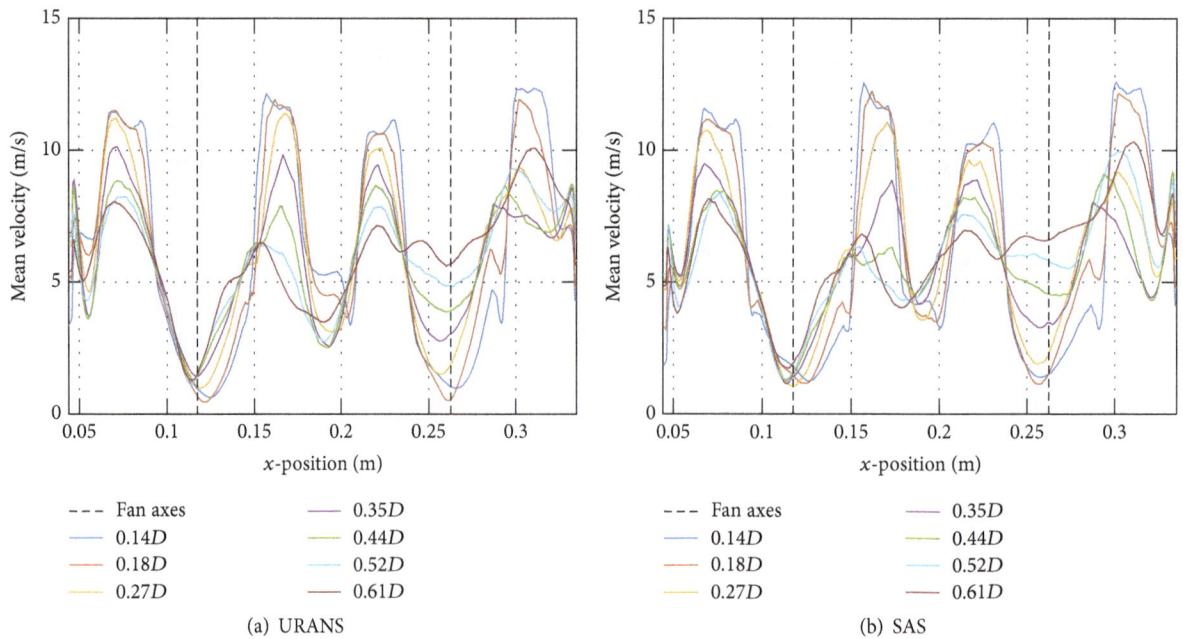

(a) URANS

(b) SAS

FIGURE 17: Mean velocity profiles along multiple axes perpendicular to both fan axes at distances of $0.14D$, $0.18D$, $0.27D$, $0.35D$, $0.44D$, $0.52D$, and $0.61D$ to the fan outlet for URANS and SAS simulation.

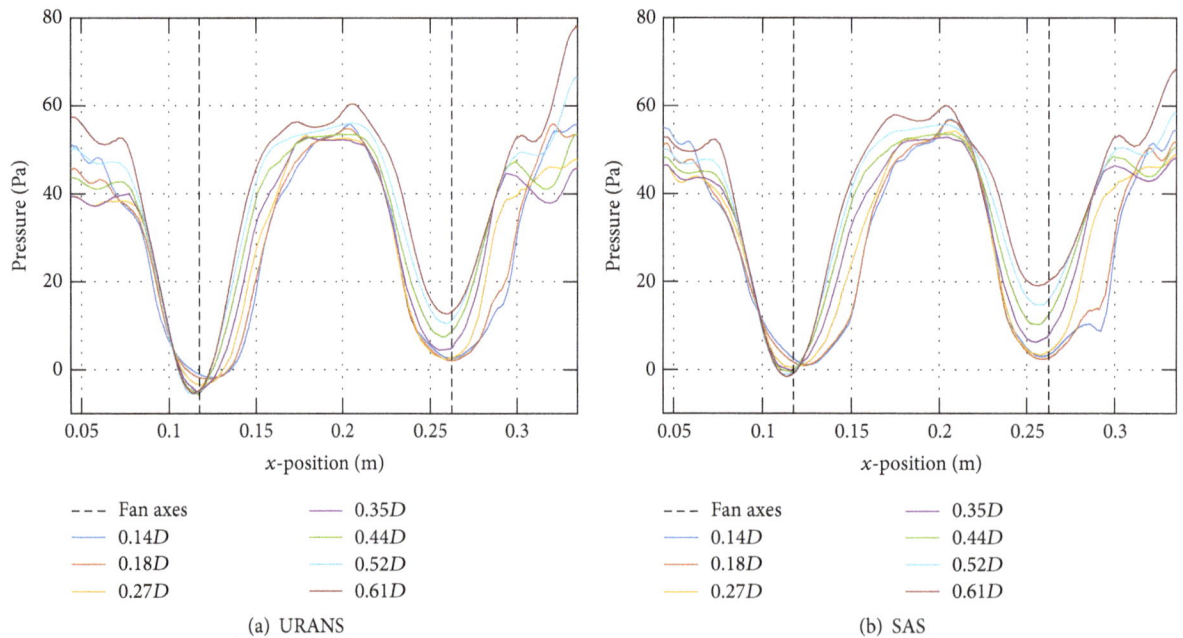

(a) URANS

(b) SAS

FIGURE 18: Mean pressure profiles along multiple axes perpendicular to both fan axes at distances of $0.14D$, $0.18D$, $0.27D$, $0.35D$, $0.44D$, $0.52D$, and $0.61D$ to the fan outlet for URANS and SAS simulation.

fluctuations generated by random turbulence in contrast to triggered phenomena like blade passing. For the SAS model the fluctuations are approximately one to two orders of magnitude higher than for the URANS model.

The damping of the URANS model can be best demonstrated by looking at the Fourier transform of the turbulent kinetic energy in Figure 20. First of all, the turbulent kinetic

energy contained within the flow is higher for the SAS model, especially at higher frequencies. Secondly, the damping of turbulent eddies is much lower for an increasing downstream distance. With the SAS model the four observed positions show a decrease of less than one order of magnitude while the URANS model reduces the turbulent kinetic energy by four orders of magnitude.

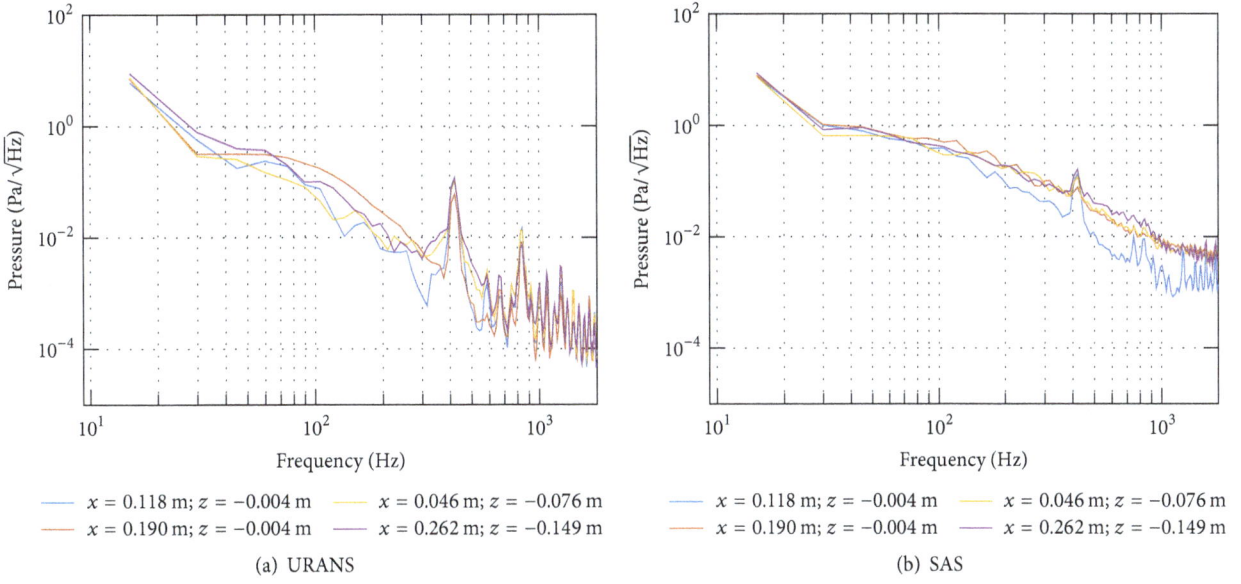

(a) URANS (b) SAS

FIGURE 19: Fourier transform of hydraulic wall pressure $0.5D$ behind the fan at four different positions $x = 0.118$ m; $z = -0.004$ m, $x = 0.190$ m; $z = -0.004$ m, $x = 0.046$ m; $z = -0.076$ m, $x = 0.262$ m; $z = -0.149$ m for URANS and SAS simulation.

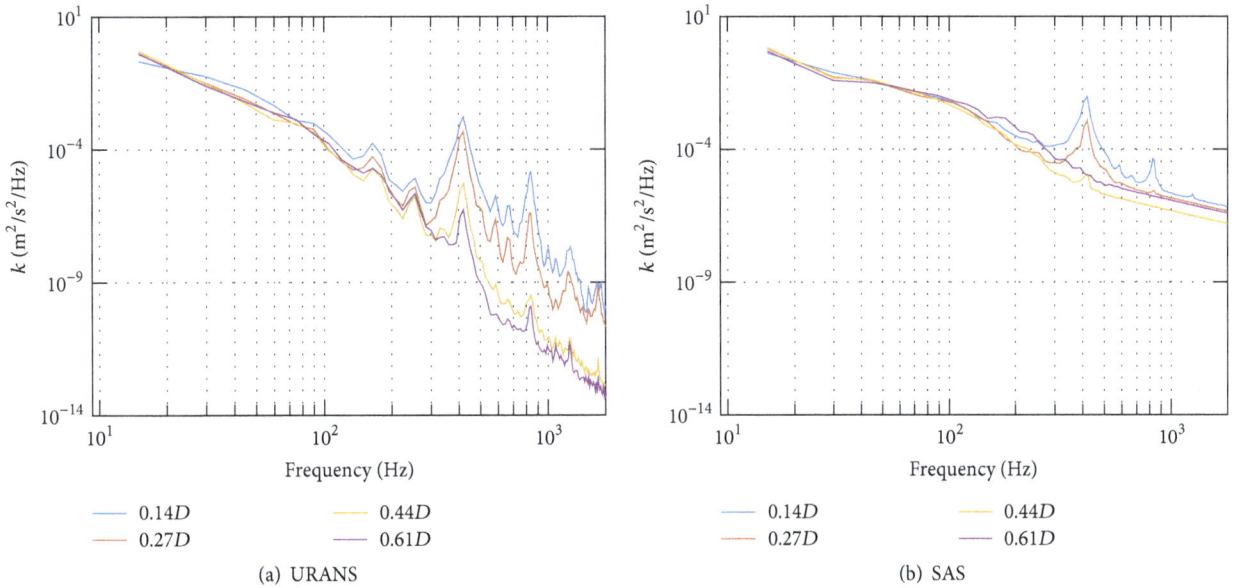

(a) URANS (b) SAS

FIGURE 20: Fourier transform of turbulent kinetic energy k in a plane containing both fan axes at diameter of $0.75D$ from one fan axis at four different positions downstream $0.14D$, $0.27D$, $0.44D$, and $0.61D$ of the fan outlet for URANS and SAS simulation.

6.3. Acoustics. In a first step, the aeroacoustic sound sources were computed and are displayed in Figure 21. While there exist acoustic formulations depending on the velocity the authors used the Δp formulation of the aeroacoustic sources. This formulation can be directly translated into the velocity formulation shown in (2).

As one would expect, the greatest sources are found inside the fans and in their wake. We find additional sources in the separation zones, where the flow enters the card slots. Sound sources are also generated by turbulence in the wake of the electronic cards' brackets. However, the components

on the electronic cards do not generate enough turbulence to transform into relevant sound sources. The difference between the sound sources obtained from the URANS and the SAS computations corresponds to the differences in the vortex structures observed in the two flow simulations through the Q-criterion (Figure 14). The SAS resolves finer vortex structures than the URANS simulation, leading to considerably greater sound sources in the entry regions of the card slots in the SAS case compared with the URANS case. Hence, the Q-criterion is a useful indicator to locate the generation of aeroacoustic sources from CFD data for

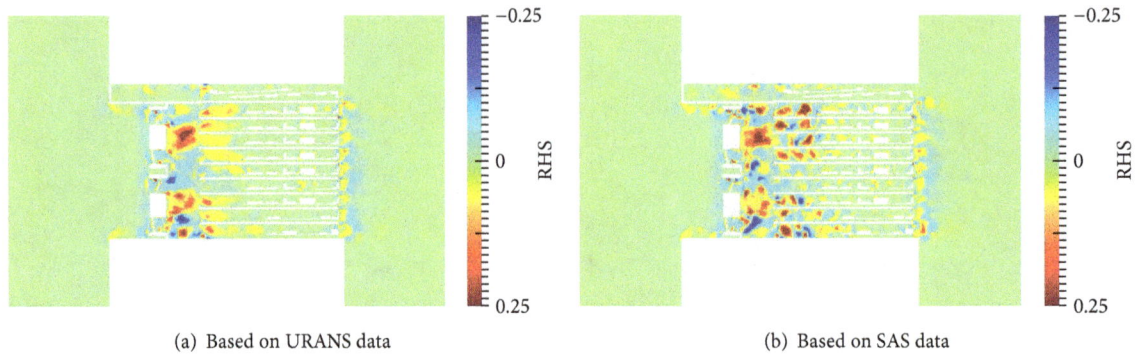

(a) Based on URANS data   (b) Based on SAS data

FIGURE 21: Finite element representation of aeroacoustic sources computed according to the right-hand side of (4); shown is the cross-section through the axes of both fans.

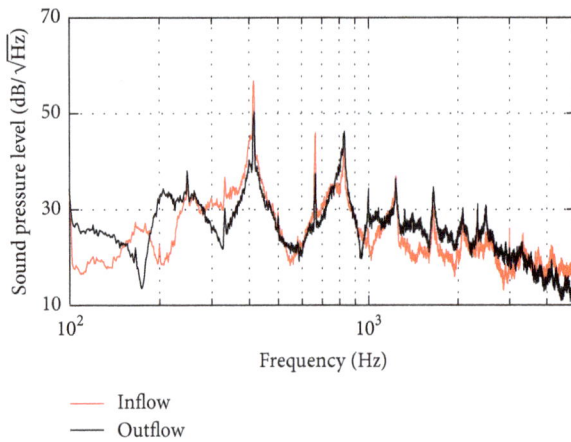

— Inflow
— Outflow

FIGURE 22: Measured sound pressure level at inflow and outflow position.

incompressible flows as it is proportional to the Lighthill sources.

The results of the acoustic measurements are presented in Figure 22 in a frequency range up to 10 kHz and are compared to the acoustic simulations in Figure 23. The fast Fourier transformations were performed with a 1 s-Hanning window for the measurement data and a 0.0667 s-Hanning window for the simulation data, resulting in frequency resolution of 1 Hz and 15 Hz, respectively. The acoustic measurements are shown to explain some of the higher frequent phenomena that are not present in the acoustic simulations due to the lower temporal resolution.

Figure 23 illustrates the acoustic spectra of the simulation and measurements at the inflow and outflow positions (see Figure 2). At a frequency of ≈ 4800 Hz as the spectra approach the Nyquist frequency, which computes as $(2 \cdot \Delta t)^{-1} = (2 \times 10^{-4} \text{ s})^{-1} = 5000$ Hz and is not shown in the spectra, the amplitudes decrease quickly due to an insufficient temporal resolution. While the simulations provide data up to this frequency, only a range up to 1 kHz should be used for evaluations at which the acoustic waves are resolved with 10 points in the temporal domain and 42 points in the spatial domain. This discrepancy in temporal and spatial

resolution can be avoided by decreasing the time step to 2–2.5×10^{-5} s resulting in an equal resolution in time and space. For the current simulation cases, damping due to insufficient temporal resolution is to be expected for frequencies higher than 1 kHz.

Supplementary acoustic and mechanical modal analyses were performed to aid explaining some of the effects observed in measurements and CAA simulations. Thereby, the first significant peak in the spectrum recorded at the outflow position of the device occurs around 215 Hz and can be attributed to an acoustic cavity resonance in the front panel slot (Figure 24(a)). The reason why this peak is only present at the outflow side is that the front panel slot is sealed at the inflow side. The next peak at 278 Hz is caused by an acoustic eigenmode that spans the settling chamber and all of the card slots (Figure 24(b)). The peaks at 417 and 833 Hz can be identified as the blade passing frequency, $f_{bp} = f \cdot z = 83.3 \text{ Hz} \cdot 5 = 417$ Hz, and its second harmonic. The prominence of the blade passing frequency and its first harmonic over the broadband background noise can be explained by its proximity to acoustic cavity resonances at 403, 416, and 868 Hz that occur in the card slots (Figure 24(c)) and the mainboard compartment (Figure 24(d)), respectively. This effect can even be proved by performing a discrete Fourier transform on the transient sound field obtained from the acoustic simulation. From the results of the Fourier transform, we pick the spectral line closest to the eigenmodes (Figure 25). Just by visual comparison with the eigenmodes depicted in Figures 24(c), 24(d), and 24(e) one can prove that the eigenmodes are actually excited by the sound originating from the fans. A close-up look at the blade reveals the generation of acoustic source at the leading edge and the wake (Figure 26). The peaks at 333, 667, and 1333 Hz correspond to the rotation-triggered interaction of the four-strutted stator with the rotor. This generates strong acoustic sources and sound pressure levels between the area of the blades and the struts as shown in Figure 27. Compared with the blade passing frequency, this phenomenon is not triggered by the number of blades. Rotor-stator interaction triggered by the blades is located at a frequency of $4 \cdot 5 \cdot 83.3 \text{ Hz} = 1666$ Hz and yields another increase in the sound emission. The local maxima at approximately 2330 and 2490 Hz belong to the

(a) Inflow region

(b) Outflow region

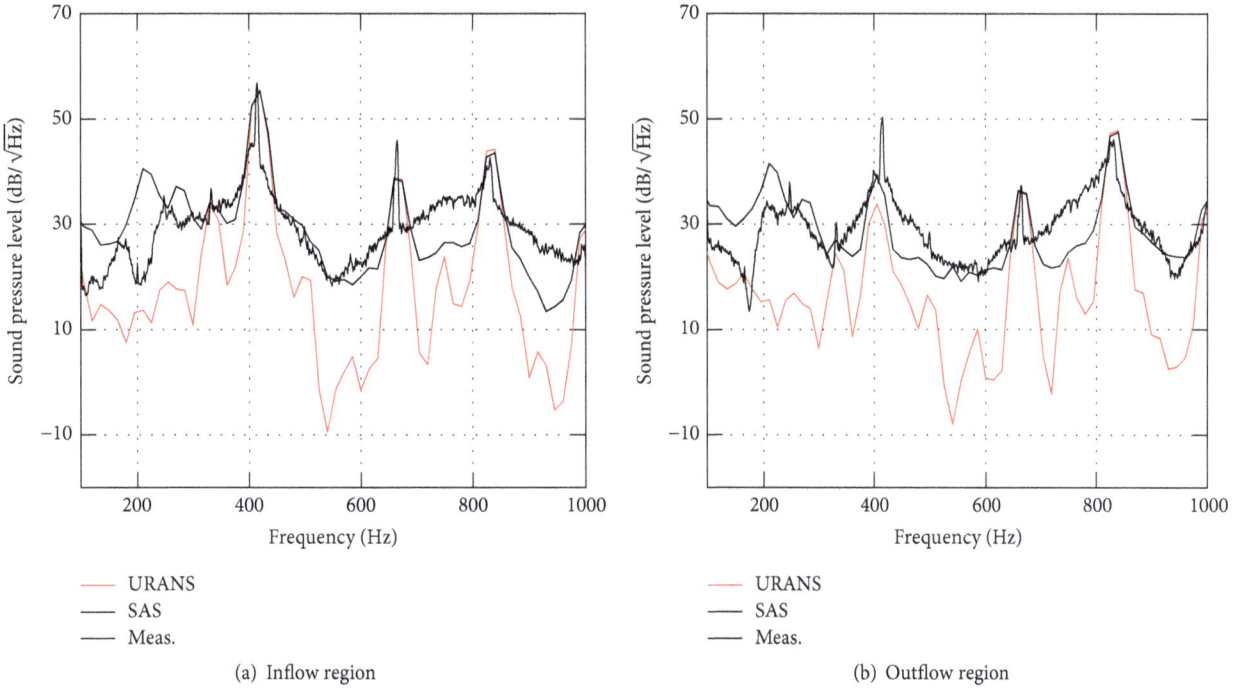

FIGURE 23: Comparison of acoustic spectra: sound pressure levels: inflow: $SPL_{meas.}$ = 66.0 dB; SPL_{URANS} = 69.8 dB; SPL_{SAS} = 70.2 dB; outflow: $SPL_{meas.}$ = 63.7 dB; SPL_{URANS} = 63.1 dB; SPL_{SAS} = 64.7 dB.

TABLE 1: Explanation of peaks in the acoustic spectra.

Frequency in Hz	Trigger	Annotation
215	Acoustic cavity eigenmode	Front panel slot; Figure 24(a)
278, 403	Acoustic cavity eigenmodes	Card slots; Figures 24(b) and 24(c)
416	Acoustic cavity eigenmode	Mainboard slot; Figure 24(d)
417, 833	Blade passing frequency	Including higher harmonics; Figure 25
333, 667, 1000, 1333	Rotor-stator interaction	Including higher harmonics; Figure 27
1666	Blade-stator interaction	—
2330	First structural eigenmode	At 2320 Hz; Figure 9(a)
2490	Second structural eigenmode	At 2484 Hz; Figure 9(b)

first and second structural eigenmodes with frequencies of 2320 and 2484 Hz, respectively (see Figure 9). These two effects can only be identified in the spectra of the measured sound pressure (Figure 22) having a higher frequency limit. A summary of the effects resulting in a tonal sound emission for simulation and measurement is given in Table 1.

Comparing the acoustic spectra of URANS-SST and SAS-SST, again the consequences of the different modeling techniques become obvious: While the URANS-SST simulation shows higher damping of less dominant turbulent structures (cf. Figures 11, 12, 13, 14, 16, 19, and 20), part of these structures

is resolved with the SAS-SST model, also resulting in a better estimation of the acoustics at those frequencies. It is evident that the scale resolving SAS case can account far better for the broadband noise in between the dominant effects like blade passing and rotor-stator interaction. The overall sound pressure levels for the frequency range between 100 and 1000 Hz are given in Figure 23. Both simulations tend to overestimate the sound pressure level at the inlet. The overall sound pressure levels show that results based on the URANS-SST simulation are closer to the measured values than those based on the SAS-SST simulation. The reason is a compensation of errors of the sound pressure at the blade passing frequency (56.8 dB) and its first harmonic (42.6 dB) with the SAS-SST model (55.3 dB and 43.6 dB, resp.) compared with URANS-SST model (55.4 dB and 44.3 dB, resp.). Both simulations underestimate the sound pressure level at the blade passing frequency and overestimate the sound pressure level at the first harmonic of the blade passing frequency. At the outlet the overall sound pressure level is an error range of ±1 dB while at the inlet the error is within ±4.2 dB. The sound pressure simulated for the rotor-stator interaction and its higher harmonics is in general too low with the SAS simulation being in slightly better agreement than the URANS case.

7. Conclusions

CFD and CAA simulations of the air-cooling of a generic electronic device were performed. The flow field was simulated using ANSYS CFX and the sound propagation was calculated using the in-house FE solver CFS++. The acoustic

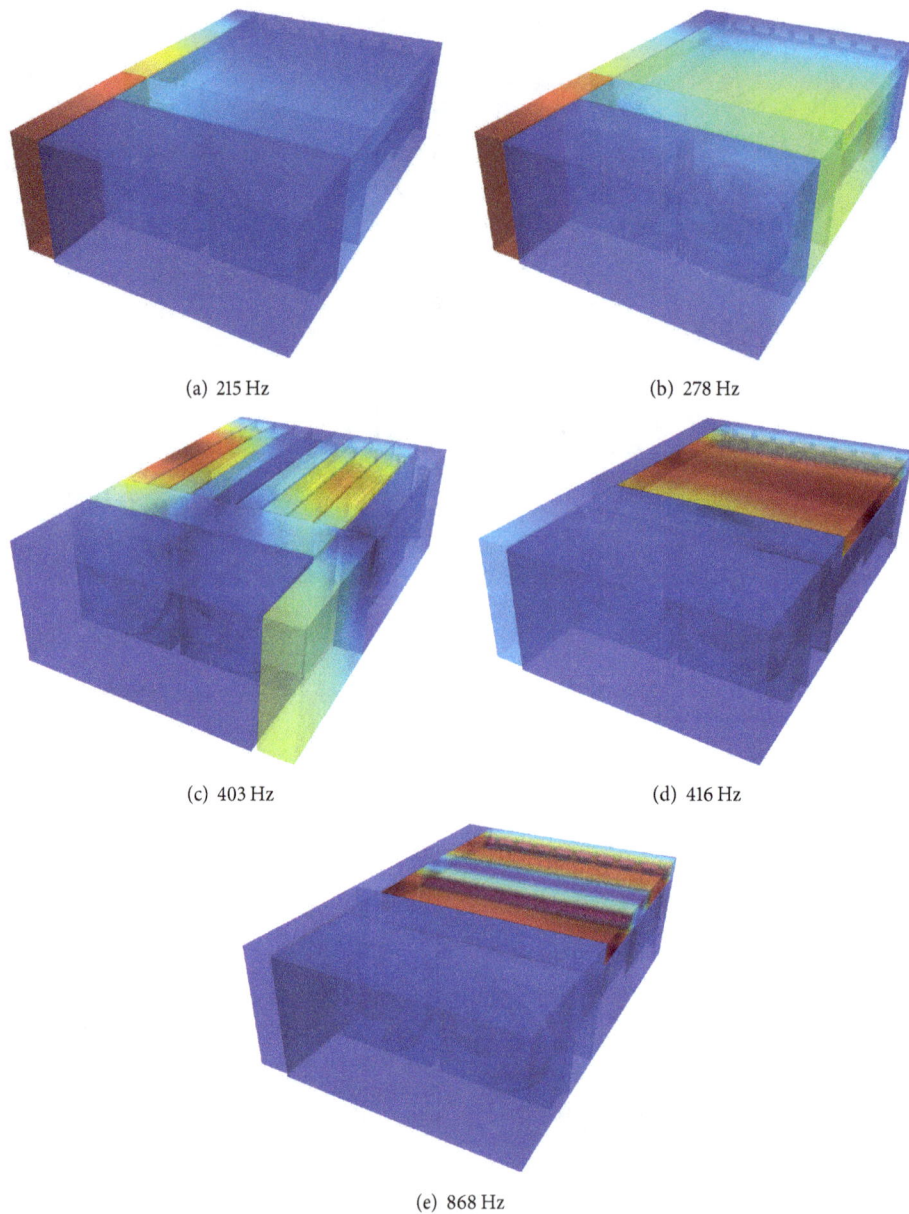

(a) 215 Hz

(b) 278 Hz

(c) 403 Hz

(d) 416 Hz

(e) 868 Hz

FIGURE 24: Acoustic eigenmodes determined via modal analysis of normalized sound pressure (blue: 0; red: 1).

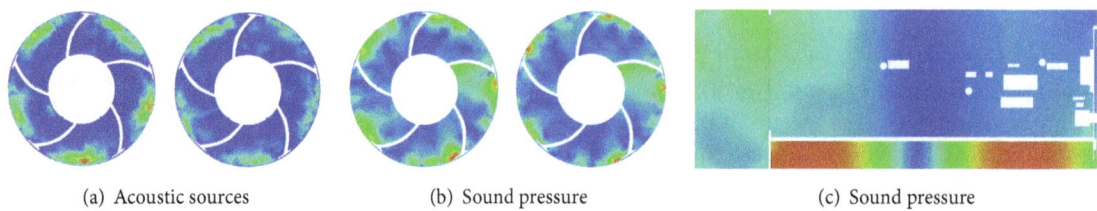

(a) Acoustic sources

(b) Sound pressure

(c) Sound pressure

FIGURE 25: Normalized discrete Fourier transform of the aeroacoustic sources computed according to the right-hand side of (4) (a) and sound pressure field computed by the CAA method based on URANS data at 837 Hz on plane perpendicular to the fan axes (b) and on a vertical midplane through the device (c) (blue: 0; red: 1).

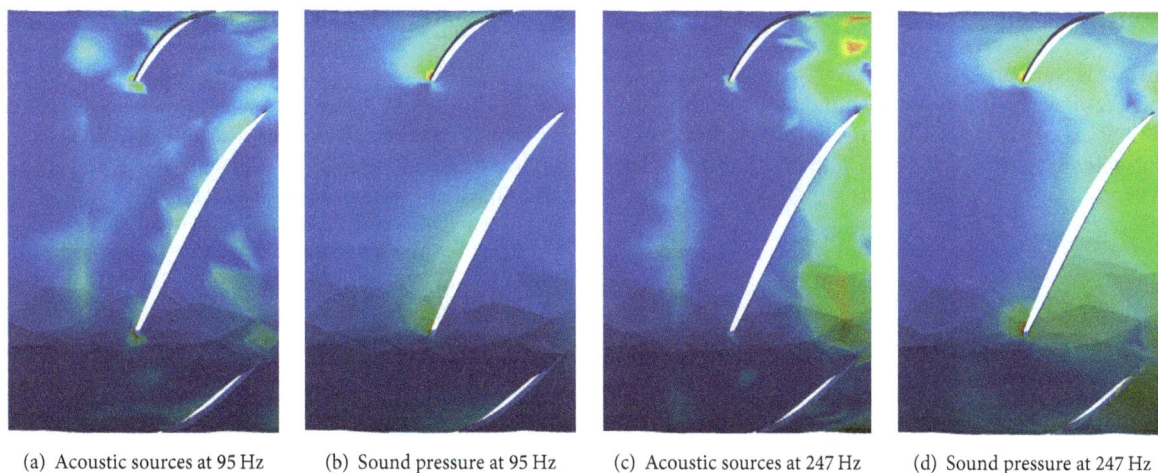

(a) Acoustic sources at 95 Hz   (b) Sound pressure at 95 Hz   (c) Acoustic sources at 247 Hz   (d) Sound pressure at 247 Hz

FIGURE 26: Normalized discrete Fourier transform of the aeroacoustic sources computed according to the right-hand side of (4) and sound pressure field computed by the CAA method based on URANS data at 95 Hz (a, b) and 247 Hz (c, d) on a cylindrical plane concentric to the fan axis at a radius of 35 mm (blue: 0; red: 1).

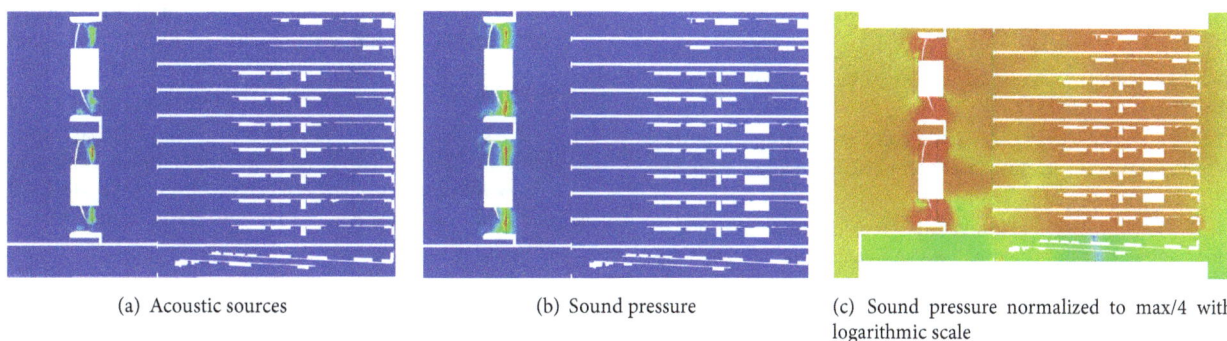

(a) Acoustic sources   (b) Sound pressure   (c) Sound pressure normalized to max/4 with logarithmic scale

FIGURE 27: Normalized discrete Fourier transform of the aeroacoustic sources computed according to the right-hand side of (4) and sound pressure field computed by the CAA method based on URANS data at 666 Hz on plane through both fan axes (blue: 0; red: 1).

results were compared with measurements and showed good agreement. In the case of the CFD simulation, two different turbulent modeling approaches were used: URANS-SST and SAS-SST. Both CFD and CAA simulations demonstrated that the SAS approach is capable of resolving turbulent scales in a more detailed fashion, provided that the spatial discretization (CFD element size) and temporal resolution (time stepping) have been chosen properly. The acoustic source terms were calculated according to Lighthill's acoustic analogy.

The acoustic results with the SAS-SST approach are in better agreement with measurements than the CAA results based on the URANS-SST approach, even though the overall sound pressure level of the URANS-SST is in better agreement with the measurement as the errors compensate each other in a positive manner. The estimated overall sound pressure levels show a maximum error of only 4.2 dB whereas the best estimation is merely 0.6 dB too low. The acoustic peaks in the spectra were linked to different effects, that is, blade passing frequency, rotor-stator interaction, blade-stator interaction, and structural and acoustic eigenmodes of the system.

Even though the results are in good agreement with the measurements, especially the reproduction of the spectra in the SAS case, a more precise estimation might be achieved by accounting for three aspects. First, as already mentioned in the CFL section of this paper, the time step should be further decreased to achieve a CFL number below 1 in the whole simulation domain, especially in the fan domains where most of the acoustic sources are generated. This will additionally account for a better temporal resolution of the acoustic simulation and hence allow resolving up to higher frequencies. Secondly, the acoustic domain should be modified to meet the same conditions as the experiments, that is, an acoustic domain that enables acoustic transmission from inlet to outlet outside of the electronic device instead of the modeled rigid wall. Thirdly, a perfectly matching layer (PML) should be introduced to avoid reflections. This will produce more accurate results especially in the frequency range below approx. 400 Hz.

Competing Interests

The authors declare that they have no competing interests.

Acknowledgments

This work was supported by the Bayerische Forschungsstiftung (BFS, Bavarian Research Foundation) within the project "FORLärm—Lärmminderung von technischen Anlagen" (Noise Reduction in Technical Equipment) under Grant no. 890-09.

References

[1] M. Kaltenbacher, *Numerical Simulation of Mechatronic Sensors and Actuators: Finite Elements for Computational Multiphysics*, Springer, Berlin, Germany, 2015.

[2] E. Baugh, "Acoustic limitations in notebook thermal design," in *Proceedings of the 10th Electronics Packaging Technology Conference (EPTC '08)*, pp. 725–730, IEEE, Singapore, December 2008.

[3] M. Nantais, C. Novak, and J. Defoe, "Graphics processing unit cooling solutions: acoustic characteristics," *Canadian Acoustics*, vol. 34, no. 3, pp. 78–79, 2006.

[4] L. Huang, "Characterizing computer cooling fan noise," *The Journal of the Acoustical Society of America*, vol. 114, no. 6, pp. 3189–3200, 2003.

[5] L. Huang and J. Wang, "Acoustic analysis of a computer cooling fan," *Journal of the Acoustical Society of America*, vol. 118, no. 4, pp. 2190–2200, 2005.

[6] J. Defoe and C. Novak, "Review of computational aeroacoustics for application in electronics cooler noise," *Canadian Acoustics*, vol. 34, no. 3, pp. 76–77, 2006.

[7] F. R. Menter, "Two-equation eddy-viscosity turbulence models for engineering applications," *AIAA Journal*, vol. 32, no. 8, pp. 1598–1605, 1994.

[8] B. E. Launder and D. B. Spalding, *Mathematical Models of Turbulence*, Academic Press, London, UK, 1972.

[9] A. Kolmogorov, "Equations of turbulent motion of an incompressible fluid," *Izvestiya Akademii Nauk SSSR, Seriya Fizicheskaya*, vol. 6, pp. 56–58, 1942.

[10] J. C. Rotta, *Turbulente Strömungen*, Teubner, Stuttgart, Germany, 1972.

[11] F. Menter and Y. Egorov, "Revisiting the turbulent scale equation," in *Proceedings of the IUTAM Symposium on One Hundred Years of Boundary Layer Research*, pp. 279–290, 2004.

[12] F. Menter and Y. Egorov, "A scale adaptive simulation model using two-equation models," in *Proceedings of the 43rd AIAA Aerospace Sciences Meeting and Exhibit*, AIAA paper 2005-1095, Reno, Nev, USA, January 2005.

[13] Y. Egorov, "Menter, development and application of SST-SAS turbulence model in the DESIDER project," in *Proceedings of the 2nd Symposium on Hybrid RANS-LES Methods*, Corfu, Greece, 2007.

[14] F. R. Menter and Y. Egorov, "The scale-adaptive simulation method for unsteady turbulent flow predictions. Part 1: theory and model description," *Flow, Turbulence and Combustion*, vol. 85, no. 1, pp. 113–138, 2010.

[15] M. J. Lighthill, "On sound generated aerodynamically. I. General theory," *Proceedings of the Royal Society of London, Series A: Mathematical and Physical Sciences*, vol. 211, no. 1107, pp. 564–587, 1952.

[16] M. Kaltenbacher, M. Escobar, S. Becker, and I. Ali, "Numerical simulation of flow-induced noise using LES/SAS and Lighthill's acoustic analogy," *International Journal for Numerical Methods in Fluids*, vol. 63, no. 9, pp. 1103–1122, 2010.

[17] S. Triebenbacher, M. Kaltenbacher, M. Escobar, and B. Flemisch, "Nonmatching grids for the coupled computation of flow induced noise," in *Proceedings of the 13th AIAA/CEAS Aeroacoustics Conference*, AIAA 2007-3512, Rome, Italy, May 2007.

5

Quantification and Analysis of Suspended Sediments Concentration Using Mobile and Static Acoustic Doppler Current Profiler Instruments

Angga Dwinovantyo,[1] Henry M. Manik,[2] Tri Prartono,[2] and Susilohadi Susilohadi[3]

[1]*Graduate School of Marine Technology, PMDSU Batch II, Bogor Agricultural University, IPB Darmaga Campus, Bogor 16680, Indonesia*
[2]*Department of Marine Science and Technology, Faculty of Fisheries and Marine Sciences, Bogor Agricultural University, IPB Darmaga Campus, Bogor 16680, Indonesia*
[3]*Marine Geological Institute, Ministry of Energy and Mineral Resources of the Republic of Indonesia, Jl. Dr. Djunjunan No. 236, Bandung 40174, Indonesia*

Correspondence should be addressed to Henry M. Manik; henrymanik@ipb.ac.id

Academic Editor: Kim M. Liew

The application of Acoustic Doppler Current Profiler (ADCP) can be used not only for measuring ocean currents, but also for quantifying suspended sediment concentrations (SSC) from acoustic backscatter strength based on sonar principle. Suspended sediment has long been recognized as the largest sources of sea contaminant and must be considered as one of the important parameters in water quality of seawater. This research was to determine SSC from measured acoustic backscattered intensity of static and mobile ADCP. In this study, vertically mounted 400 kHz and 750 kHz static ADCP were deployed in Lembeh Strait, North Sulawesi. A mobile ADCP 307.2 kHz was also mounted on the boat and moved to the predefined cross-section, accordingly. The linear regression analysis of echo intensity measured by ADCP and by direct measurement methods showed that ADCP is a reliable method to measure SSC with correlation coefficient (r) 0.92. Higher SSC was observed in low water compared to that in high water and near port area compared to those in observed areas. All of this analysis showed that the combination of static and mobile ADCP methods produces reasonably good spatial and temporal data of SSC.

1. Introduction

Measurement of suspended sediment concentration (SSC) is important in the studies of sediment transport [1]. The measurement can be done in various ways, for example, by using conventional (laboratory analysis), optical, and acoustic methods. Conventional methods using gravimetric method in the laboratory have several disadvantages; for instance, it requires numerous water samples at each point and each depth of the observation station. Although this method is the most accurate for determining SSC to date, it still has limitation: it is unable to provide time series data and spatial characteristic of SSC because it requires continuous water sampling [2, 3]. Moreover, this method is also relatively expensive and takes a long time of measurement process and sampling at all observation site, as well as optical method using laser in situ scattering and transmissometry [4]. To overcome this problem, the application of underwater acoustic technologies is used to quantify SSC, which can provide continuous and large coverage in greater detail of SSC data [5].

Utilization of acoustic instrument for determining SSC can be conducted using Acoustic Doppler Current Profiler (ADCP). This process uses extracted amplitudes from ADCP as echo intensity (EI) and converts it to SSC through calibration based on the sonar equation principle [6]. The calibration process was done by comparing the EI value to the laboratory-analyzed SSC at each layer of the water

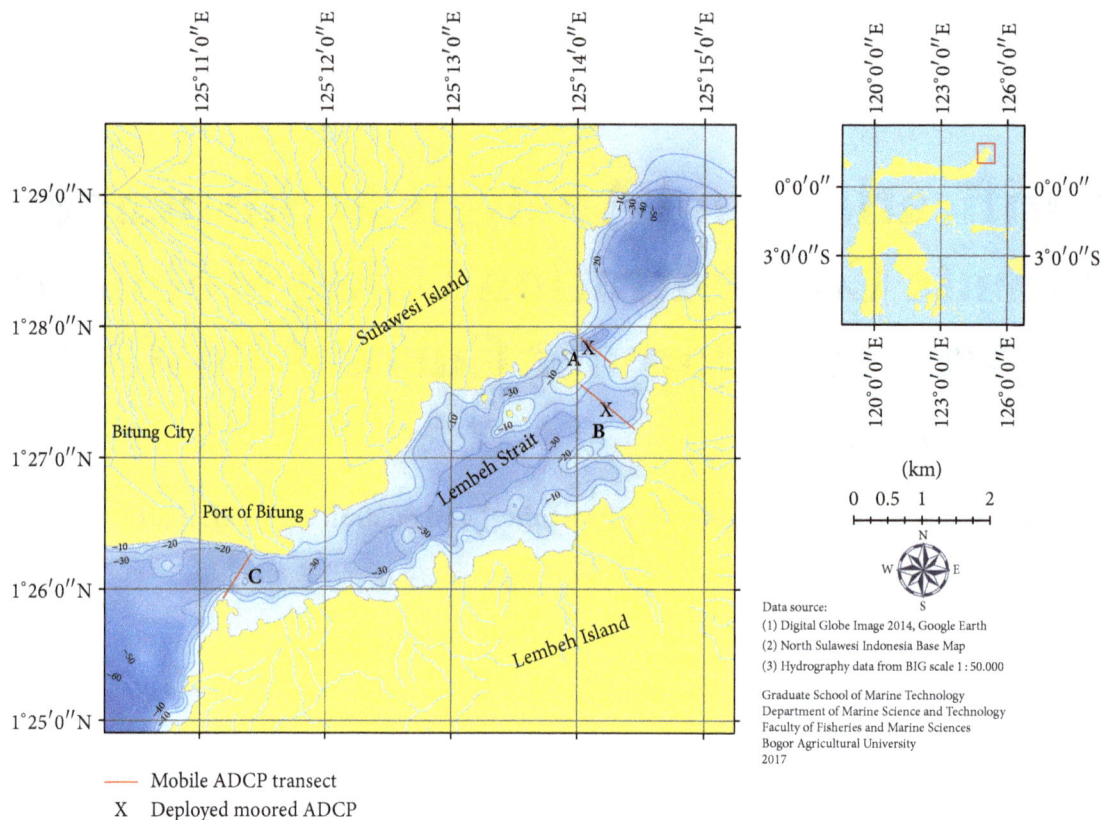

FIGURE 1: Research location in Lembeh Strait, North Sulawesi, Indonesia.

column using simple linear regression analysis. The slope and intercept obtained from linear equations were then used to convert EI values into SSC $(mg L^{-1})$. There are two acquisition methods of ADCP, static (upward-looking) and mobile (downward-looking). Usually SSC is measured statically using ADCP [7]. However, this method has a disadvantage of not being able to see the spatial distribution of SSC in a particular area. The combination of static and mobile method will provide better and more detailed information of temporal and spatial resolution data of SSC [8].

The purpose of this study was to acquire and quantify the acoustic intensity of static and mobile ADCP to determine suspended sediment concentrations. In this study, we used three ADCP instruments with various frequencies in several areas in Lembeh Strait. This study also compared the results of suspended sediment concentrations obtained from ADCP and from laboratory results, as well as comparing SSC results from static and mobile ADCP methods. Specifically, the objectives of the study focused on the calibration methods of each ADCP instrument, the estimation of total suspended solids concentration from backscatter in static and mobile ADCP, and the comparison of the effect of tides on suspended sediment concentrations in oceanography aspect. This research is important because there has not been much research done on strait area with the influence of dominant tidal currents using the combination of static and mobile ADCP methods.

2. Methods

2.1. Time and Location. The collection of water samples and field data using ADCP instruments was conducted in Lembeh Strait, North Sulawesi, Indonesia (Figure 1), in April 2016. Data processing and analysis were done at Laboratory of Ocean Acoustic Data Computation and Sonar System, Marine Acoustic and Instrumentation Division, Department of Marine Science and Technology, Faculty of Fisheries and Marine Sciences, Bogor Agricultural University, Indonesia.

Static ADCP instruments were placed in two different locations: (A) shipping lane area ($1°28'2.8800''$ LU-$125°14'6.3240''$ BT) and (B) dive site area ($1°27'31.2240''$ LU-$125°14'2.1846''$ BT), while mobile ADCP instrument was mounted on the boat and moved according to predefined cross-section: (A) shipping lane area, (B) dive site area, and (C) Port of Bitung area.

2.2. Tools and Materials. The tools and materials used in this research were static ADCP Nortek WAV 6579 with 400 kHz frequency and SonTek Argonaut-XR with 750 kHz frequency and mobile ADCP Teledyne RD Instruments Workhorse Mariner 307.2 kHz frequency (Figure 2) equipped with Differential Global Positioning System (D-GPS) C-Nav and gyroscope motion sensors. Tides data were recorded using Tidemaster Tide Gauge. For the laboratory analysis of SSC, Whatman filter paper with 47 μm diameter and vacuum

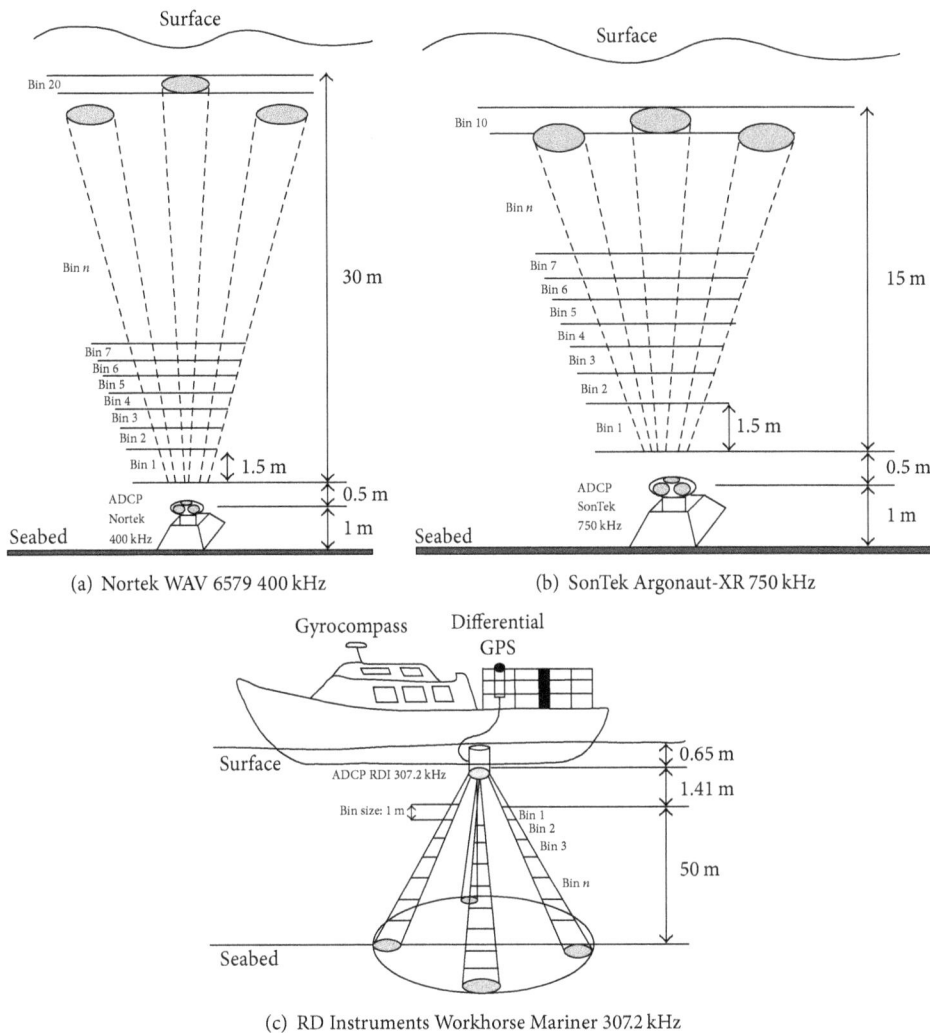

(a) Nortek WAV 6579 400 kHz

(b) SonTek Argonaut-XR 750 kHz

(c) RD Instruments Workhorse Mariner 307.2 kHz

FIGURE 2: ADCP setup: static ADCP (a) Nortek, (b) SonTek, and (c) Mobile ADCP RD Instrument.

pump were used. Water sample was taken using van Dorn Bottle Sampler. Horiba U-50 water quality checker was used as tool to measure temperature, salinity, and pH.

2.3. Data Acquisition, Processing, and Analysis

2.3.1. Experiment Setup. Both static ADCP instruments were deployed on the seabed in two different places (Figure 1) from April 3, 2016, to April 23, 2016. Vertical profiling of SSC was made using upward-looking SonTek ADCP which was deployed at (A) shipping lane area, and Nortek ADCP was deployed at (B) dive site area. These static ADCPs were mounted 50 cm above the seabed with 1.5 m resolution, maximum depth 15, and 30 m for ADCP SonTek and ADCP Nortek, respectively, and were set to 20-minute recording interval. Mobile ADCP was placed 0.65 m below the water surface, with maximum detection depth 50 m and resolution 1 m, and recorded at 1 s interval time. A previous research was conducted using an optical method of LISST-SL measuring device and was compared in stationary and moving mode to evaluate suspended sediment concentration [9].

Water sampling was conducted to measure the SSC concentration at several stations points along with the tidal time. One liter of water sample was taken using a van Dorn bottle; then it was stored in a clean polyethylene bottle. The samples were preserved by being kept in cold conditions in a closed place and hidden from sun. To compare the results between laboratory-analyzed and ADCP-based SSC, data were collected at the same depth and the same time, and to compare the results between static and mobile ADCP, data from mobile ADCP were collected at coordinate 1°28′2.8800″N-125°14′6.3240″E (location of SonTek ADCP) and at 1°27′31.2240″N-125°14′2.1846″E (location of Nortek ADCP) at a particular water sampling time. All of the experiment setup is shown in Figure 3.

2.4. Data Analysis

2.4.1. Correction of Transmission Losses. Several corrections were necessary for converting raw data recorded by ADCP to SSC, for example, changes in transmission power due to distance, sound absorption by particles and chemical

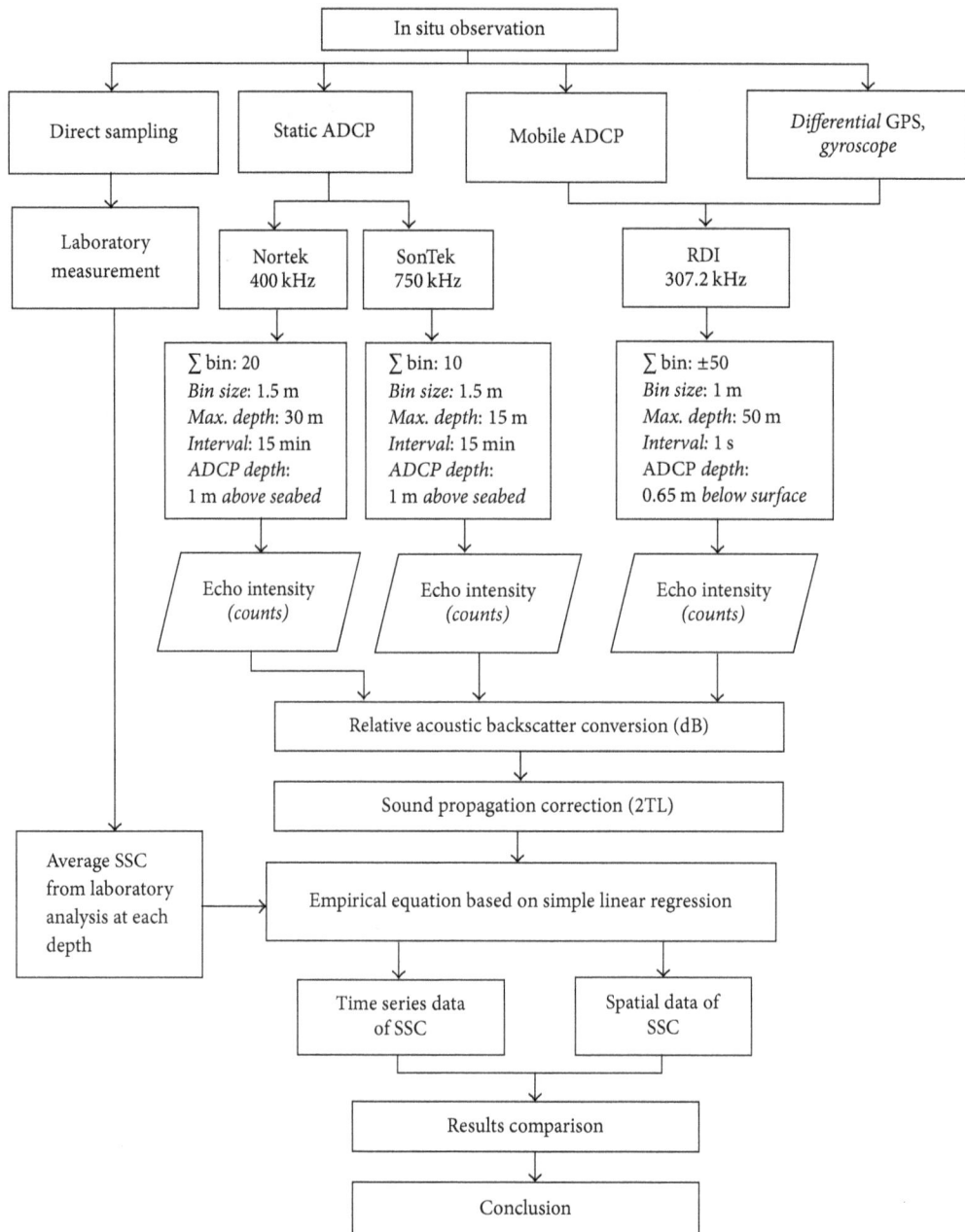

FIGURE 3: Research framework of data acquisition and processing for measuring suspended sediment concentration.

compounds in water column like magnesium sulphate and boric acid, and acoustic energy at near field. In the process of sound wave propagation, attenuation and geometrical spreading were primary factors needed to be removed due to suspended sediment or particles in the water column [10]. The problem was corrected by calculating the transmission loss (TL):

$$\text{TL} = 20 \log_{10} R + \alpha R, \qquad (1)$$

where α is absorption coefficient (dB m^{-1}) and R is the range between ADCP transducer and the measured layer of bin (m). ADCP instrument can differentiate and measure concentration of suspended sediment at different depths

(bin). The measurement on this research was performed on each bin. TL values depended on frequency and SSC because of the ADCP frequency used and SSC encountered in this research. For frequency above 100 kHz, the movement of suspended sediment by sound generated from ADCP produced viscous drag which caused transmission energy loss. The distance between ADCP transducer and the bin was measured in R for each depth:

$$R = \frac{r + 0.5 L_{\text{xmit}}}{\cos \theta}. \qquad (2)$$

From (2), r value is the distance between transducer and the center of bin or half of the bin size, L_{xmit} is the transmission length of acoustic pulse, and θ is the angle of transducer. Absorption coefficient is highly dependent on temperature, salinity, pH, and the presence of materials that can absorb sound waves. In addition, the absorption coefficient factor is also determined by the frequency of ADCP [11]. Absorption coefficient was calculated using the following formula:

$$\alpha = 0.106 \frac{f_1 f^2}{f_1^2 + f^2} e^{(\text{pH}-8)/0.56}$$

$$+ 0.52 \left(1 + \frac{T}{43}\right) \left(\frac{S}{35}\right) \frac{f_2 f^2}{f_2^2 + f^2} e^{-D/6} \quad (3)$$

$$+ 0.00049 f^2 e^{-(T/27 + D/17)},$$

where f is the frequency used in ADCP (kHz), T is the average measured temperature in water column (°C), D is the maximum depth (m), pH is the average pH, and S is the average salinity in seawater (psu). The frequency of ADCP used in the research was the most important term because absorption coefficient depended on the frequency of acoustic instruments. Since the distribution and difference of pH in seawater were not much different, the pH value was not very important in this equation. Based on (3), f_1 and f_2 values indicate absorption by specific chemicals which were boric acid and magnesium sulphate, respectively, and are calculated as

$$f_1 = 0.78 \sqrt{\frac{S}{35}} e^{T/26}$$

$$f_2 = 42 e^{T/17}, \quad (4)$$

where T is the temperature in water column (°C) and S is the salinity in seawater (psu). Both f_1 and f_2 depend on temperature, but only f_1, the boric acid frequency, depends on the salinity.

2.4.2. Echo Intensity to dB Conversion. ADCP instruments received acoustic signal as echo intensity (count). E_r symbol belongs to the noise in counts. The measured echo level values on transducer (RL) were calculated by reducing the echo intensity on each depth in count unit (E) with constant of measured RSSI amplitude seen by the ADCP in the absence of the noise in counts unit (E_r), and then the results were multiplied to echo intensity scale (K_c). The echo intensity (E) values must be much greater than the noise (E_r) since E_r may be affected by environmental noise ($E \gg E_r$). E_r value used in this research could be found by analyzing the signal received in each transducer [11] which was 40 counts. Echo level calculation was determined as

$$\text{RL} = K_c \left(E - E_r\right). \quad (5)$$

In RL value based on (5), K_c value depends on temperature variation but is relatively constant in each depth of each site. K_c values were calculated by

$$K_c = \frac{127.3}{T_e + 273}, \quad (6)$$

where K_c is a factor used to convert the amplitude counts reported by the ADCP's receiver circuit to decibels (dB per RSSI count), numerical value of 127.3 is the mean value of RSSI measurement from ADCP manufacturers, and T_e is the temperature of the ADCP electronics (°C) and add 273 to convert from degrees Celsius (°C) to Kelvin (K). Relative acoustic backscatter (RB) could be determined by adding (5) to (1). Since ADCP principle is active sonar, the transmission losses (2TL) were used in this formula. The RB values were calculated as follows:

$$\text{RB} = \text{RL} + 2\text{TL}. \quad (7)$$

The conversion process from echo intensity in counts to dB was simplified by

$$E_{\text{dB}} = K_c \left(E_{\text{count}} - E_r\right) + 40 \log_{10} R + 2\alpha R. \quad (8)$$

Equation (8) shows that E_{dB} depends on the TL. Based on the empirical formula, 1 count of static ADCP Nortek 400 kHz is equivalent to 0.45 dB, and 1 count of SonTek 750 kHz is equivalent to 0.72 dB. In mobile ADCP RDI 307.2 kHz, 1 count is equivalent to 0.43 dB [12–14]. This variation of conversion process depends on the instrument frequency.

The distance between the transducer and the first bin should be longer than the near field zone or in the far-field. The factor that can distinguish the far-field from near field is the critical range and written as follows [1]:

$$R_{\text{cr}} = \frac{\pi R_0}{4}, \quad (9)$$

where R_0 represents the distance of Rayleigh area (m) which depends on the frequency used [12]. The near field area was determined by the calculation of the correction factor (ψ) based on the equation [15] as follows:

$$\psi = \frac{\left[1 + 1.35Z + (2.5Z)^{3.2}\right]}{\left[1.35Z + (2.5Z)^{3.2}\right]}. \quad (10)$$

The contrast between R_{cr} and R_0 was symbolized as Z.

2.4.3. Conversion to Suspended Sediments Concentration (SSC). Data retrieval using ADCP over a given time span resulted in variation of relative acoustic backscatter (RB) values along with the movement of suspended particles and the amount of suspended sediment. The variation of echo intensity in dB (E_{dB}) value in water column can be attributed to the amount of suspended sediment concentration (SSC) in mg L^{-1}. This relationship is based on the following sonar equation:

$$\text{SSC} = 10^{(A + B \cdot E_{\text{dB}})}. \quad (11)$$

The values of A and B were obtained from simple linear regression of laboratory-analyzed SSC. The SSC estimation was conducted on suspended sediment samples obtained through in situ measurement using ADCP and through gravimetric analysis. The depth for water sampling had to be at the same level as the bin depth observed by ADCP [16]. The comparison between laboratory-analyzed and ADCP-based SSC was then tested using the analysis of independent sample t-test with 95% confidence interval.

2.4.4. Laboratory Analysis of SSC. Water samples were taken from 30 different depths, starting from 1 until 30 m with 1 m interval by using van Dorn bottle. The amount of water sample was 1 liter and then stored in polyethylene bottles. The bottles were later put in a cooling box packed with ice cube (<4°C) before being analyzed in the laboratory. This process was repeated 5 times in different water tidal conditions (from high water tide, ebb tide, and then low water tide). Filter paper with pore diameter 47 μm was used to filter 200 mL of water (C). Before the filtration process, the filter paper was dried at the temperature of 103–105°C for 1 hour, and then the paper was cooled in a desiccator and weighed (B). After water filtration process, the filter paper plus residue was dried again for at least one hour at 103–105°C, cooled in the desiccator, and weighed (A). Direct SSC measurement (mg L^{-1}) was calculated by [17]

$$\text{SSC} \left(\text{mg L}^{-1} \right) = \frac{(A - B) \times 1000}{C}. \tag{12}$$

2.4.5. Comparison between Static and Mobile Method. In evaluating SSC measurement using ADCP, it is important to understand the comparison between the results of static and mobile method. Since frequency is the distinguishing factor between these two methods, verification process was required to evaluate the SSC results. The comparison areas were located at A (shipping lane area) and B (dive site area) because at those stations, the data acquisition was performed using both static and mobile ADCP. This process was done by taking the same time and coordinate of ensemble data from static and mobile ADCP. All mobile ADCP data were acquired at location A on April 10, 2016, and at location B on April 11, 2016. This mobile ADCP data was corrected from tidal condition and plotted using MATLAB.

3. Results and Discussion

During the field observation in static ADCP area, it was found that the seabed substrate consisted of mud and fine sand. Lembeh Strait had strong water currents, likely caused by tidal and bathymetry morphology. In Figure 4, it can be seen that the water condition near the bottom was categorized as turbid with visibility ranging from 5 to 10 m. Based on several factors that have been described, the transmission losses correction factors (absorption and distance) became important because they were related to sound signal propagation. The factors affecting sound propagation recorded by ADCP were distance, medium, and coefficient of attenuation [18, 19].

3.1. Echo Intensity Conversion. Acoustic signal attenuation by suspended sediment (α_s) was divided into three factors: due to medium viscosity, scattering component, and diffraction caused by the loss of energy [20]. By substituting the study area environmental condition (temperature (T) = 28.55°C,

FIGURE 4: Conditions near the bottom in the deployed static ADCP area.

salinity (S) = 33.23 psu, and pH = 8.13) into (3), absorption coefficients (α) were determined as 0.0533 dB m^{-1} on ADCP Nortek 400 kHz, 0.1676 dB m^{-1} on ADCP SonTek 750 kHz, and 0.0361 dB m^{-1} on ADCP RDI 307.2 kHz. The average sound speed of the ADCP device was 1538.98 m s^{-1} and has 20° beam angle on ADCP Nortek and RDI and 25° on ADCP SonTek. Each tool had different scale factors. The value of 1 count echo intensity was equivalent to 0.72 dB count^{-1} on SonTek ADCP instrument; 0.4 dB count^{-1} on the Nortek ADCP instrument; and 0.43 dB count^{-1} on the RD Instrument. The results of echo intensity correction to the transmission losses were then converted into dB units with the result of the time series shown in Figure 5.

The backscattered values of Figure 5 indicate that the water column could be divided into three layers according to the backscattering strength: near surface, water column, and above bed. Backscattering from near the bottom was greater than that from the near surface and water column area. The color differences displayed on the echogram were the representation of the relative acoustic backscatter. Based on previous research [21], if the relative acoustic backscatter value had a high value then it was indicated that the observation area had high concentration of suspended sediment. There were various relative acoustic backscatter values obtained from each ADCP instrument. The relative acoustic backscatter values obtained from ADCP Nortek, SonTek, and RDI ranged between 53–89 dB, 60–95 dB, and 58–85 dB on the near surface; 68–107 dB, 75–110 dB, and 85–98 dB in the water column; and 82–134 dB, 100–138 dB, and 85–100 dB near the bottom of the water, respectively. The highest value was found at the depth of 25 to 30 m. The cause of acoustic backscatter differences in Figure 5 was the presence of suspended sediments in the water column. The characteristic of acoustic backscatter might reflect those of the concentration of scatterers in water column such as suspended sediment. Stronger acoustic backscatter values were due to an indication of higher suspended sediment concentrations in the detected region and vice versa. In addition, the following observation depended on the ADCP's frequency used in the acquisition process. Despite this, the high acoustic backscatter near the bottom of water was indicated by the resuspension of fine fraction of suspended sediment which resulted in negligible variation of the backscattering strength.

TABLE 1: Comparison of the results of simple linear regression analysis for three ADCP instruments used in the study. Each calibration process involved 50 data pairs at different tidal conditions in 8 hours during the ebb, flood, high water, and low water of tides.

Method	ADCP instrument	n	Slope	Intercept	r
Static	Nortek 400 kHz	50	0.0392	1.4021	0.9457
	SonTek 750 kHz	50	0.0435	1.4639	0.9346
Mobile	RDI Workhorse 307.2 kHz	50	0.0202	1.6121	0.9226

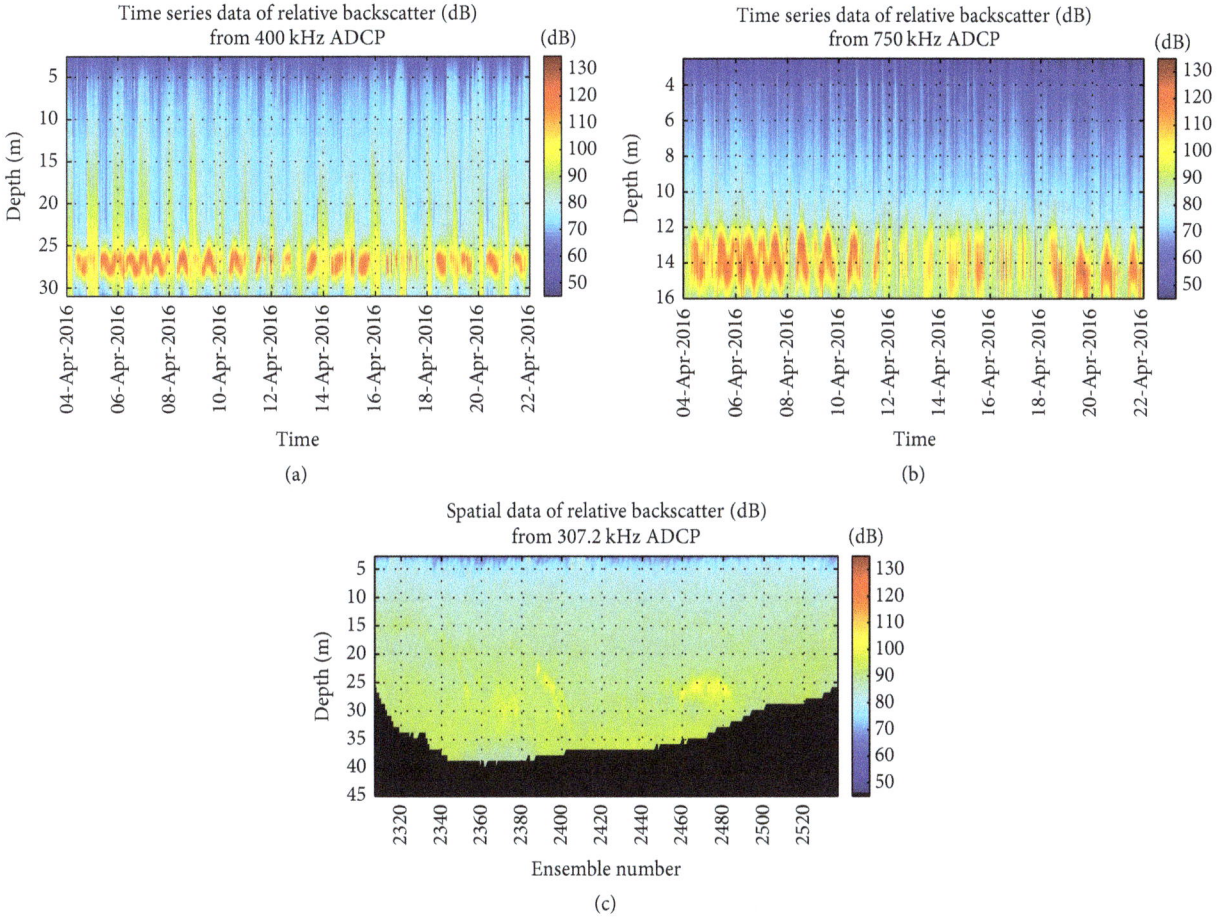

FIGURE 5: Conversion of echo intensity to relative acoustic backscatter from (a) Nortek, (b) SonTek, and (c) one of the sites near the Port of Bitung using RDI ADCP Instruments.

3.2. Calibration.

Measured ADCP backscattered data were converted to SSC data using the laboratory data as calibration and linear regression analysis. SSC data attained from laboratory analysis were then compared to relative acoustic backscatter from ADCP at the same measurement depth. Simple linear regression was used for this analysis. Water sampling was conducted over the ebb, flood, high water, and low water of tidal condition at every one hour during observation. This calibration process was used to determine slope and intercept data, given at Table 1. Figure 6 illustrates a rather high correlation between acoustic backscatter and laboratory-analyzed SSC with all correlation coefficient > 0.9 with each calibration process involving 50 data pairs.

The calibration results show that the relative acoustic backscatter values of ADCP were positively correlated to the laboratory-analyzed suspended sediment concentration. The slope and intercept values were later used to estimate the time series and spatial distribution of suspended sediment concentration. The relative acoustic backscatter values of each bin at all ensembles were then multiplied by the slope and intercept obtained from (9) and the concentration of suspended sediment was figured [22].

3.3. Results of Estimated Suspended Sediment Concentration

3.3.1. Static ADCP.

In Figure 7, it is shown that the concentration of suspended sediment near the seabed was higher than those near the surface and water column. At the dive site area (Figure 7(a)), the highest fluctuation was found at the depth

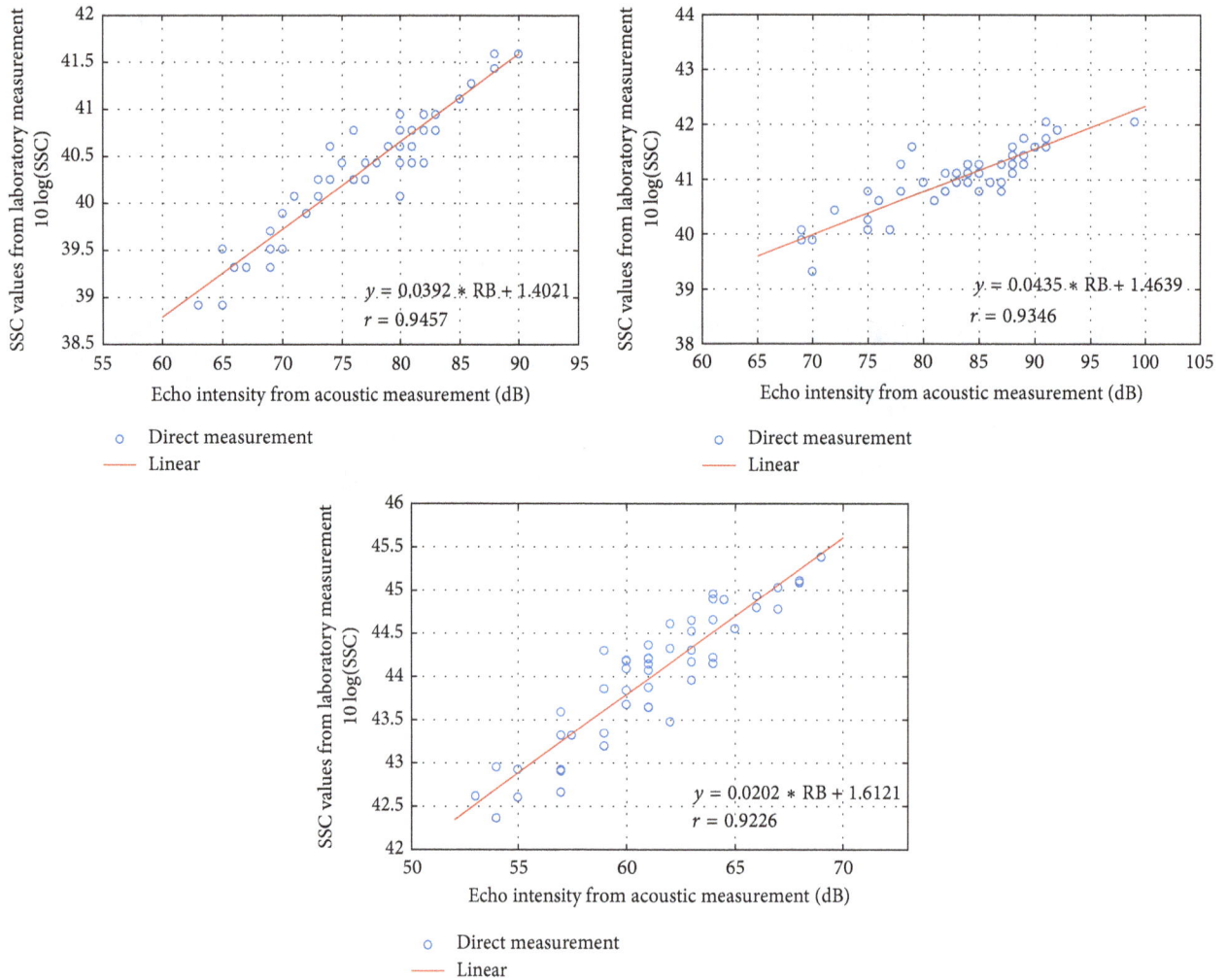

FIGURE 6: Calibration of ADCP relative acoustic backscatter data using the laboratory-analyzed SSC. Each sample was collected in every one hour at different tidal conditions.

between 25 and 30 m. The lowest concentration of suspended sediment in the observed area shown in the echogram was $45\,\mathrm{mg\,L^{-1}}$ and the highest value was $95\,\mathrm{mg\,L^{-1}}$. The SSC were different among the locations and likely depended on location. The SSC at shipping lane area was markedly higher (Figure 7(b)) than those at the other locations and the highest concentration was found at 12–16 m.

The suspended sediment material at the water column was composed by substrate material from the seabed, plankton (especially zooplankton), and other floating microorganisms and particles in the water column. Based on the observation, substrate conditions on the seabed at both locations were almost similar, composed by sandy mud and very fine sand [23]. Resuspension of sediment that primarily contributed to the SSC near bottom was possibly due to advection of bottom water. The movement of suspended material was also influenced by the tidal currents.

This area had an average tidal range of 1.2 m and the tidal pattern was mixed semidiurnal. There were correlations between SSC differences with that tidal condition at high

water, low water, ebb tides, and flood tides. The comparison between ebb and flood tides in SSC measurement shows that the concentration varied. It was indicated that the SSC values were lower during flood tide, but the SSC values were higher during ebb tides. For more details, the data in Figure 7 was magnified by only three days' observation time to prove that the SSC was affected by tidal current (Figure 8).

The effect of tidal current to the SSC values was also confirmed by tide phenomenon that occurred periodically. The indication was the tidal caused water mass movement through tidal currents, carrying suspended particles that affected the amount of suspended sediment concentration. During the low water, the SSC values were significantly higher because high amount of suspended sediment was alternatingly eroded, resuspended, and deposited. If the size of sediment was small (e.g., in sandy mud or very fine sand form), the suspended sediment would have higher concentration [23, 24]. This tidal current affected suspended sediment concentration as an energy source, causing transport of particles.

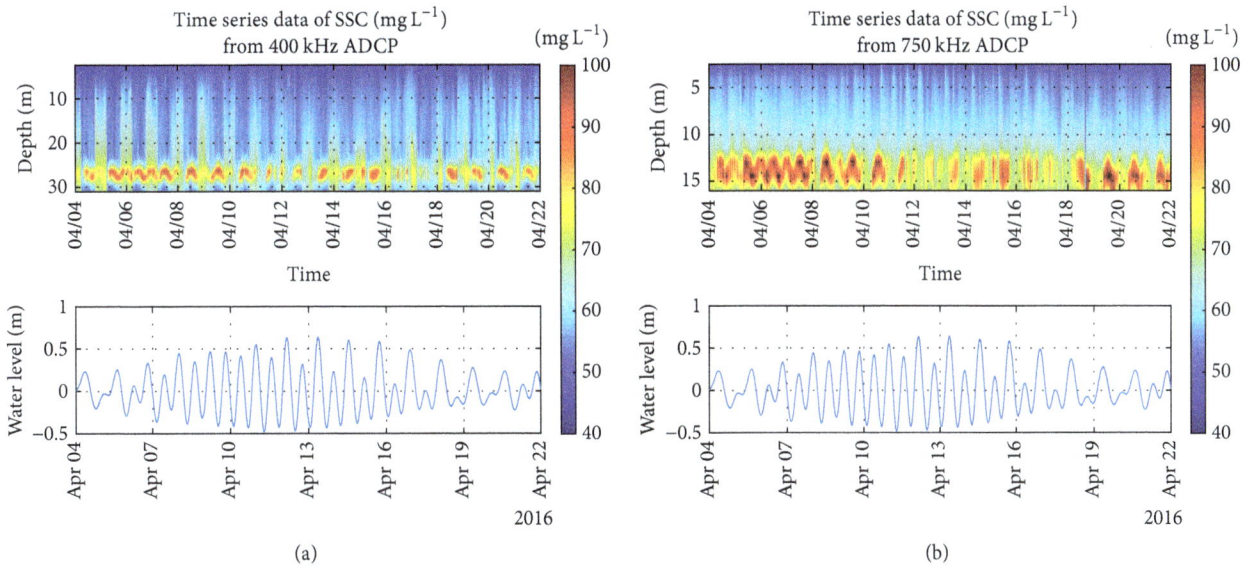

FIGURE 7: Time series of suspended sediment concentration from static ADCP at (a) dive site area and (b) shipping lane area after acoustic data processing with tidal water level (m) comparison. From this data, it is shown that the upper layer of the water column has lower concentration compared to bottom layer.

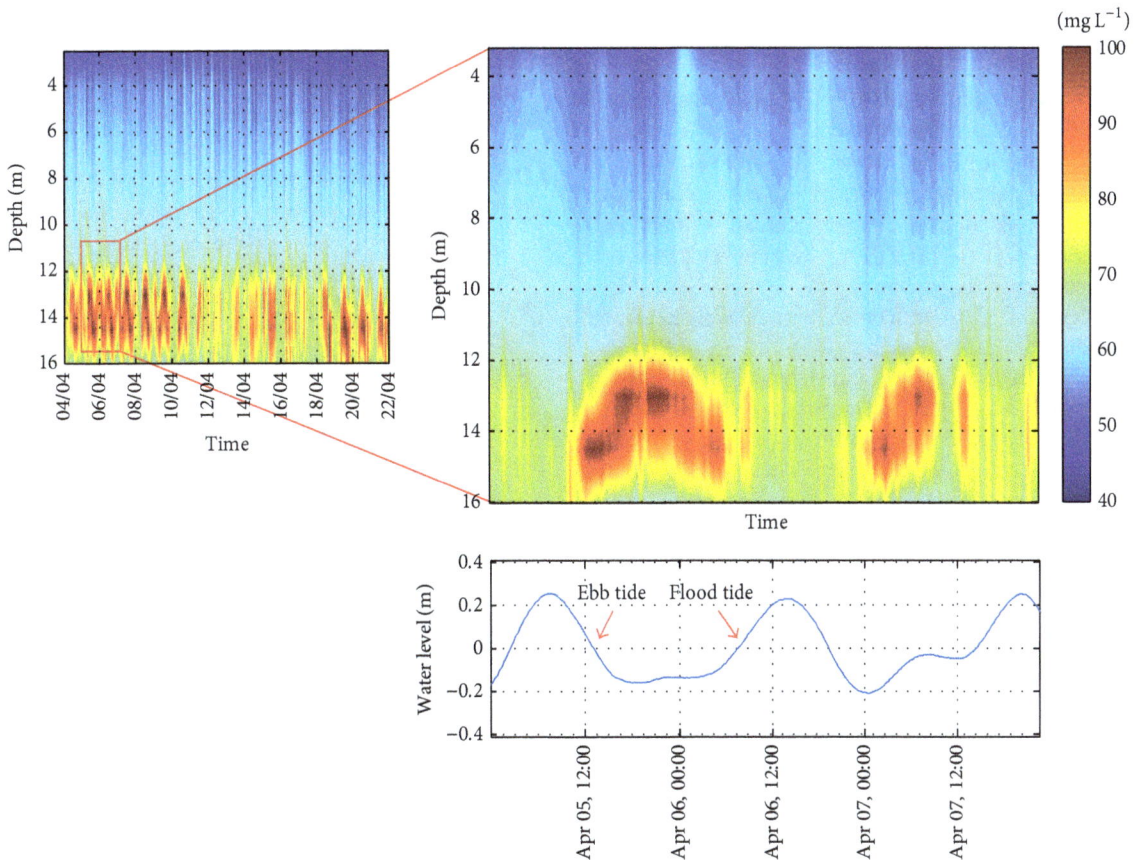

FIGURE 8: The influence of tides on suspended sediment concentration.

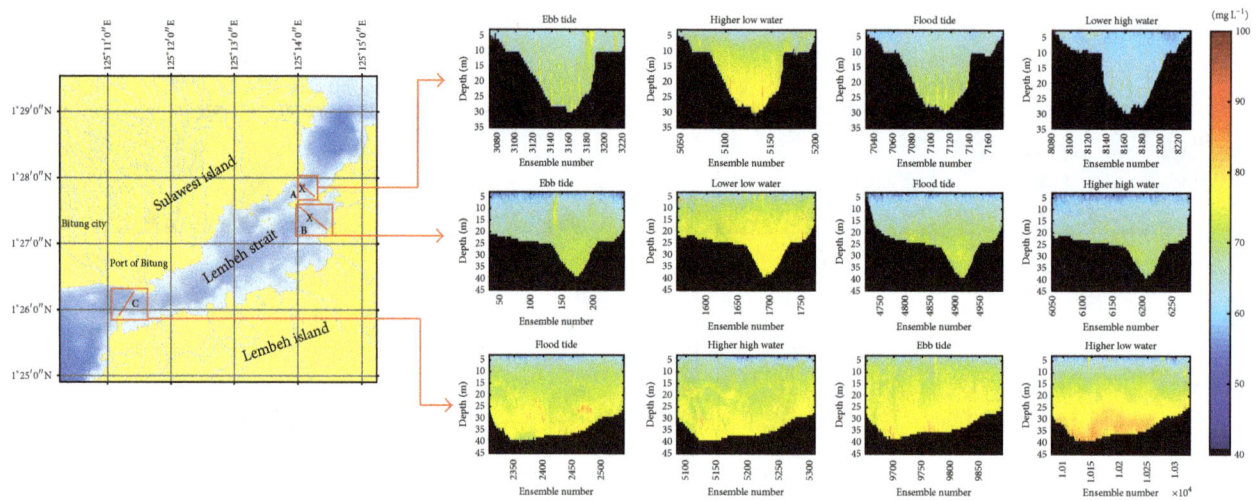

FIGURE 9: Spatial distribution of suspended sediment concentration using the mobile ADCP in three observation areas of Lembeh Strait during four different tidal conditions.

3.3.2. Mobile ADCP. For the mobile method analysis, the averaged data of the surveyed cross-section at different location and tidal time were chosen. The research has previously been performed with various detailed echo intensity at each depth. Field investigations were performed during both high and low tides and also flood and ebb tide cycles in all area. Spatial distributions of average suspended sediment concentration are shown in Figure 9.

Over the cross-section area, the deepest point was around 40 m. Suspended sediment concentration values along Lembeh Strait using mobile method ranged from 45 to 80 mg L^{-1}. The range increased during low tide, which was increased by 10 to 15 mg L^{-1} in all locations. Comparing all of the sites, spatially, the near port area had the highest concentration of suspended sediment. The reason for the higher SSC values obtained at near port during this research was likely because there was sediment input through inlet from Port of Bitung. Port activities such as dredging and disposal to the water column increased SSC [25]. At the locations of shipping lines and dive site area, the limited supply of suspended sediment from mainland or rivers led to lower concentration when compared to areas near the port.

An important aspect of tidal condition in Lembeh Strait was that the current usually flowed from the southwest to northeast during ebb tide to low water and vice versa during flood tide to high water. During high tide, the volume of water in Lembeh Strait increased and thus reduced the concentration of suspended sediment. The fluctuation patterns of suspended sediment concentration at all locations were relatively similar. Based on Figure 9, during the ebb tide to low water tide, the concentration of suspended sediment was higher, otherwise during the flood tide to high water tide, the concentration of suspended sediment was lower.

Based on static and mobile ADCP measurements, it was proved that suspended sediment variation along Lembeh Strait was affected by tidal condition. Suspended sediment concentration was affected by the types and characteristics

of suspended sediment, morphology of bathymetry, seabed sediment, frequency of ADCP instrument, and factors of tidal condition.

3.3.3. Comparison between ADCP-Based SSC and Laboratory-Based SSC. In evaluating the estimation results of the suspended sediment concentration, the correlation between the measurement results of ADCP to the measurement in the laboratory was required to be identified. Accuracy and precision were considerable factors in the utilization of ADCP for the quantification of suspended sediment, and errors were still found in the calculations [4]; for example, on Figure 10, it shows the distribution of suspended sediment concentration derived from ADCP acoustic intensity and direct measurement. By comparing the 1 : 1 line to the best fit linear regression line, the ADCP-based SSC estimates were slightly lower than the estimates based on direct measurement. Results of the acoustic method show a good qualitative agreement with statistical relationships between the SSC and direct SSC measurement method with coefficient correlation (r) > 0.8.

The result of the measurement shows that the estimated concentration of suspended sediment was slightly lower than actual measurement through laboratory analysis which was caused by several factors; for example, there were differences of particle size distribution at various depths as well as differences in sampling intervals. Limitation of ADCP in observing different grain sizes of suspended sediment has been described previously by researcher (e.g., [5]). The ADCP instrument was unable to distinguish between large particle sizes and small particle sizes that accumulate in particular volume in this research, so the measurement using ADCP will be slightly lower in estimating the suspended sediment concentration [26]. However, in other studies different results were found, that is, overestimation in the estuary because of relatively large events on net suspended sediment flux [27].

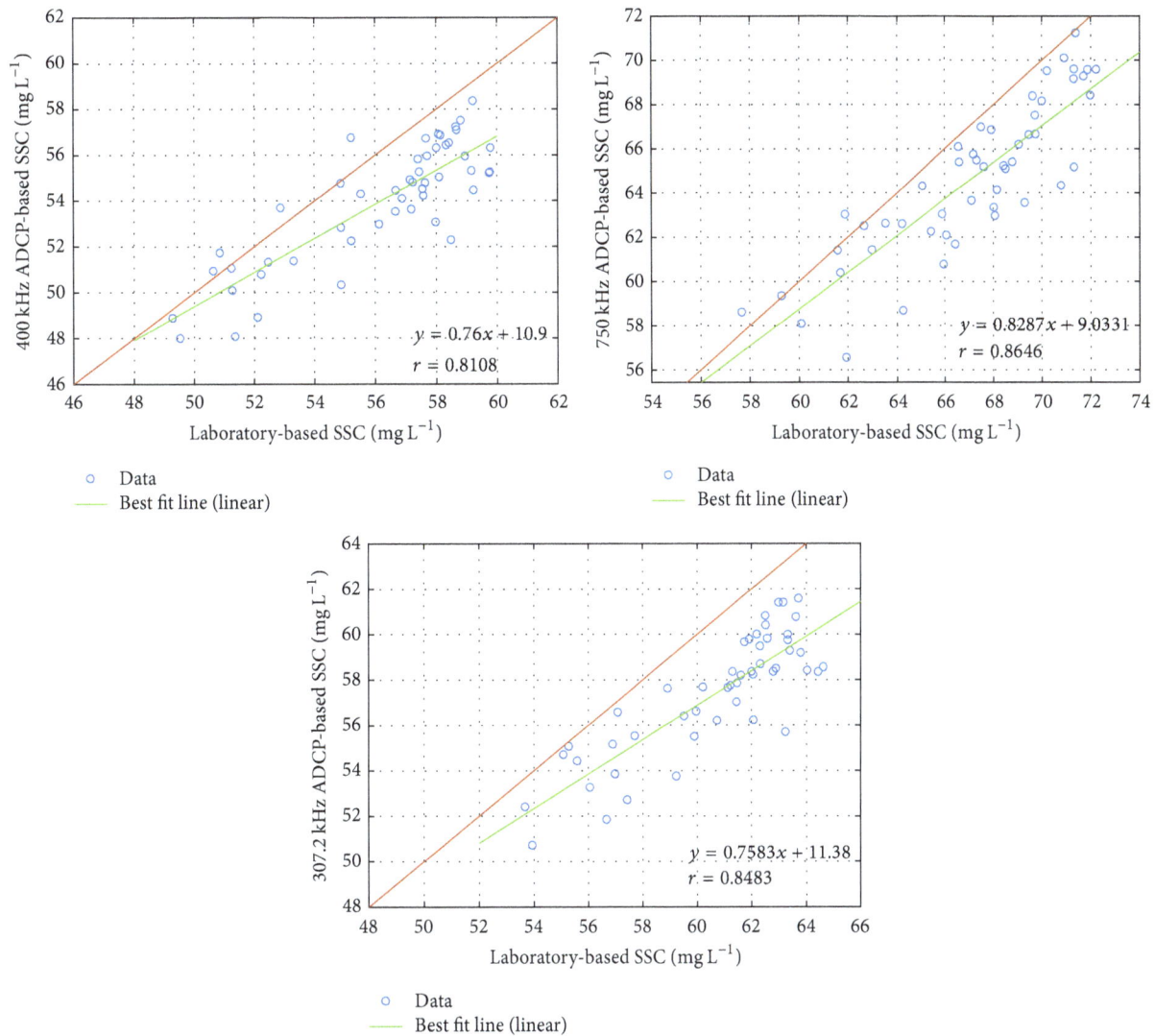

FIGURE 10: Validation of estimated suspended sediment concentrations measured by comparing all ADCP results to the results of laboratory measurements.

3.3.4. Comparison between Static ADCP and Mobile ADCP.
Based on the SSC measurement using static and mobile ADCP, it was known that statistically the results obtained from ADCP were not significantly different when compared to laboratory measurements. However, a validation of the results from ADCP was required by comparing these two methods. During data processing, the mobile and static ADCP coordinates were equated to obtain the same data at one point of the observation area. Factors of data retrieval also needed to be considered. The data in Figure 11 were obtained from the results of ensemble number selection on mobile ADCP RDI Workhorse 307.2 kHz which had the same coordinates as the static ADCP SonTek. A total of 44 ensembles were selected based on the location of the same coordinate points. Intensity data on April 10, 2016, at 10:15 to 14:05 were selected; then the data were transformed into suspended sediment concentration. Figure 11 consists of the time series data of suspended sediment concentration of the

static ADCP SonTek on April 10, 2016, for 24 hours and partially collected data from 10:00 to 14:10.

In Figure 11, it is indicated that the SSC results from the two methods had similar characteristic near the bottom of the water with the depth of 12–15 m. The suspended sediment concentration was higher than those of other depths, ranging from 75 to 83 mg L^{-1}. This strengthens the evidence that in the observation area there was stirring of the bottom sediment so at that depth the concentration was higher. At 2–10 m of depth, the suspended sediment concentration obtained from static ADCP SonTek was slightly higher than that of mobile ADCP. The results obtained from the static ADCP had blue and green colors in echogram and had concentrations ranging from 53 to 58 mg L^{-1} while the results of the mobile ADCP had darker blue color with the concentration ranging 50–55 mg L^{-1}. This was due to the higher frequency of static ADCP (750 kHz), making it more sensitive to detect smaller particles when compared to mobile ADCP (307.2 kHz).

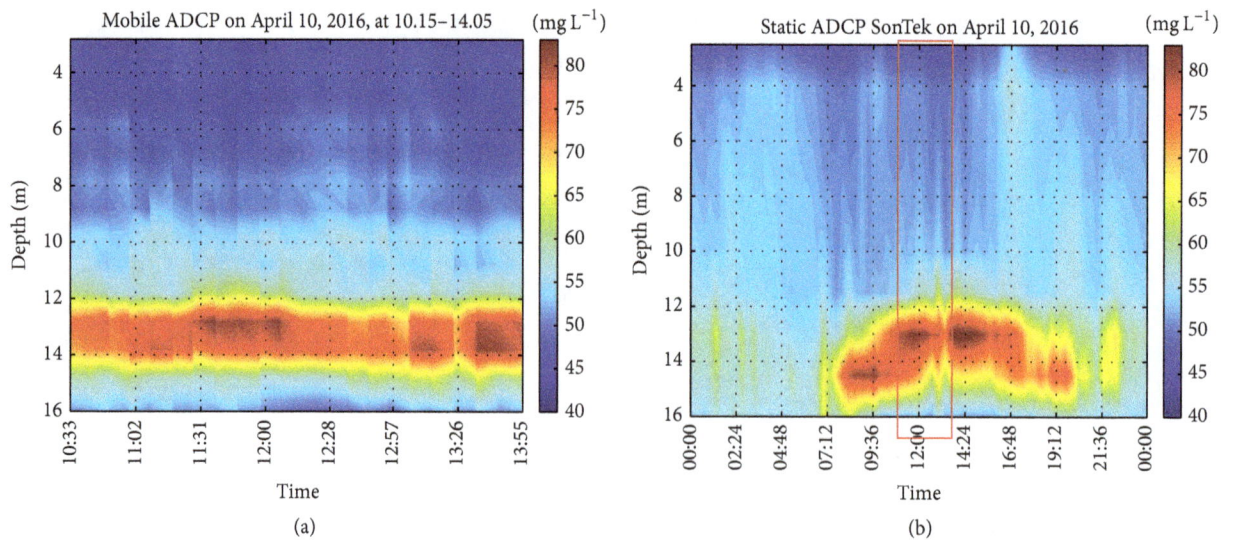

FIGURE 11: Comparison of SSC results from (a) mobile ADCP RDI 307.2 kHz and (b) static ADCP SonTek 750 kHz on April 10, 2016.

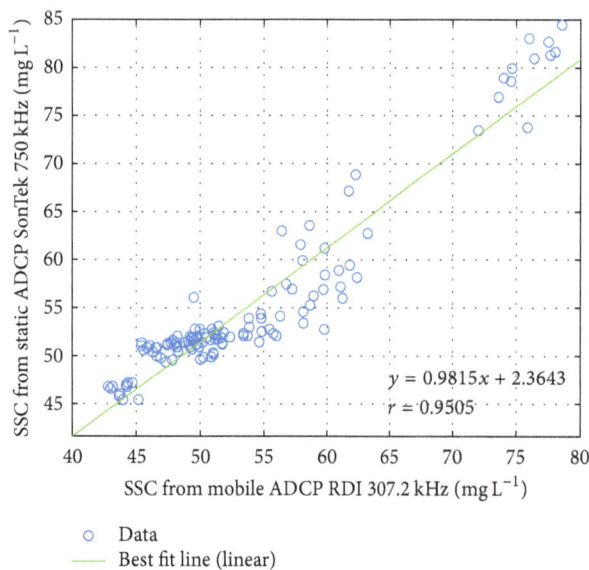

FIGURE 12: Comparison of SSC values obtained from mobile ADCP RDI 307.2 kHz and static ADCP SonTek 750 kHz at the same coordinates and time.

Because of its high sensitivity, the static ADCP was able to quantify smaller sediments, thereby affecting the intensity of the echo reversing back into the transducer to be higher. The suspended sediment concentration in Figure 11 was then extracted and plotted to analyze the error between static and mobile ADCP methods.

Based on the results of the analysis in Figure 12, the highest suspended sediment concentration indicated by reddish-yellow color was found in the 25–30 m depth range with the concentration of 70–80 mg L^{-1} (Figure 13). This result also proves that there was a suspended sediment stirring process in this area as well. At the depth of 2 to 23 m, the concentration of suspended sediments tended to be the same. This was

because the frequencies used in the static and mobile ADCP were not much different (400 and 307.2 kHz). The result of data processing of suspended sediment concentration was then extracted and plotted using MATLAB.

In determining the suitable data retrieval method using ADCP for monitoring suspended sediment concentration in a region, things to consider were data factors provided by ADCP. If the time series data was to be obtained, then static ADCP method had to be selected. However, if the distribution of sediment concentration available in an area was required, the mobile ADCP method could be used [28]. From the results of the study (Figures 11 and 13) it can be seen that both data retrieval methods using ADCP were equally good. The comparison of results obtained from static and mobile methods shows a close relationship with the correlation coefficient, with (r) 0.95. What was needed to be considered, however, was the frequency of the ADCP tool used for the detection and quantification of suspended sediment.

The plot result involved 340 data items with slope and intercept values of 0.9488 and 2.1558. The resulting correlation coefficient (r) had a value of 0.9545 (Figure 14), greater than the comparison between static ADCP SonTek and mobile ADCP RDI. This might be due to frequencies that did not vary much in the static ADCP Nortek (400 kHz) and the mobile ADCP RDI (307.2 kHz), unlike SonTek which had a frequency of 750 kHz compared to RDI's 307.2 kHz. From this finding, it can also be indicated that the results obtained from the static ADCP and the mobile ADCP measurements had a close relationship.

4. Conclusion

The intensity echo value (EI) obtained from the ADCP instrument could be converted to suspended sediment concentration based on sonar equations. The results showed that suspended sediment concentration in the Bitung Port

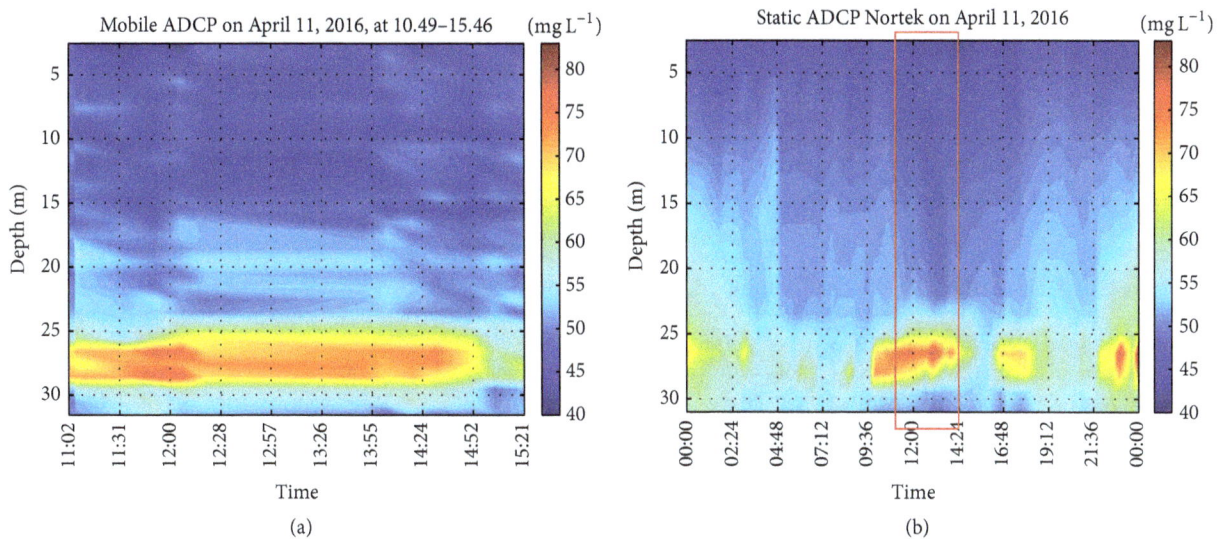

Figure 13: Comparison of SSC results from (a) mobile ADCP RDI 307.2 kHz and (b) static ADC 400 kHz on April 11, 2016.

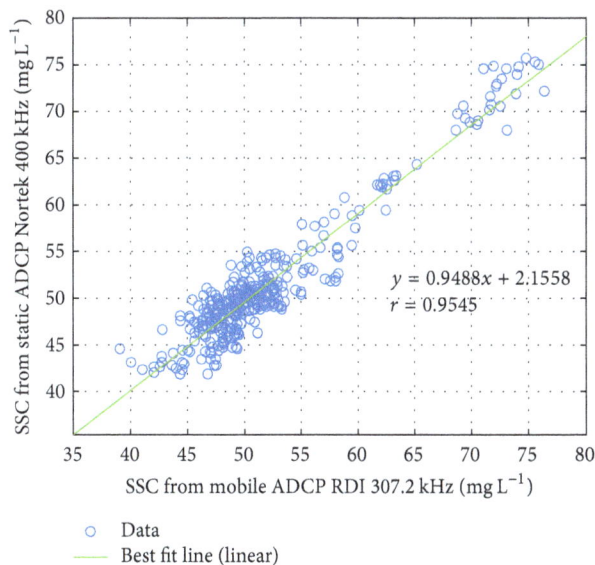

Figure 14: Comparison of SSC values obtained from mobile ADCP RDI 307.2 kHz and static ADC 400 kHz at the same coordinates and time.

area was the highest, which was caused by input from the mainland. The highest suspended sediment concentrations in all locations were located near the bottom especially in areas that had a central slope at the bottom of the strait. The combination of static and mobile ADCP methods for quantification of suspended sediments in Lembeh Strait waters provided good data and could be used as a tool in monitoring sediment transport. Tidal and low tide factors influenced the concentration of suspended sediment.

Further research is needed on the acoustic response of static and mobile ADCP based on the particle size and suspended sediment in the water column. This needs to

be performed in order to determine the concentration of suspended sediment of each size and type.

Acknowledgments

The authors would like to acknowledge the Directorate General of Higher Education (DIKTI), Indonesia, for financially supporting this research through PMDSU Research Grant Scheme (no. 330/SP2H/LT/DRPM/IX/2016 and no. 136/SP2H/LT/DRPM/IV/2017). This work is also supported by Marine Geological Institute, Ministry of Energy and Mineral Resources, Bandung, Indonesia.

References

[1] H.-B. Park and G.-H. Lee, "Evaluation of ADCP backscatter inversion to suspended sediment concentration in estuarine environments," *Ocean Science Journal*, vol. 51, no. 1, pp. 109–125, 2016.

[2] C. Tessier, P. Le Hir, X. Lurton, and P. Castaing, "Estimation of suspended sediment concentration from backscatter intensity of Acoustic Doppler Current Profiler," *Comptes Rendus - Geoscience*, vol. 340, no. 1, pp. 57–67, 2008.

[3] S. A. Moore, J. Le Coz, D. Hurther, and A. Paquier, "On the application of horizontal ADCPs to suspended sediment transport surveys in rivers," *Continental Shelf Research*, vol. 46, pp. 50–63, 2012.

[4] M. Guerrero, N. Rüther, S. Haun, and S. Baranya, "A combined use of acoustic and optical devices to investigate suspended sediment in rivers," *Advances in Water Resources*, vol. 102, pp. 1–12, 2017.

[5] A. K. Rai and A. Kumar, "Continuous measurement of suspended sediment concentration: Technological advancement and future outlook," *Measurement*, vol. 76, pp. 209–227, 2015.

[6] C. Sahin, R. Verney, A. Sheremet, and G. Voulgaris, "Acoustic backscatter by suspended cohesive sediments: Field observations, Seine Estuary, France," *Continental Shelf Research*, vol. 134, pp. 39–51, 2017.

[7] J. W. Gartner, "Estimating suspended solids concentrations from backscatter intensity measured by acoustic Doppler current profiler in San Francisco Bay, California," *Marine Geology*, vol. 211, no. 3-4, pp. 169–187, 2004.

[8] D. Felix, I. Albayrak, and R. M. Boes, "Continuous measurement of suspended sediment concentration: Discussion of four techniques," *Measurement*, vol. 89, pp. 44–47, 2016.

[9] S. Haun, N. Rüther, S. Baranya, and M. Guerrero, "Comparison of real time suspended sediment transport measurements in river environment by LISST instruments in stationary and moving operation mode," *Flow Measurement and Instrumentation*, vol. 41, pp. 10–17, 2015.

[10] Poerbandono and T. Suprijo, "Modification of attenuation rate in range normalization of echo levels for obtaining frequency-dependent intensity data from 0.6 MHz and 1.0 MHz devices," *Journal of Engineering and Technological Sciences*, vol. 45, no. 2, pp. 140–152, 2013.

[11] K. L. Deines, "Backscatter estimation using broadband acoustic Doppler current profilers," in *Proceedings of the 1999 IEEE 6th Working Conference on Current Measurement*, pp. 249–253, March 1999.

[12] S. Jiang, T. D. Dickey, D. K. Steinberg, and L. P. Madin, "Temporal variability of zooplankton biomass from ADCP backscatter time series data at the Bermuda Testbed Mooring site," *Deep-Sea Research Part I: Oceanographic Research Papers*, vol. 54, no. 4, pp. 608–636, 2007.

[13] P. Ghaffari, J. Azizpour, M. Noranian, V. Chegini, V. Tavakoli, and M. Shah-Hosseini, "Estimating suspended sediment concentrations using a broadband ADCP in Mahshahr tidal channel," *Ocean Science Discussions*, vol. 8, no. 4, pp. 1601–1630, 2011.

[14] P. Gruber, D. Felix, G. Storti, M. Lattuada, P. Fleckenstein, and F. Deschwanden, "Acoustic measuring techniques for suspended sediment," in *Proceedings of the IOP Conference Series: Earth and Environmental Science*, vol. 49, no. 122003, pp. 1–11, 2016.

[15] M. N. Landers, *Fluvial Suspended Sediment Characteristics by High-Resolution, Surrogate Metrics of Turbidity, Laser-Diffraction, Acoustic Backscatter, and Acoustic Attenuation*, Georgia Institute of Technology, 2012.

[16] S. Baranya and J. Józsa, "Estimation of suspended sediment concentrations with ADCP in danube river," *Journal of Hydrology and Hydromechanics*, vol. 61, no. 3, pp. 232–240, 2013.

[17] American Public Health Association, *Standard Methods for the Examination of Water and Wastewater*, APHA, 22nd edition, 2012.

[18] M. Guerrero, R. N. Szupiany, and M. Amsler, "Comparison of acoustic backscattering techniques for suspended sediments investigation," *Flow Measurement and Instrumentation*, vol. 22, no. 5, pp. 392–401, 2011.

[19] F. Jourdin, C. Tessier, P. L. Hir et al., "Dual-frequency ADCPs measuring turbidity," *Geo-Marine Letters*, vol. 34, no. 4, pp. 381–397, 2014.

[20] M. G. Sassi, A. J. F. Hoitink, and B. Vermeulen, "Impact of sound attenuation by suspended sediment on ADCP backscatter calibrations," *Water Resources Research*, vol. 48, no. 9, pp. 1–14, 2012.

[21] A. Dwinovantyo, H. M. Manik, T. Prartono et al., "Estimation of suspended sediment concentration from Acoustic Doppler Current Profiler (ADCP) instrument: A case study of Lembeh Strait, North Sulawesi," in *Proceedings of the IOP Conference Series: Earth and Environmental Science*, vol. 54, no. 012082, pp. 1–8, 2017.

[22] V. B. Piotukh, A. G. Zatsepin, and S. B. Kuklev, "Amplitude calibration of an acoustic backscattered signal from a bottom-moored ADCP based on long-term measurement series," *Oceanology*, vol. 57, no. 3, pp. 455–464, 2017.

[23] S. Solikin, H. M. Manik, S. Pujiyati, and S. Susilohadi, "Pemrosesan Sinyal Data Sub-bottom Profiler Substrat Dasar Perairan Selat Lembeh," *Jurnal Rekayasa Elektrika*, vol. 13, no. 1, pp. 42–47, 2017.

[24] J. Xiong, X. H. Wang, Y. P. Wang et al., "Mechanisms of maintaining high suspended sediment concentration over tide-dominated offshore shoals in the southern Yellow Sea," *Estuarine, Coastal and Shelf Science*, vol. 191, pp. 221–233, 2017.

[25] X. Liu, X. L. Feng, and J. Liu, "Characteristics of suspended sediment and resuspension process in Wendeng coastal area, Shandong peninsula," *Advanced Materials Research*, vol. 807-809, pp. 1595–1599, 2013.

[26] G. W. Wilson and A. E. Hay, "Acoustic backscatter inversion for suspended sediment concentration and size: A new approach using statistical inverse theory," *Continental Shelf Research*, vol. 106, pp. 130–139, 2015.

[27] M. A. Downing-Kunz and D. H. Schoellhamer, "Seasonal variations in suspended-sediment dynamics in the tidal reach of an estuarine tributary," *Marine Geology*, vol. 345, pp. 314–326, 2013.

[28] J. J. Nauw, L. M. Merckelbach, H. Ridderinkhof, and H. M. van Aken, "Long-term ferry-based observations of the suspended sediment fluxes through the Marsdiep inlet using acoustic Doppler current profilers," *Journal of Sea Research*, vol. 87, pp. 17–29, 2014.

6

The Tenability of Vibration Parameters of a Sandwich Beam Featuring Controllable Core: Experimental Investigation

Shreedhar Kolekar,[1,2] Krishna Venkatesh,[3] Jeong-Seok Oh,[4] and Seung-Bok Choi[5]

[1]*Mechanical Engineering Department, Jain University, Bengaluru, Karnataka State, India*
[2]*Mechanical Engineering Department, Satara College of Engineering & Management Limb, Satara, Maharashtra State 415015, India*
[3]*Centre for Incubation, Innovation, Research & Consultancy, Bengaluru, Karnataka State, India*
[4]*Division of Automotive & Mechanical Engineering, Kongju National University, Cheonan-si, Chungnam 31080, Republic of Korea*
[5]*Department of Mechanical Engineering, Inha University, No. 253, Yonghyun-dong, Nam-gu, Incheon 402-751, Republic of Korea*

Correspondence should be addressed to Seung-Bok Choi; seungbok@inha.ac.kr

Academic Editor: Kim M. Liew

This study presents experimental results of the vibration parameters of a sandwich beam featuring magnetorheological (MR) fluid as core material. For simplicity, the sandwich beam is considered as a single-degree-of-freedom (SDOF) system and the governing equation is derived in time and frequency domains. Then, from the governing equation, the vibration parameters which can be controllable by external stimuli are defined or obtained. These are the field-dependent natural frequency, damping factor, loss factor, and quality factor of the sandwich beam. Subsequently, a sandwich beam incorporating with controllable MR fluid core is fabricated and tested to evaluate the vibration parameters. MR fluid is prepared using the engine oil, iron particles, and grease as an additive and it is filled into the void zone (core) of the sandwich beam. The fabricated beam is then tested at four different conditions and the vibration parameters are numerically identified at each test. It is shown that both the natural frequency and damping property can be tuned by controlling the intensity of the magnetic field applied to MR fluid domain.

1. Introduction

The development of sandwich structural systems with integrated control capabilities of modal characteristics is crucial to control unwanted vibrations and to avoid resonance problem due to external disturbances. These systems can provide higher flexural stiffness to weight ratio, lower lateral deformations, higher buckling resistance, and higher the natural frequencies. The distributed control force throughout the sandwich structures could be achieved by embedding controllable smart materials as cores or layers between two base structures. This approach can facilitate structure vibration control over a broad range of frequencies through variations in distributed stiffness and damping properties in response to applied stimuli. These structures are called smart sandwich structures like a smart structure in which both the natural frequency and damping property can be controlled by applying external fields such as voltage and current. The development of electrorheological (ER) fluid based sandwich structures was initiated by Gandhi et al. [1–3]. In this work, it has been shown that the dynamic characteristics of ER fluid based sandwich structures can be tuned showing the increment of damping ratio and natural frequencies as the electric field increases. As extension works, Choi et al. showed that the transient vibration of a flexible link robot could be effectively controlled by applying control voltage and also demonstrated that mode shapes of sandwich plate with ER fluid core could be controlled by localizing core zones [4–7]. Experiments were also performed using various ER fluid cores including corn starch, corn oil, and zeolite-silicone oil. Substantial variations in natural frequencies of sandwich beams with these cores were observed by changing the applied electric field [8, 9]. Leng et al. [10] experimentally investigated the vibration analysis of ER fluid composite sandwich beam. It was concluded that the first three modes of natural frequencies and damping factors were increased with increasing the applied electric field. Yalcintas and Coulter [11] developed an analytical model to characterize the forced vibration response of a simply supported ER sandwich beam using RKU (Ross-Kervin-Ungar) model. The numerical solutions were validated through experimental measurements. Yeh and Chen [12, 13] evaluated the variation in the stiffness and natural frequency of the sandwich plate with ER fluid by varying the applied electric field. They concluded that the resonance frequencies of the sandwich plate could be increased with increase in electric field and decreased with increase in thickness of the ER fluid core. It was also found that the thickness of the core has a significant effect on the stability of the sandwich structure system.

Since magnetorheological (MR) fluid has same characteristics as ER fluid except external stimuli, several researchers have attempted to develop sandwich structures featuring MR fluid cores. Yalcintas and Dai [14, 15] investigated dynamic responses of MR fluid adaptive sandwich beam using the energy approach and compared the responses with the structure employing ER fluid. It was concluded that the natural frequencies of MR fluid based adaptive sandwich beam could be nearly twice those of ER fluid based sandwich beam. Sun et al. [16] analytically studied the dynamic responses of a MR fluid sandwich beam using the energy approach and the results are validated by experimental measured data. Oscillatory rheometry techniques were used to carry out experiments to develop the relationship between the applied magnetic field and complex shear modulus of the MR fluid. Yeh and Shih [17] studied theoretically the dynamic responses of MR material based adaptive beam under axial harmonic load using DiTaranto sandwich beam theory. Hu et al. [18] investigated the vibration characteristics of MR fluid based sandwich beam using DiTaranto sixth-order partial differential equation. It was shown that the natural frequencies and loss factors of the MRF beam were increased with increasing applied magnetic field strength. Vasudevan et al. [19] derived the governing differential equations of motion by FEM and Ritz formulations for a sandwich beam with MR fluid treatment and validated through experiments conducted on a cantilever sandwich beam. Various parametric studies were performed in terms of variations of the natural frequencies and loss factor as functions of the applied magnetic field and thickness of the MR fluid layer for various boundary conditions. Lara-Prieto et al. [20] experimentally investigated the controllability of vibration characteristics of MR fluid based sandwich beams under various magnetic field intensities. The effects of applied magnetic field at partial and full length of MR fluid sandwich beam were analyzed. The effectiveness of the linear quadratic regulator and flexible mode shape method based optimal control techniques on controlling transient and forced vibration responses of a fully and partially treated MR fluid sandwich were investigated by Vasudevan et al. [21]. The vibration response of a MR fluid sandwich plate was analyzed by Li et al. [22]. It was shown that the natural frequencies increase with increase in applied magnetic field. However, the loss factors decrease in higher modes with increase in magnetic field. Yeh [23] studied the free vibration characteristics of a magnetorheological elastomers based sandwich plate. The loss factor and the

natural frequencies of the sandwich plate were evaluated under various magnetic fields. Rajamohan et al. [24] studied to find the properties and also vibration response of a partially treated multilayer MR fluid beam and governing equations have been derived for partially treated multilayer prototype beam using finite element and Ritz method and compared the results with experimental and Ritz method; the effects of length and locations of MR fluid layers on the properties of the beam are investigated under different magnetic field conditions and demonstrated upon the boundary conditions and mode of vibration to be controlled for the effective vibration suppression has been derived. Rajamohan et al. [25] investigated governing equations for a partially treated MR fluid layer using FEM and Ritz approach, two different configurations of a partially treated MRF sandwich beam are considered, and the parametric studies were performed to investigate the influence of intensity of an external magnetic field and location and length of MR fluid layers on the dynamic characteristics of the structure with different boundary conditions. Rajamohan et al. [26] worked on governing equations for nonhomogeneous multilayer MR beam which were derived under nonhomogeneous conditions using FEM and Ritz formulation; the beam is formed using three different types of MR fluid and has various shear modulus properties and results showed that natural frequency at higher modes could be controlled by locating the MR fluid layers at desired locations. The natural frequency at higher modes could be increased with decreasing the length of MR fluid layer and it confirms that amplitude of vibration could be easily reduced using controllable MR fluid having different shear modulus located at the desired location and applied to more critical parts to realize more efficient vibration control. Rajamohan and Natarajan [27] worked on the dynamic behaviour of a rotating MRF sandwich beam using FEM and Ritz approach; various parametric studies were performed to study the effect of magnetic field on natural frequency and loss factors. The effect of thickness of MR fluid on natural frequency and effect of rotational speed and hub radius on natural frequencies corresponding to all the modes of vibration of rotating MR sandwich beam increased significantly with the increase in applied magnetic field intensity. Momeni et al. [28] investigated MRF sandwich beam using both experimental and simulation processes. FEM model is used to simulate vibration response under random loading and FEM approach is validated with experimental one and shows that as the magnetic field increases correspondingly the natural frequency for the sandwich beam increases. Walikar et al. [29] worked on engine oil based MR fluid using nickel as magnetisable particle and oleic acid as a surfactant with variation in concentration of nickel particles and found that effect of different magnetic field and concentration of magnetisable particles increase the natural frequency of the beam and amplitude of vibration decreases. Joshi [30] worked on vibration control of cantilever sandwich beam using laboratory prepared MR fluid and observed the variations in vibration amplitude and shifts in magnitude of resonance natural frequency. So the variations usually decreases in vibration amplitude and loss factors and increase in natural frequency as electric/magnetic field increases. However the variations in above parameters were more effective in MR adaptive structures compared with ERF structures.

Despite many research works on sandwich structures having controllable cores such as ER and MR fluids, the study on the vibration parameters which characterize vibration motions of sandwich structures is considerably rare. It is noted that in order to define or explain the vibration parameters such as loss factor a specific and simple model which governs vibration motions needs to be adopted. Many of previous works on smart sandwich structures provide the vibration parameters which are directly obtained from experimental tests without the specific definition. Consequently, this work presents criteria to evaluate the vibration parameters of smart sandwich beams considering a single-degree-of freedom (SDOF). After defining the vibration parameters from the governing equation of the SDOF, a sandwich beam with controllable MR fluid as core was fabricated and tested under free and forced vibration conditions at different magnetic fields. Then, the field-dependent vibration parameters such as natural frequency and loss factor are obtained and compared at four different conditions: empty sandwich beam, MR fluid sandwich beam at 0 T, MR fluid sandwich beam at 0.1 T, and MR fluid sandwich beam at 0.2 T, respectively. It is shown that the vibration parameters heavily depend on the magnetic intensity.

2. Vibration Parameters of SDOF Model

As mentioned in Introduction, several advantages can be achieved by applying smart sandwich structures due to the controllability of core materials. Some of advantages are as follows.

(i) Control of Vibration Amplitude at Resonance. Damping can be used to control the excessive resonance vibrations which may cause high stresses leading to the permanent failure. It should be used in conjunction with other appropriate measures to achieve the most satisfactory approach for random excitations and it is not possible to detune the system and design to keep random stresses with acceptable limit without ensuring that the damping in each mode at least exceeds a minimum specified value. This is a case for sonic fatigue of aircraft fuselage, wing, and control surface panels when they are excited by the jet noise and boundary layer turbulence induced excitations.

(ii) Noise Control. Damping is useful for the control of noise radiation from vibrating surfaces or control of noise transmission through a vibrating surface.

(iii) Damping Phenomenon. The damping is nothing but the energy dissipation in a vibrating structures. The energy which is dissipated in vibrating structures usually depends upon physical mechanisms that exist in the active structures and the physical mechanisms are very complicated physical processes and it is very difficult to analyze the system. The type of damping phenomenon that existed in the structures and usually depends upon the mechanism, which predominates under the given situation, is very essential. In a true physical

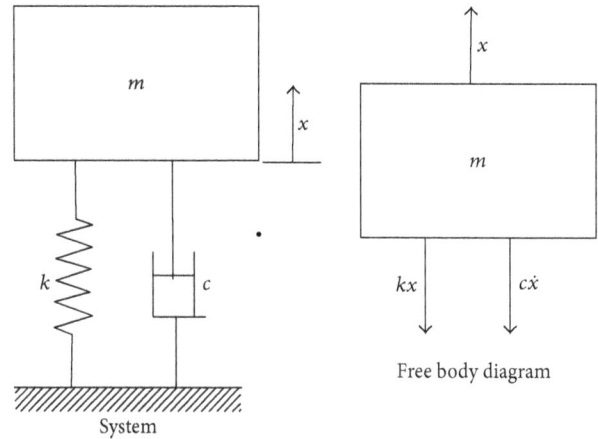

FIGURE 1: Single degree-of-freedom system.

situation, the development of a mathematical equation of motion for the vibrating structure with a physical damping mechanisms is very significant. In the year 1970, Scanlan [31] has found the mathematical damping model which does not give much information.

In order to define vibration parameters of smart sandwich structures featuring controllable core materials, the SODF shown in Figure 1 is adopted because spring-mass-damper model is an oversimplification of the most real structures. In a free vibration system under the undamped case, the vibration response of the SDOF system will never die out. The easiest approach to introduce a dissipation will take place in viscous dashpot system as shown in the figure. The damping force (F_d) which is directly proportional to instantaneous velocity is given by

$$F_d = c\dot{x}, \qquad (1)$$

where c is called a dashpot or viscous damping constant. The loss factor (η) which measures damping phenomenon and is defined as the sinusoidal excitation of the system to the corresponding sinusoidal response of the system is as follows:

$$\eta = \frac{c\,|\omega|}{k}, \qquad (2)$$

where k is the stiffness of the system. The above equation is similar to the equation for the viscoelastic systems developed by the Ungar and Kerwin's [32]. Eq. (2) shows a linear dependence between loss factor to driving frequency and inversely proportional to the stiffness of the system. This kind of frequency dependence has been discussed by Crandall in the year 1970 [33], but in actual practices it is not possible this form and in such a case often resorts to an equivalent ideal dashpot system. The theoretical objections to the approximately constant value of damping over a range of frequency, as can be observed in aeroelasticity problems, have been raised by Naylor in the year 1970 [33]. From (2), the frequency- dependent dashpot system is given by

$$c\,|\omega| = \frac{k\eta\,(\omega)}{|\omega|}. \qquad (3)$$

From Figure 1, the frequency domain representation of equation of motion can be written as follows:

$$\left[-m\omega^2 + i\omega c\,(\omega) + k\right] X\,(i\omega) = F\,(i\omega), \qquad (4)$$

where $X(i\omega)$ is the response function and $F(i\omega)$ is the excitation function. The viscous damping or dashpot has frequency dependence. Substituting (3) into (4) yields the following:

$$\left[-m\omega^2 + k\left\{1 + i\eta\,(\omega)\,\mathrm{sgn}\,(\omega)\right\}\right] X\,(i\omega) = F\,(i\omega), \qquad (5)$$

where $\mathrm{sgn}(\omega)$ is the signum function. For the "time domain" representations of (3) and (4) are expressed as follows:

$$m\ddot{x} + c\,(\omega)\,\dot{x} + kx = f. \qquad (6)$$

Then, by assuming the response function as $x(t) = x_0 \cos(\omega t - \delta)$ for the harmonic function at the frequency (ω) and also the phase lag (δ), the relationship between forcing function to the excitation can be related by

$$F = kx + \frac{k\eta}{|\omega|}\frac{dx}{dt}, \qquad (7)$$

where k is the stiffness. $\eta = \tan\delta$ is called the loss factor for inertial and stiffness properties. The phase angle δ lies between $0°$ and $90°$, and the loss factor η lies between 0 and ∞. The relationship exists between η and δ values.

On the other hand, the governing equation of the system shown in Figure 1 is given by

$$m\frac{d^2x}{dt^2} + kx + \frac{k\eta}{\omega}\frac{dx}{dt} = F_0 \cos\omega t. \qquad (8)$$

For the harmonic response function in a steady state condition, transients at any start have usually died out. If the stiffness (k) and loss factor (η) depend upon frequency for real materials, then the maximum amplitude at the resonance frequency (ω_r) can be written as $\omega_r = \sqrt{k/m}$. Amplification factor (A) is nearly equal to $1/\eta(\omega r)$ and the loss factor is given by

$$\eta = \frac{1}{A} = \frac{\Delta\omega}{\omega_r}, \qquad (9)$$

where $\Delta\omega$ is the separation of the frequencies; its response is $1/\sqrt{2}$ times the peak response. In summary, $\eta = 1/A = \Delta\omega/\omega_r$; $\Delta\omega$ is the frequencies separation response which is equivalent to $(1/\sqrt{2})X$. The peak response which is known as half power bandwidth is equal to $\eta = 1/A = \Delta\omega/\omega_r$.

Now, the following evaluation (vibration) parameters are defined as criteria for vibration characteristics of smart sandwich structures.

Natural Frequency or Velocity Resonant Frequency (ω_n). The calculation of natural frequencies is of major importance in the study of vibrations. Because of friction and other resistances vibrating systems are subjected to damping to some degree due to dissipation of energy. If the damping

is small, it has very little effect on natural frequency of the system.

(i) *Natural frequency* $(\omega_n) = \dfrac{2\pi}{T}$ rad/sec $\qquad (10)$

(ii) *Time period* $(T) = \dfrac{1}{\omega}$ sec. $\qquad (11)$

(iii) Damping Factor (ζ). The half power point method is another method for finding the damping factor form graphical representation and it provides fairly accurate results. Suppose that the natural frequency (f_n) can be measured at the peak amplitude $(X_{\max})$, and f_1 and f_2 are two values of natural frequency of the unit of Hz. Then, the damping factor can be obtained by

$$\zeta = \frac{f_2 - f_1}{2f_n}. \qquad (12)$$

(iv) Peak Amplitude. The maximum displacement of a vibrating body, in vibration study amplitude, should be minimized and natural frequency is to be increased.

(v) Logarithmic Decrement (Δ). Whenever the damping system is stroked by the impulse force, the response is exponentially decreased and from this the logarithmic decrement is obtained by

$$\Delta = \ln\frac{x_1}{x_2} = \ln\frac{x_n}{x_{n+1}}. \qquad (13)$$

It is noted that the above definition is only useful for viscous and hysteretic type of damping within limits because the ratios are equal for the viscous cycles. On the other hand, using forcing function and excitation function expressed in Fourier transform the solution of (4) can be expressed as follows:

$$x\,(t) = \frac{1}{2\pi}\int_{-\infty}^{\infty}\frac{Fe^{j\omega t}d\omega}{k} - m\omega2 + jk\eta. \qquad (14)$$

The above equation consists of two parts; one is real and another is an imaginary function, and then the following solution is obtained:

$$x\,(t) = \frac{F}{\pi}\int_{0}^{\infty}\frac{\left(k - m\omega^2\right)(\cos\omega t) + k\eta\sin\omega t}{\left(k - m\omega^2\right)^2 + (k\eta)^2}d\omega. \qquad (15)$$

The values of stiffness (k) and loss factor (η) are not to be constants for the real systems in a certain wide frequency range. For small values of loss factor (η), the accuracy is the best and the following equations are obtained:

$$x\,(t) = \frac{F}{\sqrt{km}}e^{-1/2\eta t\sqrt{k/m}}\sin t\sqrt{\frac{k}{m}}. \qquad (16)$$

$$\Delta = \frac{\pi\eta}{2} \qquad (17)$$

(vi) Resonance Frequencies. The peak values of the displacement, velocity, and acceleration response of a system

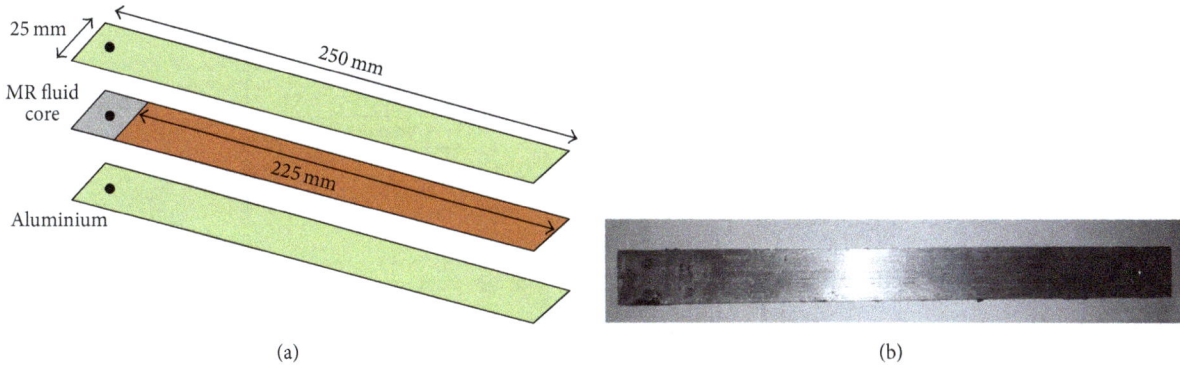

FIGURE 2: Fabricated sandwich beam with MR fluid core; (a) schematic, (b) photograph.

undergoing forced, steady state vibration occur at slightly different forcing frequencies. Since a resonance frequency is defined as the frequency for which the response is maximum, a simple system has three resonance frequencies if defined only generally. The natural frequency is different from any of the resonance frequencies. There is a relationship between damped natural frequency and undamped natural frequency using the damping ratio.

Now, the following vibration parameters to evaluate smart sandwich structures are achieved as follows:

(i) *Damped Frequency*: $\omega_d = \omega_n \sqrt{1 - \zeta^2}$ (18)

(ii) *Displacement Resonant Frequency*:

$$Df = \omega_n \sqrt{1 - 2\zeta^2}$$ (19)

(iii) *Acceleration Resonant Frequency*:

$$Af = \frac{\omega_n}{\sqrt{1}} - 2\zeta^2$$ (20)

(iv) *Loss Factor*: $\eta = 2\zeta$ (21)

(v) *Quality Factor* $= \frac{1}{2}\zeta$ (22)

(vi) *Specific damping capacity*: $Sc = 4\pi\zeta$. (23)

3. Design and Manufacture of Sandwich Beam

A simple sandwich beam is selected which is capable of withstanding load primarily by resisting bending. Beams are traditionally descriptions of building or civil engineering structural elements, but smaller structures such as truck or automobile frames, machine frames, and other mechanical and structural systems contain beam structure that are designed and analyzed in a similar fashion. So this simple mechanical model can be implemented in to more complex structures. Here in this research work the controllable capabilities of MR fluid in adaptive structures were analyzed in real time. In order to evaluate the criteria for vibration characteristics of sandwich structures, a sandwich beam with MR fluid core was fabricated and tested in this work. Figure 2 presents the schematic diagram and photograph

TABLE 1: Dimensions of each layer.

SI number	Type of layer	Dimensions of layer
(01)	Base structure	$250 \times 25 \times 3$ mm
(02)	Core layer (h_2)	$225 \times 20 \times 3$ mm

TABLE 2: Properties of beam and MR fluid.

SI number	Type of layer	Properties
(01)	Base structure (aluminium)	Density = 2700 kg/m³ Young's modulus = 70 Gpa
(02)	Core Layer (MR fluid)	Density = 2781 kg/m³

of the proposed smart sandwich beam tested in this work. MR fluid sandwich beam consists of constrained layer, base structures, and MR fluid core between upper and lower base structures which are made of aluminium. The dimensions of each layer are given in Table 1 and properties of the sandwich beam with MR fluid core are given in Table 2. The beam top and bottom base layer has length of 250 mm and width of 25 mm, while MR fluid core length is 200 mm with the width of 20 mm. The thickness of each aluminium layer and core layer is 3 mm. At the clamped end of the plate a small 45 mm extra plate is fixed in between two aluminium strips; other surfaces of the beam are sealed with silicon rubber; for filling up of MR fluid in a cavity a small hole of 5 mm size diameter was drilled in each side of the beam. From one end of the of the beam MR fluid is injected using hypodermic syringe; this allowed air bubbles to escape from other side of the hole and finally two holes were sealed and allowed to dry it.

3.1. Synthesis of MR Fluid. MR fluid used in this work was made according to the following steps. *Step 1.* Pour the desired amount of carrier oil into the beaker. Stir it for one hour using mechanical stirrer. *Step 2.* Take a small quantity of grease and mix it with carrier oil and stir it in a mechanical stirrer for 2-3 hours until the grease particles will be completely dissolved and suspended in carrier oil. After complete mixing slowly pour the iron particles in mixture and again stir it for 6–8 hours. *Step 3.* Prepared MR fluid

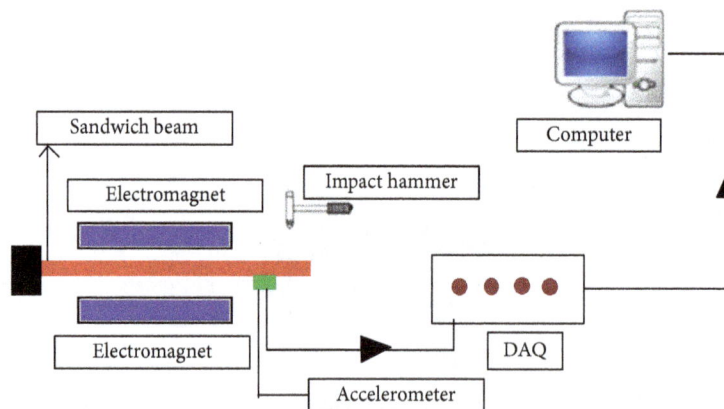

FIGURE 3: Schematic Experimental test rig for the MRF sandwich beam.

TABLE 3: Properties of SAE grade 20W-40.

SI number	Properties	Values
(01)	Density at 20°C	Density = 876 kg/m^3
(02)	Kinematic viscosity at 100°C cSt	13.5–15.5
(03)	Viscosity index	110
(04)	Flash point	200°C

is ready for the use. In this research work low viscosity silicon oil is used for the preparation of MR fluid in the laboratory. The composition is summarized as follows: carrier oil-Engine oil (SAE Grade 20W-40) of 70% and Magnetic particles-Electrolytic Iron powder (Industrial Metal Powders Pune India) of 30% Additives, Commercial Grease of 10%. The sample is denoted as SAE Grade 20W-40 (Mobil Diesel Special Multi Grade Diesel Engine oil) and the principal properties are given in Table 3.

3.2. Experimental Setup. The main elements which were employed in this work include MR fluid sandwich beam, accelerometer, exciter, power amplifier, Data acquisition system (DAQ), Electromagnets, Display unit, and Impulse hammer. *(i) Accelerometer.* Place the accelerometer on the free end of the beam by applying gum solution on it and connect it to the data acquisition system (DAQ). It senses the analogous displacement of the beam and sends it to the data acquisition system (DAQ). Here in this research work Kistler model 9722A2000, an accelerometer which has the sensitivity of 10.84 mV/g is used. *(ii) Exciter.* It is used to give desired excitation to the beam; the power is given to the exciter by controller which is connected with the computer to select the excitation parameter. The different types of excitation can be generated by the exciter by using power amplifier Ex: Sine swept sine, rectangular, triangular, and so on. In case of forced vibration we can use swept sine signal in which user has to select the initial and final frequency and swept rate. Here in this research work LDSV101-permanent magnet shaker is used in this experiment (sine peak force 8.9 N).

It is extensively used in industries and it is suitable for the analysis of dynamic behaviour of the structures and materials *(iii) Power Amplifier.* It is connected to the exciter to give particular excitation frequency to the exciter. Here in this research work TECHRON 5507 is used as power amplifier. The high level of self-excitation is achieved by increasing the gain of the input signal and applied to the exciter. *(iv) Data Acquisition System.* It takes vibration signal from accelerometer and encodes in digital form. It receives the voltage signal from the accelerometer and calibrates the data into equivalent accelerometer scale through its sensitivity and sends it to the computer by using software. These data are analyzed as a time history (displacement-time) and frequency domain (i.e., FFT). Here NIC DAQ 9174 is designed for the small portable and mixed measurement test system. And it consists of high output speed USB communication NI signal streaming technology and also lab view (laboratory Virtual Instrumentation Engineering work bench) logging software. *(v) Electromagnets.* Electromagnets are used to generate magnetic field around the sandwich beam; here magnetic field was applied up to 0.2 T. *(vi) Impulse Hammer.* This looks like an ordinary hammer. On hitting the impact hammer on cantilever structure an equal and opposite force is sensed by the beam. This generates vibrations in the sandwich beam. *(vii) Display Unit.* Data acquisition system is connected to the computer and it shows the results.

3.3. Experiment Procedure

3.3.1. Free Vibration Study. Free vibrations can be defined as oscillations about a system's equilibrium position that occurs in the absence of an external excitation. (i) Place the sandwich beam to hold in one side and other end is free, fix the accelerometer on the beam by applying the gum solution on it, and connect it to the accelerometer with data acquisition system connected to the computer with the help of USB connector. (ii) Check if the connections are properly arranged or not. (iii) Open the lab view software in computer. (iv) Hammer the beam with the help of hammer under different field conditions. (v) Response curves were obtained. Figure 3 shows a schematic representation of free vibration study.

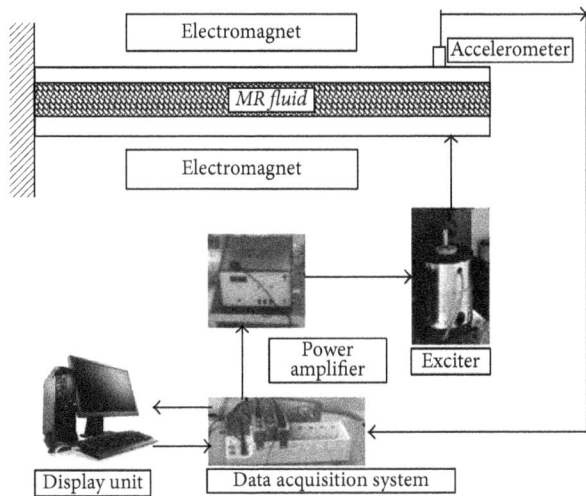

FIGURE 4: Experimental test rig for the MRF sandwich beam.

3.3.2. Forced Vibration Study. The vibration that takes place under the excitation of external forces is called forced vibration. If excitation is harmonic, the system is forced to vibrate at excitation frequency. If the frequency of excitation coincides with one of the natural frequencies of the system, a condition of resonance is encountered and dangerously large oscillations may result, which results in failure of major structures, that is, bridges, buildings, or airplane wings. MR fluid sandwich beam was clamped at the one end and the other end is free, that is, cantilever condition. Figures 4 and 5 show the schematic representation and photograph of an experimental apparatus which integrates with electromagnets, accelerometer, exciter, power amplifier, and signal analysis. The equipment used in the experiment test rig includes lab view programming and data acquisition (DAQ) board which is small and portable. The accelerometer is to sense the displacement of the beam due to the excitation by the magnet shaker. The power amplifier delivers accurate high power levels with complete self-protection for dependable operation. Initially, the empty sandwich beam is fixed to the rig using gum solution and connect in to DAQ board which is connected to computer using USB connector. The power is supplied to the exciter by increasing gain of amplifier whenever the excitation is given to the exciter having frequency range of 20 to 100 Hz. Subsequently, the sandwich beam with MR fluid core is prepared and tested by applying the magnetic field up to 0.2 T. It is noted that the parameters like amplitude and exciting frequency are set to be same for all tests. Experiment Procedures are summarized as follows. (i) Clamp one end of the sandwich beam and other end is free. (ii) Place accelerometer at the free end of the sandwich beam using gum solution to measure vibration response and connect it to the data acquisition system DAQ board system to personnel computer using personal computer. (iii) During setting of swept sine parameter make sure that vibration measurement software of the time domain is greater than the total time of excitation. (iv) Make setting to generate swept sine from exciter to excite sandwich beam. (v) Start the experiment by giving force signal to the exciter using the power amplifier and allow the beam to vibrate. (vi) Record all data obtained from accelerometer in the form of variation response with time by varying by using lab view software. (vii) Repeat the experiment for different conditions.

4. Experimental Results and Discussions

The test conditions of the smart sandwich beam are classified as follows: (i) empty beam, (ii) with MR fluid core at magnetic field $B = 0$, (iii) with MR fluid core at magnetic field $B = 0.1$ T, and (iv) with MR fluid core at magnetic field $B = 0.2$ T. Figure 6 presents the frequency responses of the proposed smart sandwich achieved at three different conditions. It is clearly observed from this figure that both the natural frequency and the peak value are significantly changed by the intensity of the magnetic field. Specifically, the natural frequency in the absence of the magnetic field is identified as 33.665544 Hz and this is increased up to 40.331989 Hz by applying the magnetic field of 0.2 T. The vibration parameters discussed in the previous section have been identified in this test and the results are summarized in Table 4. It is seen that the damping factor of the sandwich beam without the field is evaluated as 0.013719 and it is reduced to 0.009219 by applying the magnetic field of 0.2 T. Table 5 presents the detailed variation of the vibration parameters of four different cases. The results presented in this work clearly indicate that the vibration parameters of smart sandwich beam such as natural frequency and loss factor can be adaptively tuned as a function of the magnetic field. This tuning capability can be extended to more advanced control capability of vibration characteristics in real time. This can provide several benefits in vibration environment of flexible structures such as resonance avoidance.

5. Conclusion

In this work, principal criteria for the evaluation of vibration characteristics of sandwich structures with controllable core materials were discussed and investigated by undertaking an experimental work on the sandwich beam with controllable MR fluid core. After adopting the sandwich beam as a SDOF system, the vibration parameters such as natural frequency, damping ratio, loss factor, and quality factor were derived and defined. Subsequently, in order to evaluate the vibration parameters, a sandwich beam featuring MR fluid consisting of engine oil and iron particles was fabricated and tested under four different conditions. It has been identified that the natural frequency of the empty beam is higher than the case with MR fluid core in the absence of the magnetic field. This is due to the increment of the mass of MR fluid mainly attributed by iron particles. It has been also seen that the natural frequency and quality factor of the smart sandwich beam are increased as the field intensity increases, while the loss factor is decreased as the magnetic field increases. This directly means that both the natural frequency and damping property of the smart sandwich beam can be adaptively controlled by integrating an appropriate control strategy. This kind of salient benefit can provide several

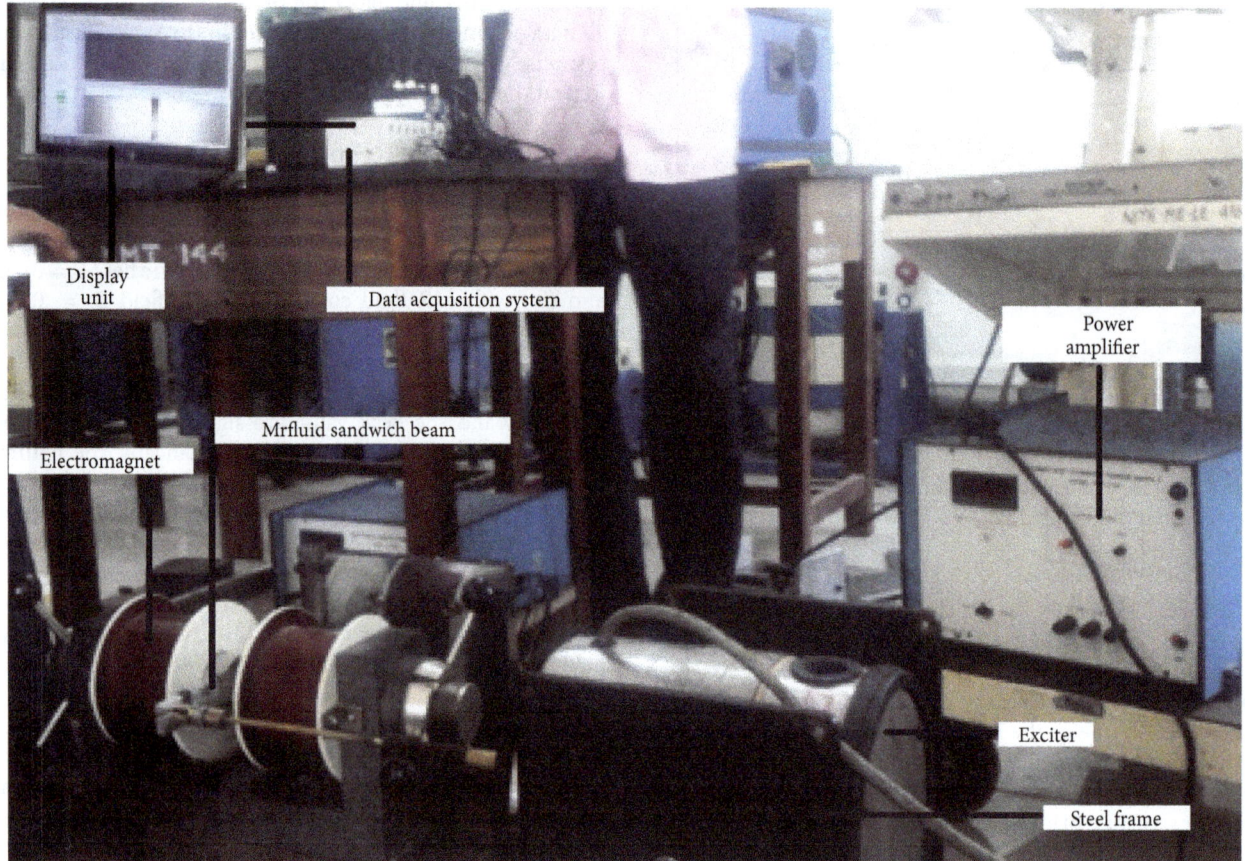

FIGURE 5: Photograph of experimental test rig for the MRF sandwich beam.

TABLE 4: Vibration parameters of sandwich beam with MR fluid core for free vibration system.

(a)

SI number	Magnetic field condition	Natural frequency (ω_n) (Hz)	Time period Sec	Peak amplitude (mm)	Damped frequency (ω_d) (Hz)	Damping factor (ζ)
(01)	Without MR fluid	24.99	0.2514	13.82	24.96	0.046
(02)	With MR fluid 0T	25.01	0.2512	13.21	24.98	0.038
(03)	0.1 T	25.90	0.2425	13.19	25.88	0.037
(04)	0.2 T	25.99	0.2417	12.92	25.97	0.035

(b)

SI number	Loss factor (η)	Logarithmic decrement (Δ)	Quality factor (Q)	Specific damping capacity (Sc)	Displacement resonant frequency (Df)	Acceleration resonant frequency (Af)
(01)	0.092	0.2893	10.86	0.578	24.93	25.04
(02)	0.076	0.2400	13.15	0.477	24.98	25.05
(03)	0.074	0.3231	13.51	0.464	25.86	25.93
(04)	0.070	0.2248	14.28	0.439	25.95	26.02

TABLE 5: Vibration parameters of sandwich beam with MR fluid core for forced vibration system.

(a)

SI number	Magnetic field condition	Natural frequency (ω_n) (Hz)	Time period Sec	Peak amplitude (mm)	Damped frequency (ω_d) (Hz)	Damping factor (ζ)
(01)	Without MR fluid	32.99	0.19	4.50744	32.99	0.014587
(02)	With MR fluid 0 T	33.66	0.18	2.97828	33.66	0.013719
(03)	0.1 T	34.99	0.17	0.928956	34.99	0.010319
(04)	0.2 T	40.33	0.15	0.566438	40.33	0.009219

(b)

SI number	Loss factor (η)	Logarithmic decrement (Δ)	Quality factor (Q)	Specific damping capacity (Sc)	Displacement resonant frequency (Df)	Acceleration resonant frequency (Af)
(01)	0.029174	0.045826412	34.27	0.1833	32.99	33.00
(02)	0.027438	0.043099509	36.44	0.1723	33.66	33.67
(03)	0.020638	0.032418094	48.45	0.1296	34.99	35.00
(04)	0.018438	0.028962342	54.23	0.1158	40.33	40.33

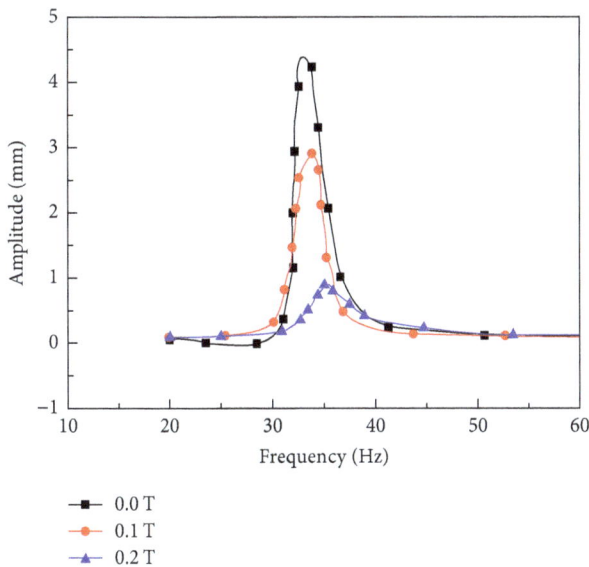

FIGURE 6: Amplitude versus Frequency.

advantages in vibration environment of flexible sandwich structures subjected to wide frequency spectrum of external disturbances. For example, the resonance problem under the forced vibration can be avoided by shifting the natural frequency and the settling time of the transient vibration can be reduced a lot.

It is finally remarked that optimal design of smart sandwich structures needs to be further explored to maximize the capability of the vibration parameters to the positive direction. Moreover, advanced control strategy to guarantee the structural stability and desired vibration parameters needs to

be developed by integrating controllable core materials such as MR fluid.

References

[1] M. V. Gandhi, B. S. Thompson, S. B. Choi, and S. Shakir, "Electro-rheological-fluid-based articulating robotic systems," *Journal of Mechanisms, Transmissions, and Automation in Design*, vol. 111, no. 3, pp. 328–336, 1989.

[2] M. V. Gandhi, B. S. Thompson, and S. B. Choi, "A new generation of innovative ultra-advanced intelligent composite materials featuring electro-rheological fluids: an experimental investigation," *Journal of Composite Materials*, vol. 23, no. 12, pp. 1232–1255, 1989.

[3] M. V. Gandhi, B. S. Thompson, and S. B. Choi, "A proof-of-concept experimental investigation of a slider-crank mechanism featuring a smart dynamically tunable connecting rod incorporating embedded electro-rheological fluid domains," *Journal of Sound and Vibration*, vol. 135, no. 3, pp. 511–515, 1989.

[4] S. B. Choi, *Control of single-link flexible manipulators fabricated from advanced composite laminates and smart materials incorporating electro-rheological fluids [Ph.D. dissertation]*, Department of Mechanical Engineering, Michigan State University, East Lansing, Mich, USA, 1990.

[5] S. B. Choi, B. S. Thompson, and M. V. Gandhi, "An experimental investigation on smart laminated composite structures featuring embedded electro-rheological fluid domains for vibration-control applications," *Composites Part B: Engineering*, vol. 2, no. 5-7, pp. 543–559, 1992.

[6] S.-B. Choi, Y.-K. Park, and J.-D. Kim, "Vibration characteristics of hollow cantilevered beams containing an electro-rheological fluid," *International Journal of Mechanical Sciences*, vol. 35, no. 9, pp. 757–768, 1993.

[7] S.-B. Choi, B. S. Thompson, and M. V. Gandhi, "Experimental control of a single-link flexible arm incorporating electrorheological fluids," *Journal of Guidance, Control, and Dynamics*, vol. 18, no. 4, pp. 916–919, 1995.

[8] Y. Choi, A. F. Sprecher, and H. Conrad, "Vibration characteristics of a composite beam containing an electrorheological fluid," *Journal of Intelligent Material Systems and Structures*, vol. 1, no. 1, pp. 91–104, 1990.

[9] Y. Choi, A. F. Sprecher, and H. Conrad, "Response of electrorheological fluid-filled laminate composites to forced vibration," *Journal of Intelligent Material Systems and Structures*, vol. 3, no. 1, pp. 17–29, 1992.

[10] J. S. Leng, Y. J. Liu, S. Y. Du, L. Wang, and D. F. Wang, "Active vibration control of smart composites featuring electro-rheological fluids," *Applied Composite Materials*, vol. 2, no. 1, pp. 59–65, 1995.

[11] M. Yalcintas and J. P. Coulter, "Analytical modeling of electrorheological material based adaptive beams," *Journal of Intelligent Material Systems and Structures*, vol. 6, no. 4, pp. 488–497, 1995.

[12] J.-Y. Yeh and L.-W. Chen, "Vibration of a sandwich plate with a constrained layer and electrorheological fluid core," *Composite Structures*, vol. 65, no. 2, pp. 251–258, 2004.

[13] J.-Y. Yeh and L.-W. Chen, "Finite element dynamic analysis of orthotropic sandwich plates with an electrorheological fluid core layer," *Composite Structures*, vol. 78, no. 3, pp. 368–376, 2007.

[14] M. Yalcintas and H. Dai, "Magnetorheological and electrorheological materials in adaptive structures and their performance comparison," *Smart Materials and Structures*, vol. 8, no. 5, pp. 560–573, 1999.

[15] M. Yalcintas and H. Dai, "Vibration suppression capabilities of magnetorheological materials based adaptive structures," *Smart Materials and Structures*, vol. 13, no. 1, pp. 1–11, 2004.

[16] Q. Sun, J.-X. Zhou, and L. Zhang, "An adaptive beam model and dynamic characteristics of magnetorheological materials," *Journal of Sound and Vibration*, vol. 261, no. 3, pp. 465–481, 2003.

[17] Z.-F. Yeh and Y.-S. Shih, "Dynamic characteristics and dynamic instability of magnetorheological material-based adaptive beams," *Journal of Composite Materials*, vol. 40, no. 15, pp. 1333–1359, 2006.

[18] B. Hu, D. Wang, P. Xia, and Q. Shi, "Investigation on the vibration characteristics of a sandwich beam with smart composites-MRF," *World Journal Modelling Simulation*, vol. 2, pp. 201–206, 2006.

[19] R. Vasudevan, R. Sedaghati, and S. Rakheja, "Vibration analysis of a multi-layer beam containing magnetorheological fluid," *Smart Materials and Structures*, vol. 19, no. 1, Article ID 015013, 2010.

[20] V. Lara-Prieto, R. Parkin, M. Jackson, V. Silberschmidt, and Z. Kęsy, "Vibration characteristics of MR cantilever sandwich beams: Experimental study," *Smart Materials and Structures*, vol. 19, no. 1, Article ID 015005, 2010.

[21] R. Vasudevan, R. Sedaghati, and S. Rakheja, "Optimal vibration control of beams with total and partial MR-fluid treatments," *Smart Materials and Structures*, vol. 20, no. 11, Article ID 115016, 2011.

[22] Y. H. Li, B. Fang, F. M. Li, J. Z. Zhang, and S. Li, "Dynamic analysis of sandwich plates with a constraining layer and a magnetorheological fluid core," *Polymer Composite*, vol. 19, no. 4-5, pp. 295–302, 2011.

[23] J.-Y. Yeh, "Vibration analysis of sandwich rectangular plates with magnetorheological elastomer damping treatment," *Smart Materials and Structures*, vol. 22, no. 3, Article ID 035010, 2013.

[24] V. Rajamohan, S. Rakheja, and R. Sedaghati, "Vibration analysis of a partially treated multi-layer beam with magnetorheological fluid," *Journal of Sound and Vibration*, vol. 329, no. 17, pp. 3451–3469, 2010.

[25] V. Rajamohan, V. Sundararaman, and B. Govindarajan, "Finite element vibration analysis of a magnetorheological fluid sandwich beam," in *Proceedings of the International Conference on Design and Manufacturing (IConDM '13)*, vol. 64, pp. 603–612, July 2013.

[26] V. Rajamohan and M. Ramamoorthy, "Dynamic characterization of non-homogeneous magnetorheological fluids based multi-layer beam," *Applied Mechanics and Materials*, vol. 110–116, pp. 105–112, 2012.

[27] V. Rajamohan and P. Natarajan, "Vibration analysis of a rotating magnetorheological fluid sandwich beam," in *Proceedings of the International Conference on Advanced Research in Mechanical Engineering (ICARME '12)*, pp. 978-93, Tirupati, India, May 2012.

[28] S. Momeni, A. Zabihollah, and M. Behzad, "Experimental works on dynamic behavior of laminated composite beam incorporated with magneto-rheological fluid under random excitation," in *Proceedings of the 3rd International Conference on Mechatronics and Robotics Engineering*, pp. 156–161, Paris, France, Feburary 2017.

[29] C. A. Walikar, S. Kolekar, R. Hanumantharaya, and K. Raju, "A study on vibration characteristics of engine oil based magnetorheological fluid sandwich beam," *Journal of Mechanical Engineering and Automation*, vol. 5, no. 3, pp. 84–88, 2015.

[30] S. B. Joshi, "Vibration study of magnetorheological fluid filled sandwich beams," *International Journal of Applied Research in Mechanical Engineering*, vol. 2, no. 2, pp. 100–104, 2012.

[31] R. H. Scanlan, "Linear damping models and causality in vibrations," *Journal of Sound and Vibration*, vol. 13, no. 4, pp. 499–503, 1970.

[32] E. E. Ungar and J. Kerwin, "Loss factors of viscoelastic systems in terms of energy concepts," *The Journal of the Acoustical Society of America*, vol. 34, pp. 954–957, 1962.

[33] V. D. Naylor, "Some fallacies in modern damping theory," *Journal of Sound and Vibration*, vol. 11, no. 2, pp. 278–280, 1970.

7

Physical and Acoustical Properties of Corn Husk Fiber Panels

Nasmi Herlina Sari,[1] I. N. G. Wardana,[2] Yudy Surya Irawan,[2] and Eko Siswanto[2]

[1]*Department of Mechanical Engineering, Faculty of Engineering, Mataram University, Nusa Tenggara Barat, Indonesia*
[2]*Department of Mechanical Engineering, Faculty of Engineering, Brawijaya University, East Java, Indonesia*

Correspondence should be addressed to Nasmi Herlina Sari; n.herlinasari@unram.ac.id

Academic Editor: Benjamin Soenarko

This research focuses on the development of a sustainable acoustic material comprising natural fibers of corn husk that were alkali modified by 1%, 2%, 5%, and 8% NaOH. The morphology and the acoustical, physical, and mechanical properties of the resulting fibers were experimentally investigated. Five different types of sample were produced in panel form, the acoustical properties of which were studied using a two-microphone impedance tube test. The porosity, tortuosity, and airflow resistivity of each panel were investigated, tensile tests were conducted, and the morphological aspects were evaluated via scanning electron microscopy. The sound absorption and tensile properties of the treated panels were better than those of raw fiber panels; the treated panels were of high airflow resistivity and had low porosity. Scanning electron micrographs of the surfaces of the corn husk fibers revealed that the different sound absorption properties of these panels were due to roughness and the lumen structures.

1. Introduction

Currently, sound-absorbing materials mostly comprise synthetic and waste products, such as foams, recycled rubber, glass wool, and polyester fibers, which can be hazardous to human health, disruptive at workplaces, and harmful to the environment [1]. As such, it would be preferable if natural materials could be used instead. Natural fibers, with their porous cell structure and relatively low density, are becoming increasingly popular because they are renewable, nonabrasive, cheaper, and available in abundance and pose lower health risks during handling and processing [2, 3]. Some applications (e.g., construction and furniture) have begun to use sound-absorbing panels comprising cellulose fibers from leaves, rice, hemp, coconut, or ramie [4–7]. However, such natural fibers tend to absorb a considerable amount of moisture. In addition, they are adversely affected by heat and microbes and are prone to decomposition [8]. Fortunately, several studies have reported that chemical treatment of the surfaces of natural fibers can reduce these disadvantages [9–11]. In particular, the effects of alkali treatment of natural fibers used in polymer composites have been discussed in relation to the physical and mechanical properties of such materials [12–14]. However, to date, few studies have

investigated the effects of alkali treatment on the acoustic properties of the surface of natural fibers [3].

The acoustic properties of absorbent fibrous materials have been extensively studied. Delany and Bazley [15] investigated a range of such materials in relation to their characteristic impedances and propagation coefficients. Attenborough [16] used flow resistivity to study predictive models of the acoustic characteristics of rigid fibrous absorbent soils and sands. Biot [17, 18] formulated theories about the propagation of high- and low-frequency stress waves in porous elastic media containing a compressible viscous fluid. Bies and Hansen [19] characterized the acoustic performance of porous materials for common applications that are based on the measurement of airflow resistance. These previous studies showed that a better understanding of the microstructure and physical parameters of a material could help in developing high-performance acoustic materials.

This study investigates the acoustic and nonacoustic properties of the panels of corn husk fiber (CHF), which is a renewable bioresource and is thus biodegradable. Panels of CHF are compared with and without alkali treatment of the fibers. The concentration of alkali (NaOH) is varied to analyze the changes in the mechanical properties and surface morphology of the fibers.

2. Materials and Methods

2.1. Material. The main material used in this study was waste corn husk that had an average length and width of 240–245 mm and 110–135 mm, respectively. The selection process was aimed at maintaining the uniformity of the selected CHFs.

2.2. Fiber Bundle Extraction. The corn husks were soaked in water for 16 days to undergo a process of microbacterial degradation. They were then washed thoroughly in fresh water and combed with a plastic brush in order to remove residue particles from the fibers surfaces and align the fibers equally. The retained inner layer was a bundle of fibers that could be separated for further use. Raw CHFs were retrieved, cleaned, and dried naturally in the air.

2.3. Alkaline Treatment of CHF. The CHFs were soaked in 1%, 2%, 5%, or 8% NaOH for 2 h in a standard atmosphere of 29°C and 64% relative humidity and subsequently rinsed five times with mineral water in order to remove NaOH from the fiber surfaces. They were dried in natural sunlight to remove any residual moisture and were then stored in plastic wrap. Subsequently, they were stored in a dry box with 40% humidity.

2.4. Fabrication of CHF Panels. Samples of raw and treated CHF were prepared and weighed using a digital analytical balance to gain fiber mass of 4.5×10^{-3} kg. Subsequent, the samples formed in a round mold with a diameter of 29 mm and thickness of 20 mm. A compaction of 37 kPa was used to form panel samples that were 29 mm × 20 mm (diameter × thickness). Five different samples were used for acoustical and porosity tests. The bulk density of each sample was measured by the ratio of the total mass m of the sample (Kg) and its volume v (m^3) as $\rho_{bulk} = m/V$ [5]. Because the dimension of a fiber lumen is in microns, it is not detected in the measurement volume. The greater density of the sample, that is, more fibers with the same thickness, was detected to have pores size decreased. However, this could form a more complicated internal path (tortuosity) which can cause greater energy loss. A photograph of the corn husk panel samples is shown in Figure 1.

2.5. Physical Properties

2.5.1. Porosity. The connected porosity of the CHF panels was nonacoustically measured using water saturation, as illustrated by Vašina et al. [20] All the samples were dried at 105°C for one day. Subsequently, they were weighed before being left in a vacuum vessel to saturate under water (density of water $\rho_w = 1000$ kg/m^3). After 24 h, they were carefully removed and weighed again. The porosity was computed using $\varepsilon = V_w/V_s$, where V_w is the volume of the sample occupied by water and V_s is the total volume of the sample. The volume of water can be calculated using the following: $V_w = (m_{wet} - m_{dry})/\rho_w$, where m_{wet} and m_{dry} are the wet and dry masses (kg) of the sample, respectively.

2.5.2. Airflow Resistivity. The airflow resistivity was based on the ASTM D-1564-1971 test. The flow resistivity was calculated using [21, 22]

$$\sigma = \frac{\left(6.8 \cdot \eta \cdot (1 - \varepsilon)^{1.296}\right)}{\left(d^2 \cdot \varepsilon^3\right)}, \tag{1}$$

where η represents the viscosity of air (1.84×10^{-5} Pa·s), ε represents the porosity, and d represents the radii of the fibers.

2.5.3. Tortuosity. The following empirical formula was used to calculate tortuosity (φ) in terms of porosity (ε) as follows [3]:

$$\varphi = 1 + \frac{(1 - \varepsilon)}{2\varepsilon}. \tag{2}$$

2.6. Mechanical Properties. Single fibers were separated manually from the fiber bundles of raw and treated CHFs. The tensile strength and Young's modulus were determined using a Tensilon RTG 1310 universal testing machine with a load cell of 10 kN. All the fiber samples were tested after conditioning the samples for 24 h in a standard testing atmosphere of 28°C and 70% relative humidity. The sample length was 25 mm, and a crosshead speed of 2 mm/min was used for tensile testing, according to ASTM D-3379-75 [23]. In total, 15 samples were tested for each alkali treatment condition and the average and standard deviation values were reported. Prior to each test, the mean diameter of the fibers was measured to an accuracy of 0.001 mm using a Mitutoyo digital micrometer.

2.7. Sound Absorption Measurement. The acoustic properties of the material formed from the fibers were measured using a two-microphone transfer-function method, according to ISO 10534-2/ASTM E-1050-98 standards. The testing apparatus was part of complete acoustic material testing system, Brüel & Kjær, as it is shown in Figure 2. A small tube setup was employed to measure different acoustical parameters in the frequency range of 100 Hz–6.4 kHz. At one end of the tube, a loudspeaker was situated to act as a sound source and the test material was placed at the other end to measure sound absorption properties. For precise fitting of samples into the measurement tube, an aluminum rod has a length of 40 mm and diameter of 29 mm and it was utilized to push the material into a preadjusted depth. Two acoustic microphones (type 4187, Brüel & Kjær) were located in front of the sample to record the incident sound from the loudspeaker and the reflected sound from the material. The recorded signals in the analyzer in terms of the transfer function between the microphones were processed using Brüel & Kjær material testing software to obtain the absorption coefficient of the sample under test. Each set of the experiments was repeated three times in order to have average measurements.

2.8. Scanning Electron Microscope. The surface morphologies of the raw and treated CHFs were observed using FEI Inspect S50 scanning electron microscope with a field emission gun. An accelerating voltage of 10 kV was used to obtain SEM

FIGURE 1: Photographs of the test panel sample. (a) Raw; (b) 1% NaOH; (c) 2% NaOH; (d) 5% NaOH; and (e) 8% NaOH.

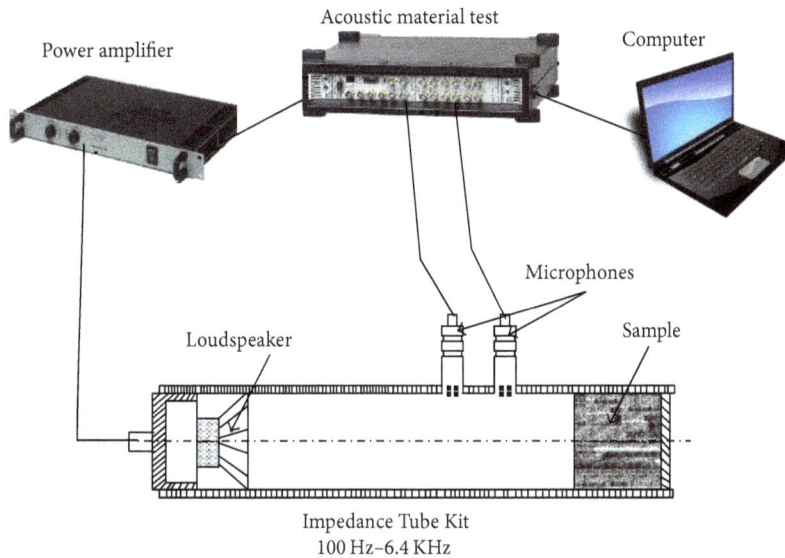

FIGURE 2: Impedance tube kit (type 4206, Brüel & Kjær).

images of the surfaces of the raw and alkali-treated CHFs. Before testing, the samples were sliced and mounted on SEM stubs with double-sided adhesive tape. To make the samples conductive, they were gold sputtered for 5 min to a thickness of approximately 10 nm under a pressure of 0.1 torr and a current of 18 mA. Micrographs were recorded at different magnifications to ensure clear images.

3. Results and Discussion

3.1. Physical Properties of CHF Panels. Large differences were observed in the physical properties of the CHF samples because of their different microstructures as a result of the chemical treatment of the raw fibers. This diversity is very interesting because it can provide considerably different porous microstructures and thus different acoustic properties. The values of porosity, tortuosity, and airflow resistivity are listed in Table 1.

The alkali treatment is increasing the airflow resistivity and decreasing the porosity of the panel. The porosity values show that the raw fibers are more porous than treated CHFs. The airflow resistivity and tortuosity of the treated samples are presumably higher than those of the raw sample. The CHF sample treated with 8% NaOH has a higher resistivity and lower porosity than the CHF samples treated with 1%, 2%, and 5% NaOH.

All the samples present an open pore structure wherein the pores are interconnected. This is one of the most important factors for noise absorption because such a structure increases airflow resistivity and thus the dissipation of the wave energy in the pores. In these samples, the multiscale fiber structure with lumina inside the fiber bundle has pores whose size can differ by many orders of magnitude (Figures 8(a)–8(j)).

3.2. Tensile Strength Properties. The tensile properties and the modulus of elasticity are compared for the raw and treated CHFs, as shown in Figures 3 and 4. The alkali treatment conducted by varying the concentration of NaOH from 1% to 8% increased the tensile strength and the modulus of elasticity. The tensile strength of a raw fiber was 160.49 ± 17.12 MPa. Under treatment, this increased to between 230.90 ± 41.85 MPa and 368.25 ± 78.97 MPa. The modulus of elasticity of a raw fiber was 4.57 ± 0.54 GPa. Under treatment, this increased to between 7.09±0.52 GPa and 15.87±1.87 GPa. These enhancements are related to a decrease in the fiber diameter, as shown in Figure 5. For CHF samples treated with 1% NaOH, the modulus of elasticity is higher than that for those treated with 2% and 5% NaOH, contributing to the sound absorption of the samples. The fibers treated with 8% NaOH had the best tensile performance, which is attributed to having the lowest diameter of 0.124 ± 0.017 mm.

TABLE 1: Physical properties of samples.

Material (CHFs)	Thickness (t) (mm)	Density ($Kg \cdot m^{-3}$)	Porosity (ε)	Airflow resistivity (σ) ($Pa \cdot s \cdot m^{-2}$)	Tortuosity (φ)
Raw	20	344	0.88 ± 0.02	1375 ± 332	1.06 ± 0.01
Treated (1% NaOH)	20	438	0.87 ± 0.01	1885 ± 93	1.07 ± 0.01
Treated (2% NaOH)	20	566	0.86 ± 0.00	2540 ± 44	1.08 ± 0.01
Treated (5% NaOH)	20	584	0.82 ± 0.01	5572 ± 157	1.11 ± 0.01
Treated (8% NaOH)	20	615	0.77 ± 0.01	11.118 ± 462	1.15 ± 0.01

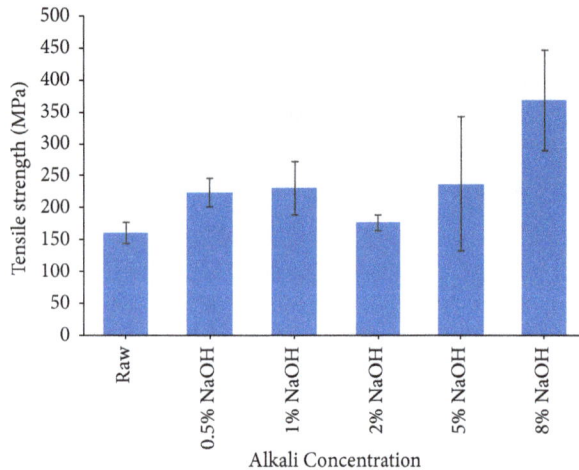

FIGURE 3: Ultimate tensile strengths of raw and treated of corn husk fiber (CHF) bundles.

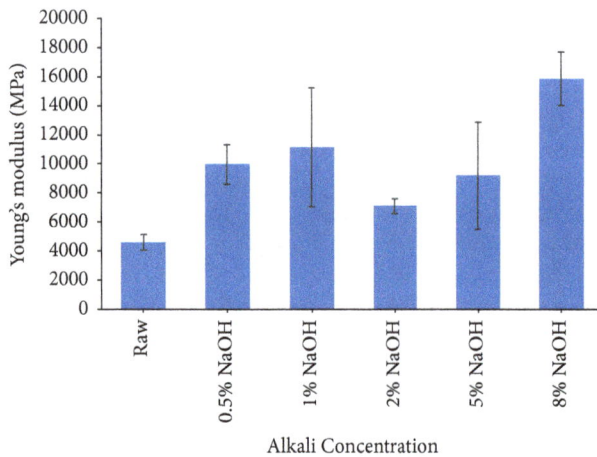

FIGURE 4: Moduli of elasticity of raw and treated CHF bundles.

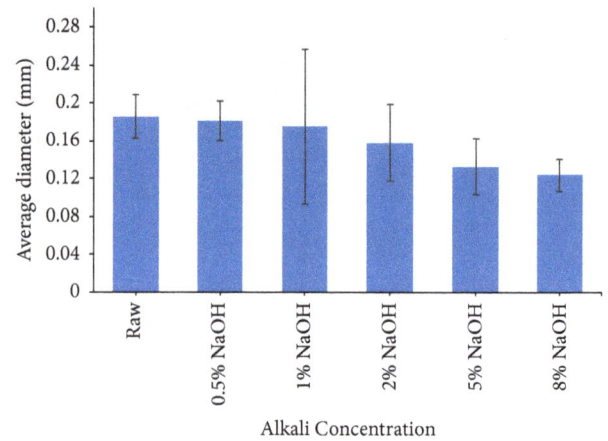

FIGURE 5: Variation in the diameters of CHF bundles.

Obi Reddy et al. [24] and Suryanto et al. [25] reported similar investigations but with different fibers. They stated that the increase in tensile strength is indeed due to the reduction in fiber diameter, which in turn is due to the loss of the hemicellulose and lignin parts of the cellulose and the associated moisture content of fiber brought about by the alkaline treatment. Hossain et al. [26] stated that the chemical treatment increases the aspect ratio and surface roughness of the fibers, thereby increasing their tensile strength. This may also explain why the CHFs treated with 8% NaOH had the lowest diameter and highest tensile strength.

3.3. Sound Absorption Analysis of CHF Panel. Cellular lumina form numerous interconnected pores. When a sound wave strikes a fibrous sound absorber, it causes the fibers of the absorbing material to vibrate, generating a tiny amount of heat because of internal friction between the fibers. The sound waves propagate vibration energy through the air spaces in the individual lumina inside the fiber. A portion of this sound energy is converted into heat in the lumina, which is then absorbed by the surrounding walls. Subjecting a sample to a pressure of 37 kPa does not change its physical properties because of the mechanical strength of the CHFs in a range 160.49–368.25 MPa. Therefore, there is no change in the surface impedance of a sample as a result of compaction.

Figure 6 shows the variation of absorption coefficient with frequency for NaOH-treated fiber panels. Increasing the NaOH concentration from 1% to 8% improves the sound absorption of the CHF panels. For the raw fibers, the maximum sound absorption coefficient is 0.93 (over 1.6–3 kHz), whereas, for the alkali-treated samples, it is 0.98–0.99 (over 1.6–3.25 kHz). This suggests that the alkali treatment changes the fiber elasticity and reduces the fiber diameter, thereby increasing the airflow resistivity and decreasing the porosity of the panel (Table 1). Moreover, for a constant fiber volume fraction, fibers having smaller diameter are more numerous than those having a larger diameter. Therefore, because the number of fibers per unit area increases as the fiber diameter

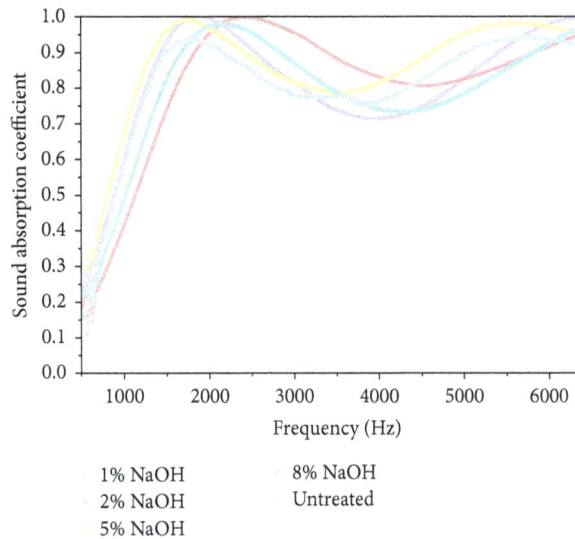

FIGURE 6: Sound absorption coefficients of CHF panels.

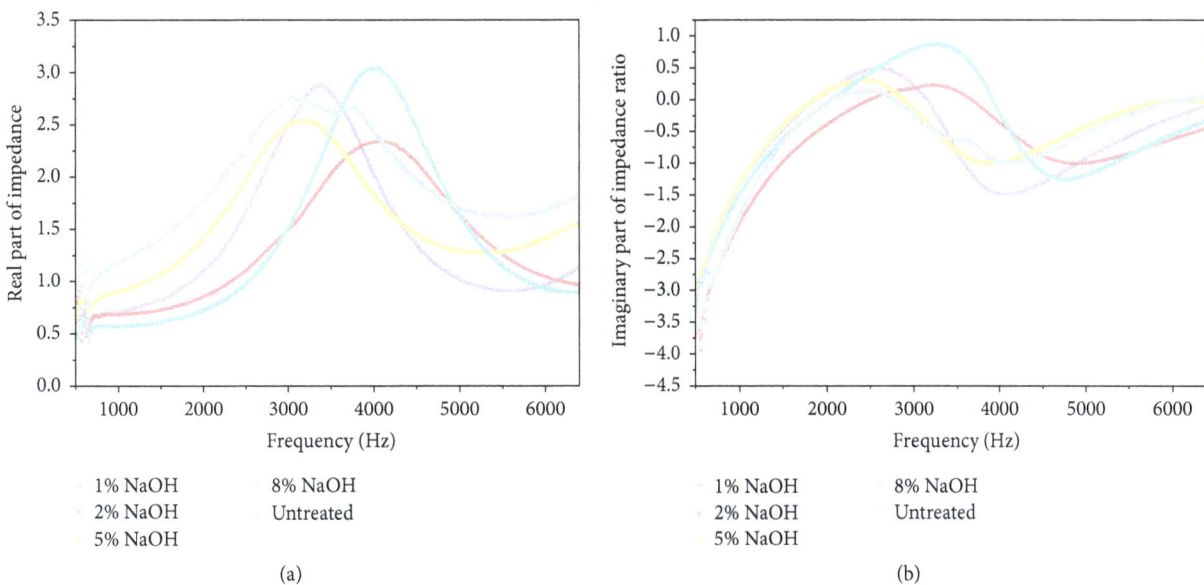

FIGURE 7: (a) Real part and (b) imaginary part of the impedance ratio of samples.

decreases, the additional thermal energy is dissipated more rapidly due to the increased frictional surface area. The sound absorption coefficient is therefore correspondingly higher than those of the raw sample. Presumably, the larger the air cavities and lumina inside the fiber, the higher the wavelength of the sound that is absorbed, which therefore lowers the sound absorption frequency (over 1.6–3.25 kHz). In contrast, because of the tightly arranged structure of the raw fibers, the incident sound waves are reflected more. Chen et al. [27] stated that pectin, lignin, hemicellulose, and other low-molecular-weight materials can form a dense layer on the surface of fibers, thereby enhancing reflection. This would also explain why the sound absorption coefficient of raw fibers is lower than those of treated fibers, even though the structural patterns of the two types of fiber are similar.

For the samples treated with 5% and 8% NaOH, although the diameters of the treated fibers are smaller than those of the raw fibers, the treated fibers have a higher Young's modulus. In such circumstances, the increase in airflow resistance and tortuosity due to treatment is accompanied by an increase in stiffness. Therefore, movements of the sound are difficult to pass through the samples. This leads to a decrease in the absorption performance of the samples. However, for the high-frequency range above 5 kHz, the treated samples show better sound absorption performance. This is related to the random distribution of fibers in the panel, which enhances the absorption of sound waves. For a thickness of 20 mm and a fiber weight of 4.5×10^{-3} kg, the absorption coefficient is 0.8 in the frequency range of 1.3–6 kHz, which is better than that for other fibers such as paddy waste fibers. Putra et al.

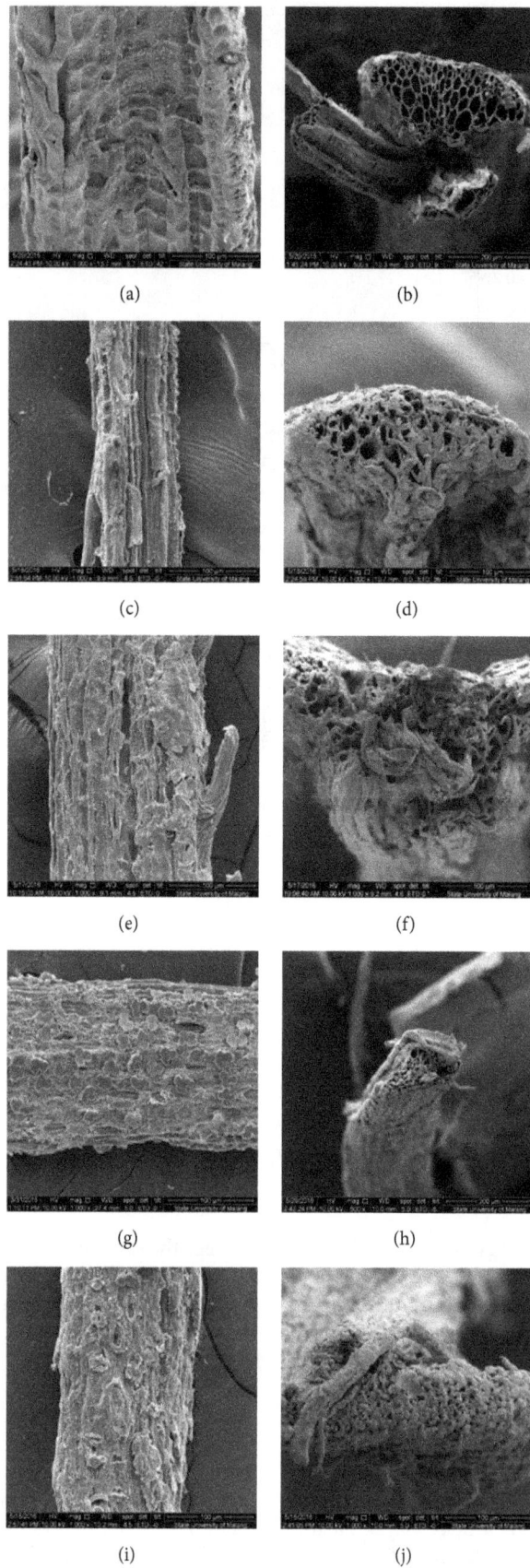

FIGURE 8: Scanning electron microscope (SEM) images of CHF: (a, b) raw and treated fibers, (c, d) 1% NaOH, (e, f) 2% NaOH, (g, h) 5% NaOH, and (i, j) 8% NaOH.

[5] showed that paddy waste fibers with a thickness of 20 mm and a fiber weight of 5×10^{-3} kg had a sound absorption coefficient of 0.7 in the frequency range of 2–5 kHz. It proves that the CHF panel can be a good alternative and a sustainable acoustic material. Furthermore, sound absorption at lower frequencies (over 1.6–3.25 kHz) is desirable for automotive applications because this frequency range corresponds to noise from the wind, tires, road, conversation, and engine running, thereby making CHF a promising candidate for automotive interior sound absorption.

Figures 7(a) and 7(b) show the real and imaginary parts, respectively, of the acoustic impedance obtained from different samples. The real part is the resistance associated with energy losses, and the imaginary part is the reactance associated with phase changes. In this case, we see a better performance for the panels treated with 2% and 5% NaOH. Increasing the alkali treatment concentration reduces the impedance values, which in turn increase the fraction of wave energy that can be transmitted into the material.

3.4. Morphology Analysis. The scanning electron micrographs of the raw and treated CHF surfaces are shown in Figure 8. As shown in Figure 8(a), the surfaces of raw CHFs have impurities and shallow grooves. The only crevices that are apparent are minute ones. However, from observations of the fiber cross sections, we noticed a fiber structure comprising dense hollows (called *lumina*) inside the fiber bundles (Figure 8(b)).

An average lumen diameter of 1.2 μm was calculated using the ImageJ software. This hollow structure (lumen) has a certain effect on the empty space for sound absorption. The sound waves propagate vibration energy through the air cavities and lumina inside the fiber. The unique lumen structure helps natural fibers to absorb sound.

The scanning electron micrographs show significant changes in surface morphology after applying NaOH treatment (Figures 8(c)–8(j)). As the percentage of NaOH on the CHF surfaces is increased, more hemicellulose and lignin are lost, which increases the sound absorption. The alkali treatment removes impurities, causing the fibers to separate. This leads to rougher surfaces and hence causes fibrillation, as shown in Figures 8(c), 8(e), 8(g), and 8(i), which enhances the mechanical and acoustical properties. From observations of the cross sections of the treated fibers, we observed that the average diameter of the lumina inside the fiber decreases (Figures 8(d), 8(f), 8(h), and 8(j)). For example, fibers treated with 5% NaOH have an average lumen diameter of 2.51 μm. Reduction in lumen diameter is certain to improve the acoustical properties of the panel.

There are numerous micropores in the porous structure of a single fiber bundle. When a sound wave impinges on the porous structure of the fibrous panel, the air motion and compression in the lumina caused by the sound vibration can cause friction on the lumen walls, thereby inhibiting movement of the air that is close to the lumen walls. Because of the frictional and viscous forces, a considerable fraction of the sound energy is converted into heat, which attenuates the acoustic energy. Heat loss due to heat exchange between the air in a lumen and the lumen walls can also cause sound energy attenuation. The random distribution of the fibers in the fibrous absorber panel allows more sound waves to impinge on the lumina of the fiber bundle, thereby strengthening the sound absorption. These special lumen structures and their distribution are the main reasons for the acoustical absorptivity.

4. Conclusions

The use of corn husk fiber for sound absorption was reported herein. The material is natural and renewable and is a waste product of corn husk processing that poses no harm to human health. All the samples studied could absorb sound. The alkali treatment of the fibers with NaOH had an effect of decreasing the fiber diameter, thereby improving the mechanical and acoustic properties of the CHF panels. The sound absorption approached 100% in the low-frequency range of 1.6–3.250 kHz for fiber samples that were treated with 2% and 5% NaOH concentrates. Panels of treated fibers were better in absorbing sound than the panels of raw fibers. In addition, the alkali treatment helped to maintain the panel shape and protect it the panel shape of fiber damage.

We note that the CHF test samples used throughout this work were not combined as polymer materials. Future work could involve studying the effect of fiber volume fraction on the sound absorption properties of CHF composite materials.

Competing Interests

The authors declare that there is no conflict of interests regarding the publication of this paper.

References

[1] R. Zulkifli, Zulkarnain, and M. J. M. Nor, "Noise control using coconut coir fiber sound absorber with porous layer backing and perforated panel," *American Journal of Applied Sciences*, vol. 7, no. 2, pp. 260–264, 2010.

[2] A. P. Jorge and C. J. Malcolm, "Recent trends in porous sound-absorbing materials," *Sound and Vibration*, vol. 44, no. 7, pp. 12–18, 2010.

[3] S. Fatima and A. R. Mohanty, "Acoustical and fire-retardant properties of jute composite materials," *Applied Acoustics*, vol. 72, no. 2-3, pp. 108–114, 2011.

[4] M. Küçük and Y. Korkmaz, "The effect of physical parameters on sound absorption properties of natural fiber mixed nonwoven composites," *Textile Research Journal*, vol. 82, no. 20, pp. 2043–2053, 2012.

[5] A. Putra, Y. Abdullah, H. Efendy, W. M. F. W. Mohamad, and N. L. Salleh, "Biomass from paddy waste fibers as sustainable acoustic material," *Advances in Acoustics and Vibration*, vol. 2013, Article ID 605932, 7 pages, 2013.

[6] U. Berardi and G. Iannace, "Acoustic characterization of natural fibers for sound absorption applications," *Building and Environment*, vol. 94, pp. 840–852, 2015.

[7] S. Ersoy and H. Küçük, "Investigation of industrial tea-leaf-fibre waste material for its sound absorption properties," *Applied Acoustics*, vol. 70, no. 1, pp. 215–220, 2009.

[8] M. S. Salit, "Tropical natural fibres and their properties," in *Tropical Natural Fibre Composites*, Engineering Materials, pp. 15–38, Springer Singapore, Singapore, 2014.

[9] P. Saha, S. Manna, S. R. Chowdhury, R. Sen, D. Roy, and B. Adhikari, "Enhancement of tensile strength of lignocellulosic jute fibers by alkali-steam treatment," *Bioresource Technology*, vol. 101, no. 9, pp. 3182–3187, 2010.

[10] M. F. Rosa, B.-S. Chiou, E. S. Medeiros et al., "Effect of fiber treatments on tensile and thermal properties of starch/ethylene vinyl alcohol copolymers/coir biocomposites," *Bioresource Technology*, vol. 100, no. 21, pp. 5196–5202, 2009.

[11] D. Shanmugam and M. Thiruchitrambalam, "Static and dynamic mechanical properties of alkali treated unidirectional continuous Palmyra Palm Leaf Stalk Fiber/jute fiber reinforced hybrid polyester composites," *Materials & Design*, vol. 50, pp. 533–542, 2013.

[12] H. Ku, H. Wang, N. Pattarachaiyakoop, and M. Trada, "A review on the tensile properties of natural fiber reinforced polymer composites," *Composites Part B: Engineering*, vol. 42, no. 4, pp. 856–873, 2011.

[13] P. Pantamanatsopa, W. Ariyawiriyanan, T. Meekeaw et al., "Effect of modified jute fiber on mechanical properties of green rubber composite," *Energy Procedia*, vol. 56, pp. 641–647, 2014.

[14] A. R. S. Neto, M. A. M. Araujo, R. M. P. Barboza et al., "Comparative study of 12 pineapple leaf fiber varieties for use as mechanical reinforcement in polymer composites," *Industrial Crops and Products*, vol. 64, pp. 68–78, 2015.

[15] M. E. Delany and E. N. Bazley, "Acoustical properties of fibrous absorbent materials," *Applied Acoustics*, vol. 3, no. 2, pp. 105–116, 1970.

[16] K. Attenborough, "Acoustical characteristics of rigid fibrous absorbents and granular materials," *The Journal of the Acoustical Society of America*, vol. 73, no. 3, pp. 785–799, 1983.

[17] M. A. Biot, "Theory of propagation of elastic waves in a fluid-saturated porous solid. II. Higher frequency range," *The Journal of the Acoustical Society of America*, vol. 28, no. 2, pp. 179–191, 1956.

[18] M. A. Biot, "Theory of propagation of elastic waves in a fluid-saturated porous solid. I. Low-frequency range," *The Journal of the Acoustical Society of America*, vol. 28, pp. 168–178, 1956.

[19] D. A. Bies and C. H. Hansen, "Flow resistance information for acoustical design," *Applied Acoustics*, vol. 13, no. 5, pp. 357–391, 1980.

[20] M. Vašina, D. C. Hughes, K. V. Horoshenkov, and L. Lapčík Jr., "The acoustical properties of consolidated expanded clay granulates," *Applied Acoustics*, vol. 67, no. 8, pp. 787–796, 2006.

[21] F. P. Mechel, *Formulas of Acoustics*, Springer, Berlin, Germany, 2nd edition, 2008.

[22] R. Maderuelo-Sanz, A. V. Nadal-Gisbert, J. E. Crespo-Amorós, and F. Parres-García, "A novel sound absorber with recycled fibers coming from end of life tires (ELTs)," *Applied Acoustics*, vol. 73, no. 4, pp. 402–408, 2012.

[23] A. Alawar, A. M. Hamed, and K. Al-Kaabi, "Characterization of treated date palm tree fiber as composite reinforcement," *Composites Part B: Engineering*, vol. 40, no. 7, pp. 601–606, 2009.

[24] K. Obi Reddy, C. Uma Maheswari, M. Shukla, J. I. Song, and A. Varada Rajulu, "Tensile and structural characterization of alkali treated Borassus fruit fine fibers," *Composites Part B: Engineering*, vol. 44, no. 1, pp. 433–438, 2013.

[25] H. Suryanto, E. Marsyahyo, Y. S. Irawan, and R. Soenoko, "Morphology, structure, and mechanical properties of natural cellulose fiber from mendong grass (*Fimbristylis globulosa*)," *Journal of Natural Fibers*, vol. 11, no. 4, pp. 333–351, 2014.

[26] M. K. Hossain, M. R. Karim, M. R. Chowdhury et al., "Comparative mechanical and thermal study of chemically treated and untreated single sugarcane fiber bundle," *Industrial Crops and Products*, vol. 58, pp. 78–90, 2014.

[27] D. Chen, J. Li, and J. Ren, "Study on sound absorption property of ramie fiber reinforced poly(L-lactic acid) composites: morphology and properties," *Composites Part A: Applied Science and Manufacturing*, vol. 41, no. 8, pp. 1012–1018, 2010.

8

Sound Radiation Characteristics of a Rectangular Duct with Flexible Walls

Praveena Raviprolu, Nagaraja Jade, and Venkatesham Balide

Department of Mechanical & Aerospace Engineering, Indian Institute of Technology Hyderabad, Kandi, Telangana 502285, India

Correspondence should be addressed to Venkatesham Balide; venkatesham@iith.ac.in

Academic Editor: Marc Thomas

Acoustic breakout noise is predominant in flexible rectangular ducts. The study of the sound radiated from the thin flexible rectangular duct walls helps in understanding breakout noise. The current paper describes an analytical model, to predict the sound radiation characteristics like total radiated sound power level, modal radiation efficiency, and directivity of the radiated sound from the duct walls. The analytical model is developed based on an equivalent plate model of the rectangular duct. This model has considered the coupled and uncoupled behaviour of both acoustic and structural subsystems. The proposed analytical model results are validated using finite element method (FEM) and boundary element method (BEM). Duct acoustic and structural modes are analysed to understand the sound radiation behaviour of a duct and its equivalence with monopole and dipole sources. The most efficient radiating modes are identified by vibration displacement of the duct walls and for these the radiation efficiencies have been calculated. The calculated modal radiation efficiencies of a duct compared to a simple rectangular plate indicate similar radiation characteristics.

1. Introduction

The most commonly used duct cross sections are rectangular, flat oval, and circular in heating, ventilation, and air-conditioning (HVAC) applications. Ducts transport the conditioned air from an air-handling unit (AHU) to an occupied space. Similarly, the sound produced from an AHU propagates to the receiver (a room or space) in longitudinal and transverse direction of the duct. The radiated noise from the duct walls traveling in the transverse direction is called "breakout noise." Breakout noise is one of the common paths of sound transmission in HVAC thin wall ducts. Cummings [1] discussed the role of various duct cross section geometries on the breakout and break-in noise. Further, Cummings provided research review progress for the past two decades on "duct breakout noise."

Of all the geometries, rectangular ducts had maximum breakout noise at lower frequencies, which means minimum wall transmission loss. Breakout noise causes the structural-acoustical coupling between the flexible duct wall structure and the acoustic domain inside the duct volume. Airborne and structure-borne sounds contribute to duct breakout noise, which is more dominant at low frequencies [1]. The hydrodynamic force and the acoustic pressure waves excite the duct wall and cause structure-born noise. These forces and pressure waves are associated with the flow and sound propagation through the duct, respectively. The excitation due to hydrodynamic force generates lower frequency vibration, which could be a focus for fatigue analysis.

An acoustic wave excites the duct walls strongly and induces vibrations. The produced vibrations of the structure will induce acoustic pressure inside the duct. This phenomenon continues under coupling and it is efficient at the strongly coupled modes. Coupling depends on the acoustic, structural natural frequency and spatial distribution of mode shapes. Transfer factor identifies the strongly coupled structural and acoustical modes [2]. Vibration displacement of the flexible structure due to acoustic excitation produces sound radiation outside the duct volume. The radiated sound power depends on vibration velocity, surface area, and radiation efficiency [3]. In the present paper, mean flow effects are not considered because of lower Mach numbers in the HVAC ducts.

Analytical methods are available in the literature for calculating the sound radiation from different ducts, such as a finite-length line source, equivalent finite-length cylindrical radiator, equivalent plate model, and finite and boundary element methods [1–5]. Cummings estimated the breakout noise from a rectangular duct using line source model and equivalent cylinder model. Astley and Cummings discussed a finite element scheme and an experimental setup to study the acoustic transmission through walls of a rectangular duct. Venkatesham et al. developed an analytical solution for prediction of breakout noise from the rectangular duct [4] and a plenum with four compliant walls [5] based on an equivalent plate model for sound radiation. In this model, the acoustic pressure and the vibration velocity vectors were expressed in terms of uncoupled acoustical and structural subsystems. The radiated sound power was expressed in terms of radiation impedance matrix and average velocity vector. Quadruple integrals are involved in the radiation impedance matrix. Venkatesham et al. [5] discussed a procedure for solving quadruple integrals based on the coordinate transformation. The same method is extended in the current manuscript to determine rectangular duct sound radiation characteristics.

Modal radiation efficiency calculation helps in finding efficient sound radiation modes in the total sound power radiated from a duct. Wallace [6] discussed an analytical formula for calculating radiation efficiencies at low frequencies from a baffled rectangular plate. Ran Lin and Pan [7] studied the vibration and sound radiation characteristics of box structures. It provides a basis for understanding the vibration energy flows between the panels of the box, grouping of various modes, radiation efficiencies, and the various kinds of sound sources.

In this paper, the uncoupled structural modes of the rectangular duct and acoustical cavity modes are calculated by the analytical method and validated with the numerical results. These rectangular duct modes are classified into four different groups and are similar to box structures as discussed in [7]. In these four groups, the most efficient radiating modes are estimated based on the symmetries between the panel pairs and the net volume displacements in a particular mode. The modal radiation efficiencies of different groups of a rectangular duct are estimated and compared analytically and numerically. These modal radiation efficiencies of the rectangular duct of the four groups are compared to that of simple rectangular plate. This comparison shows a similarity between duct sound radiation behaviours in terms of plate modes. As a part of the study, the total sound power radiated from the duct walls is estimated by using finite element method (FEM) and boundary element method (BEM). The sound radiation behaviour of a duct is also studied to understand its equivalence with monopole and dipole sources. The total radiation efficiencies of coupled acoustic and structural modes are calculated. These results are used for validation of the proposed analytical model.

2. Theoretical Formulation

The outline of the formulation of an analytical model of sound radiation from the rectangular duct is shown in Figure 1. The objective here is to calculate the total sound power radiation and radiation efficiency from the flexible duct walls. Assump-

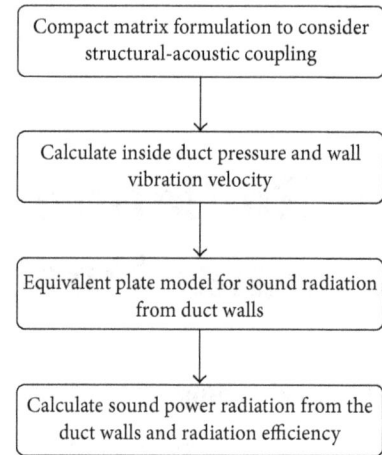

FIGURE 1: Flow chart of theoretical formulation methodology.

tions made in this model are (i) strong coupling amongst inside duct volume and flexible duct wall area; (ii) weak coupling between flexible wall area and outside environment; (iii) that coupled behaviour can be expressed in terms of a finite number of uncoupled acoustic and structural modes.

The proposed model has two stages. In the first stage, inside duct pressure and wall vibration velocity, due to an acoustic excitation, are calculated by considering the structural-acoustical coupling and impedance-mobility approach. In this approach, the coupled response between the structural and acoustical domains is represented in the form of uncoupled acoustic and structural subsystems natural frequencies and mode shapes. In the second stage, an equivalent plate model with duct wall vibrations as boundary condition is developed to predict the characteristics of sound radiation. Further details are discussed in Section 2.2.

2.1. Pressure Field inside the Duct and Wall Vibration Velocity. Pressure field inside the duct in terms of uncoupled acoustic mode shapes is given as

$$p(\mathbf{x},\omega) = \sum_n \psi_n(\mathbf{x}) a_n(\omega) = \boldsymbol{\psi}^T \mathbf{a}. \tag{1}$$

The compliant duct wall vibration velocity is given as

$$w(\mathbf{z},\omega) = \sum_m \Phi_m(\mathbf{z}) b_m(\omega) = \boldsymbol{\phi}^T \mathbf{b}, \tag{2}$$

where $\psi_n(\mathbf{x})$ is an uncoupled acoustic mode shape function of the duct, $a_n(\omega)$ is the complex amplitude of the nth acoustic pressure mode, $\Phi_m(\mathbf{z})$ is an uncoupled vibration mode shape function of the duct, and $b_m(\omega)$ is the complex amplitude of the mth vibration velocity mode.

Complex amplitude of the nth acoustic mode under structural and acoustic excitation is given as [8]

$$a_n(\omega) = \frac{\rho_0 c_0^2}{V} A_n(\omega)$$

$$\cdot \left(\int_V \psi_n(\mathbf{x}) s(\mathbf{x},\omega) dV + \int_{S_f} \psi_n(\mathbf{y}) w(\mathbf{y},\omega) dS \right), \tag{3}$$

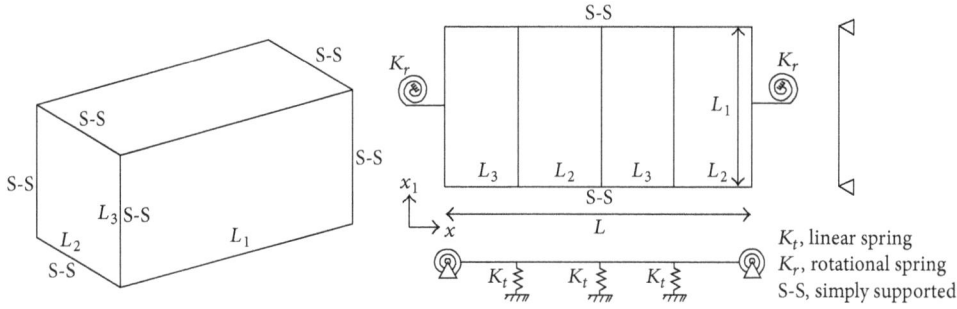

FIGURE 2: Equivalent plate representation of a rectangular duct with simply supported boundary condition.

where ρ_0 and c_0 denote the density and speed of sound in air, respectively. Function $s(\mathbf{x}, \omega)$ denotes the acoustic density function (volume velocity per unit volume) in the duct volume V and $w(\mathbf{z}, \omega)$ denotes normal velocity of a surrounding flexible structure for surface area S_f.

The complex vibration velocity amplitude of the mth mode can be expressed as [8]

$$b_m(\omega) = \frac{1}{\rho_s h S_f} B_m(\omega)$$
$$\cdot \left(\int_{S_f} \Phi_m(\mathbf{z}) f(\mathbf{z}, \omega) \, dS - \int_{S_f} \Phi_m(\mathbf{z}) p(\mathbf{z}, \omega) \, dS \right), \quad (4)$$

where ρ_s and h denote density of the material and thickness of the duct wall, respectively. $f(\mathbf{z}, \omega)$ is the force distribution on the surface of the duct wall. $p(\mathbf{z}, \omega)$ is the inside acoustic pressure distribution on the surface of the duct wall.

Uncoupled acoustical mode shapes and natural frequencies of a rectangular duct are calculated by using the formula given in [8]. Similarly, an equivalent plate model is used to calculate the uncoupled structural mode shapes and the natural frequency of rectangular duct.

Modal acoustic pressure vector $\mathbf{a}$ given in (1) and (3) has been expressed in terms of impedance and mobility as

$$\mathbf{a} = \mathbf{Z_a} (\mathbf{q} + \mathbf{q_s}). \quad (5)$$

$\mathbf{Z_a} = \mathbf{A}\rho_0 c_0^2 / V$ is the uncoupled acoustic modal impedance matrix and $\mathbf{q}$ is the modal source strength vector and $\mathbf{q_s} = \mathbf{Cb}$, where $\mathbf{C}$ is the matrix representing coupling coefficient $C_{n,m}$ between nth acoustic mode and mth structural mode. Coupling represents the spatial distribution of an acoustic mode on the flexible surface.

Modal structural vibration velocity vector $\mathbf{b}$ given in (2) and (4) can be expressed in matrix form as

$$\mathbf{b} = \mathbf{Y_s} (\mathbf{g} - \mathbf{g_a}). \quad (6)$$

$\mathbf{Y_s} = \mathbf{B}/\rho_s h S_f$ represents uncoupled structural mobility matrix and $\mathbf{g}$ is the generalized modal force vector and $\mathbf{g_a} = \mathbf{C^T a}$.

Using (5) and (6), the complex amplitude of acoustic and structural modes is given as

$$\mathbf{a} = \left(\mathbf{I} + \mathbf{Z_a CY_s C^T} \right)^{-1} \mathbf{Z_a} \left(\mathbf{q} + \mathbf{CY_s g} \right), \quad (7)$$

$$\mathbf{b} = \left(\mathbf{I} + \mathbf{Y_s C^T Z_a C} \right)^{-1} \mathbf{Y_s} \left(\mathbf{g} - \mathbf{C^T Z_a q} \right). \quad (8)$$

When $(\mathbf{I} + \mathbf{Z_a CY_s C^T})^{-1} \sim \mathbf{I}$ or $(\mathbf{I} + \mathbf{Y_s C^T Z_a C})^{-1} \sim \mathbf{I}$ or coupling matrix is zero matrix then (7) and (8) calculate uncoupled system responses. Selection of the strongly coupled acoustic and structural modes is done based on a transfer factor [2] which is given as

$$T_{n,m} = \left(1 + \frac{\left(\omega_n^2 - \omega_m^2 \right) \rho_s h S_f V}{4 \rho_0 c_0^2 C_{n,m}^2} \right)^{-1}. \quad (9)$$

$T_{n,m}$ can be interpreted as a transfer factor for nth acoustic mode coupled to mth structural mode.

2.2. Equivalent Plate Representation. In this method, an unfolded equivalent plate representation is used to model the rectangular duct. Figure 2 shows the rectangular duct and its equivalent plate representation of dimensions $L \times L_1$, respectively, where L is the duct perimeter and L_1 is the duct length. The folded joint shown in Figure 2 is modelled as a rotational spring (K_r) and creases of two adjacent plate panels are modelled as linear springs (K_t). A simply supported (S-S) boundary condition is applied to equivalent plate boundaries along the axial direction (i.e., along the x-axis at $x_1 = 0$ and L_1). Rayleigh-Ritz method is used to calculate the duct's uncoupled structural mode shapes and natural frequencies. The description of Rayleigh-Ritz model of an equivalent plate to calculate the uncoupled natural frequency of various axial boundary conditions is given in [9].

2.3. Calculation of Sound Power and Radiation Efficiency. The sound power radiated from the duct wall is expressed in terms of modal amplitudes of vibration velocity ($\mathbf{b}$) and radiation impedance matrix [$\mathbf{Z}$]. It is given as follows:

$$W_{\text{rad}} = \frac{1}{2} \mathbf{b^H} \text{Re} [\mathbf{Z}] \, \mathbf{b}, \quad (10)$$

where the superscript **H** is Hermitian transpose and Re indicate the real value.

$$\text{Re}\left[Z_{m_1 m_2 m'_1 m'_2}\right] = \frac{k}{2\pi}$$

$$\cdot \int_0^L \int_0^{L_1} \int_0^L \int_0^{L_1} \sum_{m_1}\sum_{m_2} A_{m_1 m_2} \sin\frac{m_1\pi x}{L}\sin\frac{m_2\pi x_1}{L_t} * \cdots \sum_{m'_1}\sum_{m'_2} A_{m'_1 m'_2}\sin\frac{m'_1\pi x'}{L}\sin\frac{m'_2\pi x'_1}{L_1}\frac{\sin kR}{R}dx'_1\,dx'\,dx_1\,dx, \tag{11}$$

where

$$R = \sqrt{(x-x')^2 + (x_1 - x'_1)^2}. \tag{12}$$

Equation (11) can be rewritten as

$$I_{m_1 m_2 m'_1 m'_2} = \int_0^L \int_0^{L_1}\int_0^L\int_0^{L_1}\sin\frac{m_1\pi x}{L}\sin\frac{m_2\pi x_1}{L_1}\sin\frac{m'_1\pi x'}{L}\sin\frac{m'_2\pi x'_1}{L_1}\frac{\sin kR}{R}dx'_1\,dx'\,dx_1\,dx. \tag{14}$$

The radiation impedance can be expressed in terms of Rayleigh integral as

$$\text{Re}\left[Z_{m_1 m_2 m'_1 m'_2}\right]$$
$$= \frac{k}{2\pi}\sum_{m_1}\sum_{m_2}\sum_{m'_1}\sum_{m'_2} A_{m_1 m_2} A_{m'_1 m'_2} I_{m_1 m_2 m'_1 m'_2}. \tag{13}$$

The expression of the integral I in (13) can be written as

The quadruple integral of (14) is evaluated by using an established method given in [4]. m'_1, m'_2, m_1, and m_2 are modal indices.

An analytical equation to calculate the modal radiation efficiency as a function of frequency for a given input sound source is given as [7]

$$\sigma = \frac{W_{\text{rad}}}{W_{\text{in}}}, \tag{15}$$
$$W_{\text{in}} = \rho_0 c_0 S_f \langle\mathbf{w}\rangle^2,$$

where W_{rad} is radiated sound power and W_{in} is the plane wave sound radiation power by a piston source having the same surface area of the duct structure and vibrating with the same root mean square velocity as the structure (**w**) [7]. S_f is the flexible duct wall area and is given by $L \times L_1$, and vibration velocity of the structure is given by $\langle\mathbf{w}\rangle^2$.

3. Numerical Model

Numerical models based on FEM are developed to calculate the uncoupled structural and acoustic natural frequencies. Both of these models are linked for coupled analysis. Figure 3 shows the flow chart of numerical modelling procedure to calculate sound power radiation, modal radiation efficiency, and quadratic velocity.

3.1. Uncoupled Structural Model. Figure 4(a) shows the numerical model of a duct structure. A rectangular duct of dimensions 0.3 m × 0.4 m × 1.5 m (L_1 = 1.5, L_2 = 0.3, and L_3 = 0.4) with duct wall thickness of 5 mm is modelled and meshed using SHELL 63 elements [10]. A simply supported boundary condition is applied at both ends of the duct as

shown in Figure 4(a). Aluminium material properties are applied to the structure and density ρ_s = 2770 kg/m^3, Young's modulus E = 71 GPa, Poisson's ratio υ = 0.33, and structural damping ratio = 0.01 are considered.

3.2. Uncoupled Acoustic Model. An enclosed duct volume of dimensions 0.3 m × 0.4 m × 1.5 m is modelled and meshed using SOLID185 elements [10] as shown in Figure 4(b). Acoustic medium properties are as follows: c_0 = 340 m/s, ρ_0 = 1.225 kg/m^3, and acoustic damping ratio = 0.01 are applied to acoustic volume. The boundaries of the acoustic cavity are assumed to be rigid in Figure 4(b).

3.3. Acoustic-Structural Coupled Model. Coupled model considers structural (duct) and acoustic (cavity) domains and these are solved in FEM acoustic module of LMS virtual lab-13 [11]. A constant velocity piston source excitation is applied to the inlet face of an acoustic mesh, which is shown in Figure 4(b). The coupled system of equations helps in determining structural displacements on the flexible circumferential duct walls and its pressure inside the duct.

3.4. Calculation of Sound Power Radiation. A numerical model to calculate sound radiated from the flexible duct walls is shown in Figure 5. The total sound power radiated from the duct walls is estimated using BEM acoustics exterior method. In the first step of BEM, acoustic potentials such as pressure and particle velocity are solved on the boundary mesh. In order to determine the sound radiated to the exterior of the duct, acoustic potentials are solved on the field point mesh which is a virtual surface surrounding the BE mesh. It is assumed that field point mesh is a nonreflecting surface where the wave just propagates and does not reflect. Figure 5

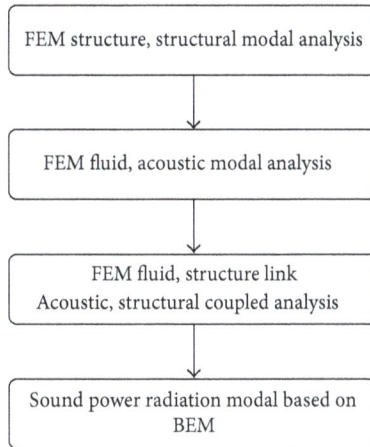

FIGURE 3: Flow chart of numerical modal to estimate sound power radiation.

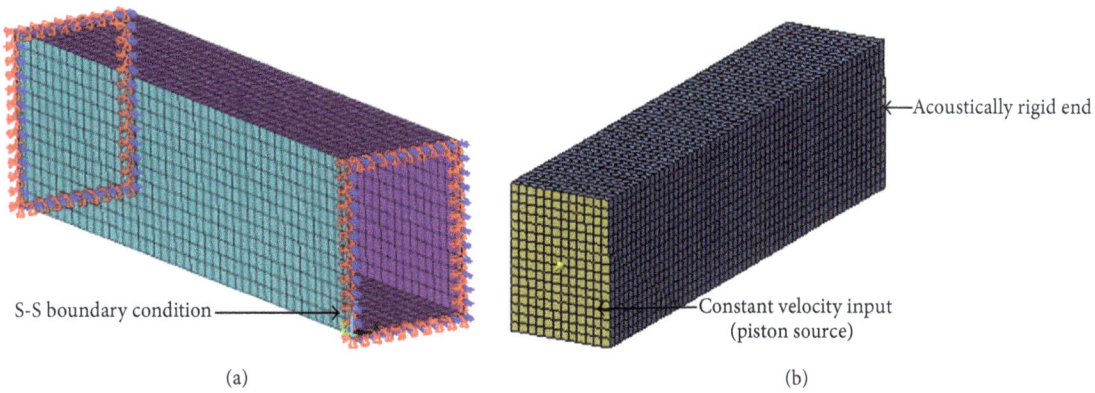

(a)

(b)

FIGURE 4: Numerical models to calculate the radiated sound power and radiation efficiency, (a) structural model and (b) acoustic model.

shows the boundary element mesh and field point mesh. The dimensions of the field point mesh are 0.4 m × 0.5 m × 1.5 m.

Modal radiation efficiency is calculated numerically by activating selected structural or acoustic modes. A theoretical study can be performed to estimate radiation efficiency by choosing different combination of structural and acoustic modes participation in sound radiation. However, two cases are discussed; that is, (i) a single acoustic mode can be coupled to multiple structural modes and (ii) single structural mode can be coupled to the multiple acoustic modes.

4. Results and Discussion

The uncoupled structural modes of the duct structure are calculated by using the proposed equivalent plate model and compared to numerical results as discussed in Section 4.1. These uncoupled modes are classified into four groups based on the net volume displacement and symmetry behaviour at a particular mode which is described in Section 4.2. The uncoupled acoustic modes are calculated using the analytical model and compared to numerical results as shown in

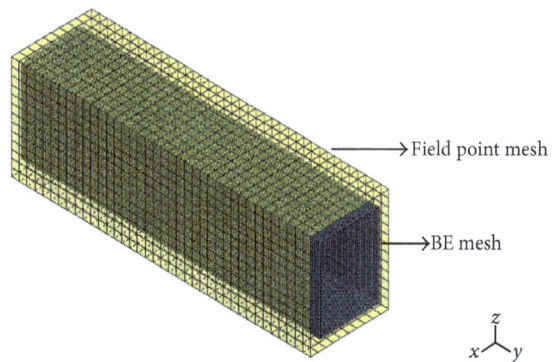

FIGURE 5: A numerical model to calculate the sound radiated from the flexible duct walls.

Section 4.3. Calculation of coupled modes and identification of strong coupling modes are based on transfer factor as discussed in Section 4.4.

TABLE 1: First ten uncoupled structural modes of a rectangular duct with a simply supported boundary condition calculated with proposed analytical and numerical model.

Mode	Analytical (Hz)	Numerical (Hz)
1	99.26	99.22
2	114.65	114.59
3	130.95	127.08
4	140.59	140.51
5	142.72	141.94
6	163.98	163.47
7	177.29	177.17
8	195.67	189.68
9	196.05	195.60
10	209.76	208.72

As the next step, modal radiation efficiencies are calculated by an equivalent plate model for all four groups and validated with the numerical results using the rectangular duct model. In the present paper, the comparisons are shown only for the strongly coupled modes (Section 4.5). In Section 4.6, duct modal radiation efficiencies calculated by the proposed method are compared to a simple rectangular plate available in literature. The behaviour of radiation efficiency when single acoustic mode coupled with multiple structural modes is studied and the results are shown in Section 4.7. The sound radiation behaviour is also studied and shown in Section 4.8.

4.1. Uncoupled Structural Modes. Table 1 shows the comparison of analytical and numerical results of the first ten uncoupled structural natural frequencies of a simply supported rectangular duct. It is observed that the analytical and numerical results are in a good agreement. An equivalent plate model is appropriate to explain the free vibration behaviour of a rectangular duct for further studies.

The structural modal analysis helps in understanding the displacement pattern of a structure subjected to a given boundary condition. The displacement pattern is expressed in terms of mode shape. Figures 6(b) and 6(d) show the developed surface of rectangular duct and its mode shape. The perimeter of the chosen duct dimensions is 1.4 m with length of 1.5 m. Figures 6(a) and 6(b) show the comparison of analytical and numerical model results of fundamental duct mode shape. Similarly, Figures 6(c) and 6(d) show the 10th mode shape. Analytical mode shapes shown in Figures 6(b) and 6(d) are equivalent plate model results. It is observed from the figure that both analytical and numerical model results have the same behaviour.

4.2. Mode Shapes Representation, Grouping and Calculating Net Volume Displacement. The structural mode shapes of the duct are analysed to understand the free vibration behaviour of the rectangular duct, symmetry behaviour, and relative change of phase between the panel pairs.

A particular notation has been followed to describe each mode shape in terms of natural frequency, the number of antinodes on panel pairs, and symmetry behaviour. The

fundamental structural mode shape shown in Figure 6(a) can be represented according to the proposed notation as $[D\_S(1,1), S(1,1)]_{99.2\,Hz}$. It can be described as follows: the first term in the bracket indicates the number of antinodes on x-panel pairs, that is, at $x = 0.5L_2$ (plate P_2) and $x = -0.5L_2$ (plate P_4) along y- and z-directions, and for the given example (1, 1) means one antinode in y- and z-directions. The second term indicates the number of antinodes on y-panel pairs, that is, at $y = 0.5L_3$ (plate P_1) and $y = -0.5L_3$ (plate P_3) along x- and z-directions. Letter "D" indicates the dominating pair which can be identified by comparing the modal displacement amplitude between the pair of plate panels. Letter "S" denotes symmetry of the mode shape behaviour of duct walls and end number is the modal frequency.

Similarly, the 10th mode shape can be represented as $[S(2,2), D\_AS(1,2)]_{209.72\,Hz}$. Here, the number of antinodes on x-panel pairs (i.e., on P_2 and P_4) is (2, 2) and the number of antinodes on y-panel pairs (that is on P_1 and P_3) is (1, 2). Dominant amplitude (D) is on y-panel pair. Letter "AS" denotes antisymmetry mode shape behaviour of duct walls. So, based on this proposed notation, the modes are categorized into four groups based on mode symmetry and antisymmetry behaviour. Group 1 represents Symmetry (S)-Symmetry (S) behaviour of x-axis pair and y-axis pair. Similarly, the other three groups are S-AS, AS-S, and AS-AS.

Net volume displacement for each uncoupled structural mode is calculated with (16). It helps to understand efficiently radiating sound modes and types of sound source

$$v_m = \sum_{n=1}^{4} \left(\sum_m \Delta v_m \right)_n, \tag{16}$$

where n represents panel number varying from 1 to 4 which are considered as four walls of the duct. Δv_m is volume displacement associated with amplitude displacement of the mth element in an nth plate, that is, each duct wall [7].

First, thirty modes of a simply supported rectangular duct are classified into four groups and net volume displacement of each mode is calculated according to (16). These results are given in Table 2. It is observed that modes in group 2, group 3, and group 4 have zero net volume displacement. The modes with odd modal index in group 1 (odd, odd) have higher net volume displacement values than the remaining modes in the group. The modes which have the symmetrical panel pairs and the largest net volume displacement in group 1 are efficient sound radiators and also identified as monopole type of sound source. Modes having (odd, even) or (even, odd) mode shapes in group 1 behave as a dipole sound source and are less efficient sound radiators. Sound radiation behavior is observed by calculating radiation efficiency and their slope.

4.3. Uncoupled Acoustic Modes. The uncoupled acoustical modes for the duct volume subjected to rigid termination boundary conditions are calculated analytically and numerically. Table 3 shows a comparison of first ten uncoupled acoustical modes of the duct. The first acoustic mode wavelength is equal to two times the largest dimension. In this case, it is 113.3 Hz in the longitudinal direction. It is observed

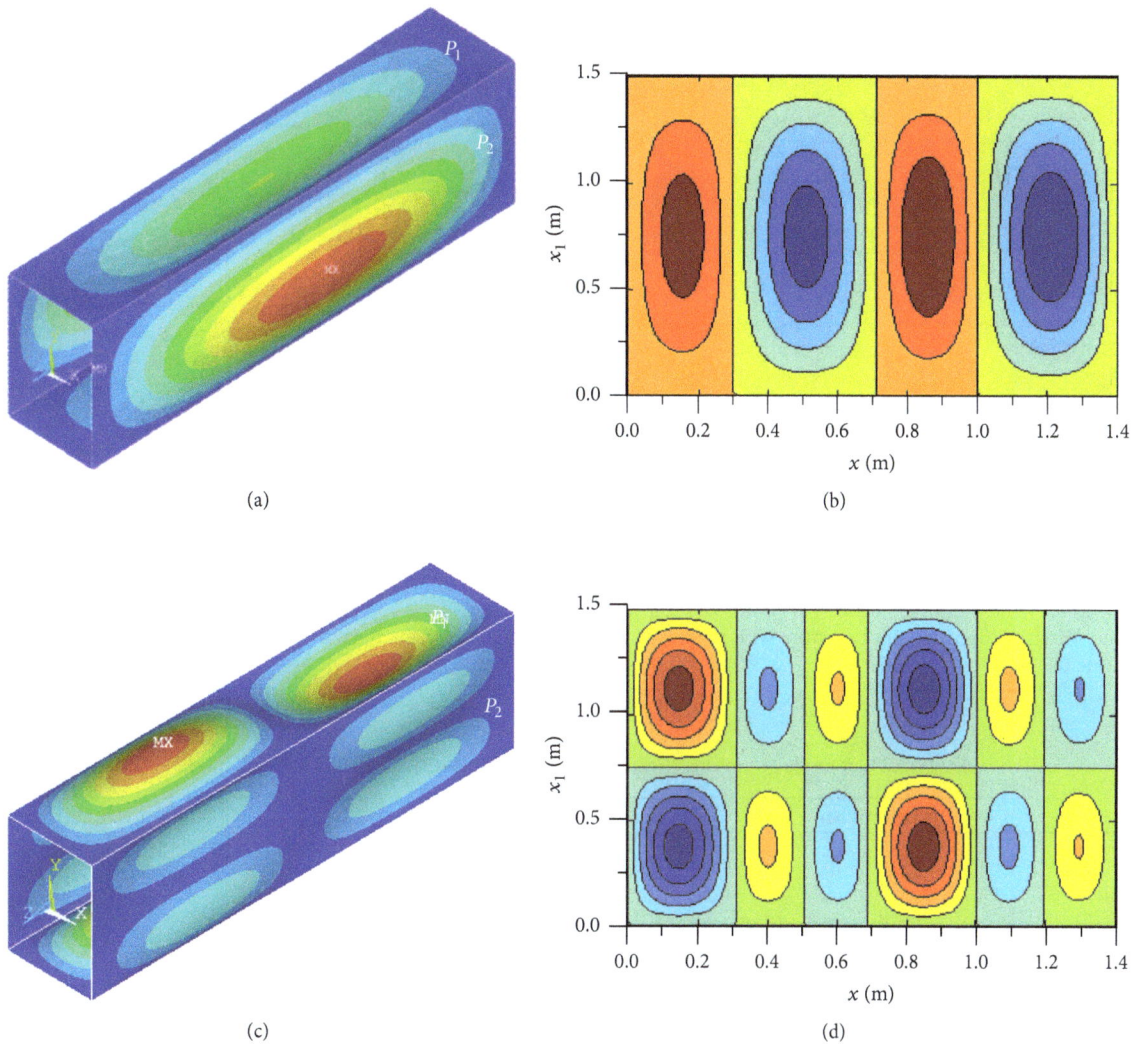

FIGURE 6: (a) Fundamental structural duct mode shape at 99.22 Hz, numerical. (b) Fundamental structural duct mode shape at 99.26 Hz, analytical. (c) Tenth duct mode shape at 208.72 Hz, numerical. (d) Tenth duct mode shape at 209.76 Hz, analytical.

that the uncoupled acoustic modes estimated analytically and numerically are in a good agreement.

4.4. Comparison of Coupled and Uncoupled Response. The sound pressure inside the duct is calculated by both analytical and numerical method for the coupled and uncoupled cases. Equation (7) from the analytical model is used to calculate the inside duct pressure. Variation of the sound pressure inside the duct at a location $(0.5 \times L_2, 0.5 \times L_3, 0.4 \times L_1)$ with respect to frequency for coupled and uncoupled analysis is shown in Figures 7(a) and 7(b). A good agreement between both numerical and analytical results is observed for inside pressure calculation as shown in Figure 7(a).

Figure 7(b) shows the sound pressure inside the duct for coupled and uncoupled conditions. It is observed from the graph that the first peak in a pressure spectrum, based on an uncoupled model occurring at 113.3 Hz, corresponds to the uncoupled acoustic mode. Similarly, two pressure peaks in the coupled model are observed at 105 Hz and 122 Hz. The

energy exchange between acoustic and structural subsystems at coupling frequency converts into a multidegree freedom system. Similarly, at 226 Hz it is divided into two coupled frequencies, each at 220 Hz and 228 Hz.

Table 4(a) gives the first thirty coupled modes which are estimated numerically by coupled modal analysis. These coupled modes contain both acoustical and structural modes. It can be observed that coupling occurs close to the acoustical modes.

Table 4(b) shows the comparison of uncoupled and coupled modes of a rectangular duct. It shows that an acoustic mode at 113.3 Hz and a structural mode at 114.59 Hz are strongly coupled, and a shift in a natural frequency leads to coupled modes at 105.36 Hz and 121.47 Hz. Similarly, the acoustic mode at 226.6 Hz is coupled with the structural mode at 224.69 Hz and leads to 219.39 and 226.30 Hz coupled modes. There also exists strong coupling of a structural mode at 345.1 Hz and an acoustical mode at 340 Hz which results in coupled modes at 343.82 Hz and 347.07 Hz.

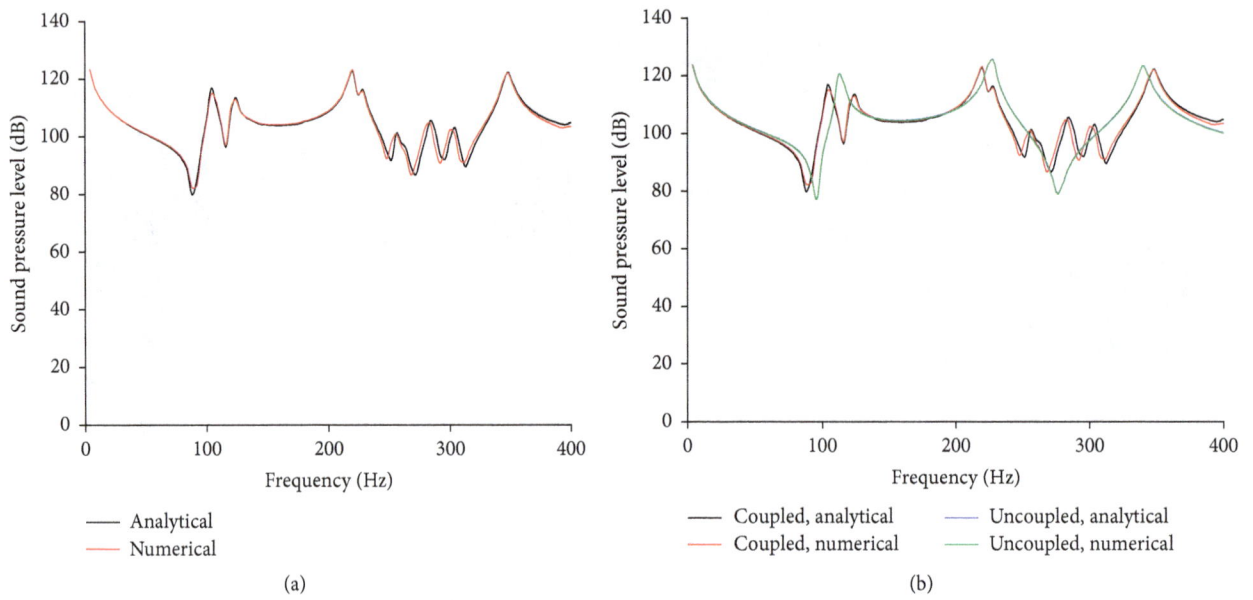

FIGURE 7: (a) Variation of inside duct pressure with respect to frequency for coupled analysis. (b) Comparison of uncoupled and coupled pressure inside the duct.

TABLE 2: Duct mode group classification and net volume displacements for each mode.

	Mode number		Net volume displacement
Group 1	1	$[D\_S(1,1), S(1,1)]_{99.2\,Hz}$	0.1182
	2	$[D\_S(1,2), S(1,2)]_{114.59\,Hz}$	0
	4	$[D\_S(1,3), S(1,3)]_{140.5\,Hz}$	0.0421
	7	$[D\_S(1,4), S(1,4)]_{177.17\,Hz}$	0
	11	$[D\_S(1,5), S(1,5)]_{224.69\,Hz}$	0.0277
	14	$[S(1,1)^*, D\_S(1,1)]_{248.96\,Hz}$	0.2630
	16	$[S(1,2)^*, D\_S(1,2)]_{259.43\,Hz}$	0
	18	$[S(1,3)^*, D\_S(1,3)]_{278.08\,Hz}$	0.0868
	19	$[D\_S(1,6), S(1,6)]_{283.08\,Hz}$	0
	21	$[S(1,4)^*, D\_S(1,4)]_{306.25\,Hz}$	0
	23	$[S(1,5)^*, D\_S(1,5)]_{345.11\,Hz}$	0.0512
	24	$[D\_S(1,7), S(1,7)]_{352.37\,Hz}$	0.0216
	29	$[S(1,6)^*, D\_S(1,6)]_{395.46\,Hz}$	0
Group 2	3	$[D\_AS(1,1), S(2,1)]_{127.08\,Hz}$	0
	5	$[D\_AS(1,2), S(2,2)]_{141.94\,Hz}$	0
	6	$[D\_AS(1,3), S(2,3)]_{163.47\,Hz}$	0
	9	$[D\_AS(1,4), S(2,4)]_{195.60\,Hz}$	0
	13	$[D\_AS(1,5), S(2,5)]_{239.22\,Hz}$	0
	20	$[D\_AS(1,6), S(2,6)]_{294.53\,Hz}$	0
	25	$[D\_S(1,7), S(2,7)]_{361.47\,Hz}$	0
Group 3	8	$[S(2,1), D\_AS(1,1)]_{189.68\,Hz}$	0
	10	$[S(2,2), D\_AS(1,2)]_{208.72\,Hz}$	0
	12	$[S(2,3), D\_AS(1,3)]_{233.06\,Hz}$	0
	17	$[S(2,4), D\_AS(1,4)]_{267.39\,Hz}$	0
	22	$[S(2,5), D\_AS(1,5)]_{312.34\,Hz}$	0
	26	$[S(2,6), D\_AS(1,6)]_{368.20\,Hz}$	0
Group 4	27	$[D\_AS(2,2), AS(2,2)]_{375.41\,Hz}$	0
	30	$[D\_AS(2,3), AS(2,3)]_{403.43\,Hz}$	0

TABLE 3: First ten uncoupled acoustic modes of a rectangular duct volume with rigid end condition.

Mode number	Analytical (Hz)	Numerical (Hz)
1	113.3	113.3
2	226.7	226.6
3	340	340
4	425	425
5	439.9	439.8
6	453.3	453.3
7	481.7	481.6
8	544.3	544.2
9	566.7	566.6
10	566.7	566.6

The fundamental acoustic mode at 113.3 Hz is closer to second structural mode at 114 Hz, and the second acoustic mode (226.6 Hz) is strongly coupled to 11th, 14th, and 17th (224 Hz, 248.96 Hz, and 279.36 Hz) structural modes. The third acoustic mode is strongly coupled to 20th structural mode at 307.82 Hz. Table 5 shows the transfer factor values for the strongly coupled acoustic modes. It is observed from the table that a strong coupling exists between an acoustic mode at 113.3 Hz and structural mode at 114.7 Hz. Transfer factor values are close to one representing a strong coupling. Generally, lesser difference in uncoupled acoustic and structural natural frequencies and good spatial matching of the mode shapes provide higher transfer factor values.

4.5. Calculation of Radiation Efficiency for Strongly Coupled Structural Modes. Based on transfer factor values, the structural mode frequencies with highest transfer factor are identified and these coupled modes are considered for sound power radiation calculation and then radiation efficiency. Figure 8 shows the radiation efficiencies of one structural mode coupled to multiple acoustic modes calculated both analytically and numerically. It shows that both analytical and numerical results are in good agreement. It is clear that equivalent plate model can be used effectively to calculate the modal radiation efficiency.

It is observed from Figure 8 that mode 14 (248.96 Hz) is the most efficient radiating mode followed by mode 1 (99.26 Hz). It is also observed that net volume displacement of mode 14 is large when compared to other modes (Table 2). This is due to substantial net volume displacement and symmetry between the panel pairs of mode 14 at 248.96 Hz. Similarly, structural modes (1, 11, 14, and 18) at 99.2 Hz, 248.96 Hz, 224.69 Hz, and 283.08 Hz have the same kind of slopes, which is a characteristic observed in (odd, odd) mode. A slope of 20 dB/decade is observed in (odd, odd) modes. Individual groups such as S-S (group 1) with (odd, odd) indices are the effective sound radiators. Structural modes (2, 21) at 114.65 Hz and 306.25 Hz are (odd, even) modes and have a slope of 40 dB/decade.

Table 6 shows the slopes of strongly coupled structural modes and comparison of slopes for analytical and numerical models. It shows that both results are in good agreement.

4.6. Comparison of Duct Radiation Efficiencies with a Plate. The radiation efficiency of a duct is compared to a plate radiation efficiency so as to understand the correlation and approximate duct sound radiation behaviour. A simply supported plate with equivalent dimensions of the duct's perimeter and length is considered for the comparison. Modal radiation efficiency for a simply supported rectangular plate can be calculated using analytical equations given in [6]. Critical frequency for a simply supported rectangular plate can be calculated using (17) as given in [6]

$$\omega_c = c_0^2 \left(\frac{\rho_s h}{D} \right)^{1/2}, \tag{17}$$

where c_o is the speed of sound, $\rho_s h$ is the surface density, and D is the flexure rigidity.

Modal radiation efficiencies for the duct are calculated for one structural mode coupled to multiple acoustic modes which exist within the critical frequency. These calculated radiation efficiencies are compared to radiation efficiency of a simply supported plate with dimensions of (1.4 m × 1.5 m). Figures 9–12 show the comparison of the duct modal radiation efficiencies with the plate modal radiation efficiency for the four different groups.

Figure 9 shows the modal radiation efficiency comparison of (1, 1) plate with the [D_S, S] group modes. All the modes in group 1 with (odd, odd) modal indices exhibit the same slope of 20 dB/decade as that of a simply supported plate (1.4 m × 1.5 m) until the critical frequency. After that, the curve becomes asymptotic and the radiation efficiency curve approaches unity.

Figure 10 shows the variation of modal radiation efficiencies corresponding to [D_AS, S] group and comparison with that of a simply supported plate of (2, 1) mode. The slope of radiation efficiency in group 2 is high when compared to group 1. Similarly, volume displacement in group 2 is less compared to group 1. A plate with equivalent dimensions has the same slope as a duct with group 2 mode shapes, but it has a higher slope when compared to the lower mode in the same group. The slope behaviour of 40 dB/decade is observed until the critical frequency, after which the values approach unity, and the curve becomes asymptotic. It is also observed that radiation efficiency values are lower for the higher modes when compared to the fundamental mode.

Figure 11 shows the variation of modal radiation efficiencies corresponding to [S, D_AS] and comparison with that of a simply supported plate of (1, 2) mode. The slope values of this group are the same as [D_AS, S] group. It exhibits that both groups have similar radiation curves. These groups have slope behaviour the same as a simply supported plate (1, 2) mode with a value of 40 dB/decade until critical frequency.

Figure 12 shows the comparison of modal radiation efficiency of [D_AS, AS] group with simply supported plate mode of (2, 2). All the modes in this group have (even, even) modal indices and exhibit similar slope to that of a simply

TABLE 4: (a) Coupled modes of a rectangular duct. (b) Comparison of uncoupled and coupled modes of a rectangular duct.

(a)

Mode number	Frequency (Hz)	Mode number	Frequency (Hz)	Mode number	Frequency (Hz)
1	100.48	11	208.23	21	282.82
2	105.36	12	219.39	22	294.39
3	121.47	13	226.30	23	301.62
4	126.61	14	232.59	24	311.92
5	140.08	15	238.81	25	343.82
6	141.47	16	254.16	26	347.07
7	162.99	17	258.02	27	352.22
8	177.12	18	260.92	28	361.28
9	189.14	19	266.91	29	367.71
10	195.22	20	282.74	30	374.89

(b)

Uncoupled mode frequency (Hz)	Mode type	Coupled mode frequency (Hz)
99.22	Structural	100.48
113.3	Acoustic	105.36
114.59	Structural	121.47
127.08	Structural	126.61
140.51	Structural	140.08
141.94	Structural	141.47
163.47	Structural	162.99
177.17	Structural	177.12
189.68	Structural	189.14
195.60	Structural	195.22
208.72	Structural	208.23
224.69	Structural	219.39
226.6	Acoustic	226.30
345.1	Structural	343.82
345	Acoustic	347.07

TABLE 5: Transfer factor values of strongly coupled acoustic modes.

Acoustic mode (Hz)	Structural mode (Hz)	Transfer factor $T_{n,m}$
113.3	114.7	0.99
	224.9	0.98
226.6	248.96	0.66
	278.08	0.57
340	306.06	0.79

supported plate (1.4 m × 1.5 m) with 60 dB/decade for group (even, even) modal indices. The slopes of these curves are the highest when compared to any other group.

Table 7 shows the comparison of slope values of all the four groups for a few selected duct modes and simply supported plate modes. Duct radiation efficiency slope values of all groups are the same as an equivalent plate. It can be observed from Figures 9–12 that the radiation efficiency curves for single structural mode coupled to multiple acoustic modes and plate mode behaviour are found to be similar.

Therefore, it can be concluded that duct radiation efficiencies of different groups can be calculated based on an equivalent plate model with minimum error. Equivalent plate model for a rectangular duct proposed in this research work is an appropriate way to predict the free vibration behaviour and sound radiation characteristic of a duct.

4.7. Calculation of Radiation Efficiency for One Acoustic Mode Coupled to Multiple Structural Modes. In order to understand the effect of coupling closer to an acoustic mode, a case of one acoustic mode coupled to multiple structural modes has been studied. Figure 13 shows the radiation efficiency of three acoustic modes such as 113, 226, and 339 Hz frequencies that are individually coupled to all structural modes with respect to frequency.

4.8. Total Radiation Efficiency and Radiated Sound Power. Figures 14 and 15 show the total radiation efficiency and sound power radiated from multiple acoustic modes when coupled to multiple structural modes. All peaks in the sound

TABLE 6: Radiation efficiency slopes of strongly coupled structural modes.

Mode	Frequency (Hz)	Analytical (dB/decade)	Numerical (dB/decade)
1	99.26	19.42	18.90
2	114.65	39.25	39.00
11	224.84	17.27	16.96
14	249.85	18.72	18.74
18	278.08	16.75	16.90
21	306.25	38.05	38.69

TABLE 7: Comparison of the radiation efficiency slopes for duct modes and simply supported plate modes.

Group	Duct mode	Duct mode analytical (dB/decade)	Duct mode numerical (dB/decade)	Simply supported plate mode (dB/decade)
1	$[D\_S(1,1), S(1,1)]_{99.2\,Hz}$	19	19	19
2	$[D\_AS(1,1), S(2,1)]_{127.08\,Hz}$	41	41	39
3	$[S(2,3), D_{AS(1,3)}]_{233.06\,Hz}$	37	39	39
4	$[D\_AS(2,2), AS(2,2)]_{375.41\,Hz}$	57	58	54

power curve are associated with the coupled modes. First peak in the total sound power curve occurs at fundamental acoustic mode 113.3 Hz. Second prominent peak in the curve occurs close to second acoustic mode at 226.6 Hz where acoustic energy is exchanged between three structural modes. Third prominent peak occurs close to third acoustic mode at 340 Hz. It is observed that second acoustic mode is more efficient in radiating sound due to strong coupling amongst three structural modes as compared to the other two acoustical modes.

Figure 16 represents pressure distribution at a point on x-panel ($x = 0.2$ m, $y = 0$ m, and $z = 0.75$ m) and y-panel ($x = 0$ m, $y = 0.25$ m, and $z = 0.75$ m), respectively. All the peaks correspond to coupled frequencies. Figure 17 represents numerical results of the sound radiation pattern of first three coupled acoustic modes at 113 Hz, 226 Hz, and 340 Hz. For the third acoustic mode, radiation pattern is similar to structural mode shape (306 Hz), where a strong coupling is observed and has a dipole behaviour.

It is observed from previous results that both analytical and numerical results are in good agreement to capture the acoustic and structural behaviour of both coupled and uncoupled subsystems. The proposed equivalent plate model for rectangular duct predicted free vibration behaviour and sound radiation characteristics. Modal grouping based on free vibration study shows that group 1 (Symmetry-Symmetry) modes with (odd, odd) indices are more effective sound radiators and thus justified based on radiation efficiency results.

5. Conclusions

Analytical and numerical models for calculating rectangular duct sound radiation characteristics have been discussed in the present manuscript. The predicted results from two

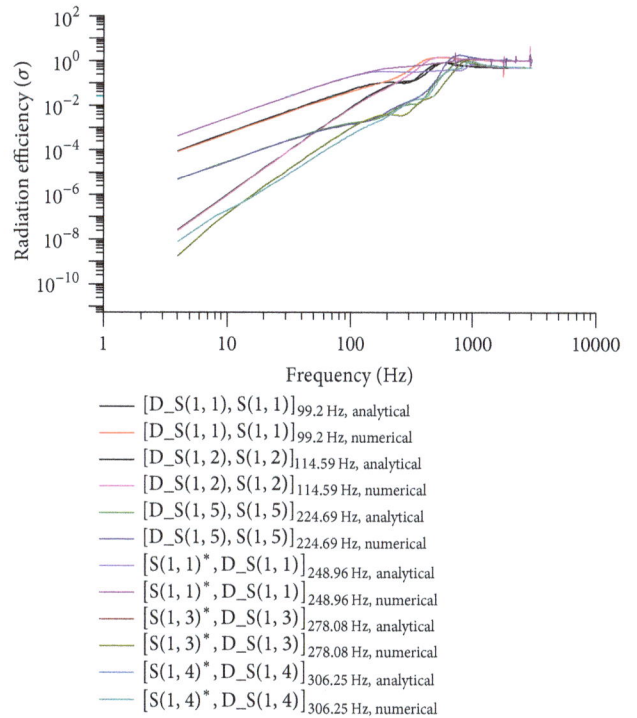

FIGURE 8: Modal radiation efficiency of the strongly coupled structural modes.

models are in good agreement. It has been verified that rectangular duct considered in this paper can be modelled as an equivalent plate for free vibration analysis. The uncoupled structural duct modes have been categorized into four groups and it was found that Symmetry-Symmetry mode group with (odd, odd) modal indices behave as efficient sound radiators. These were determined from higher net volume

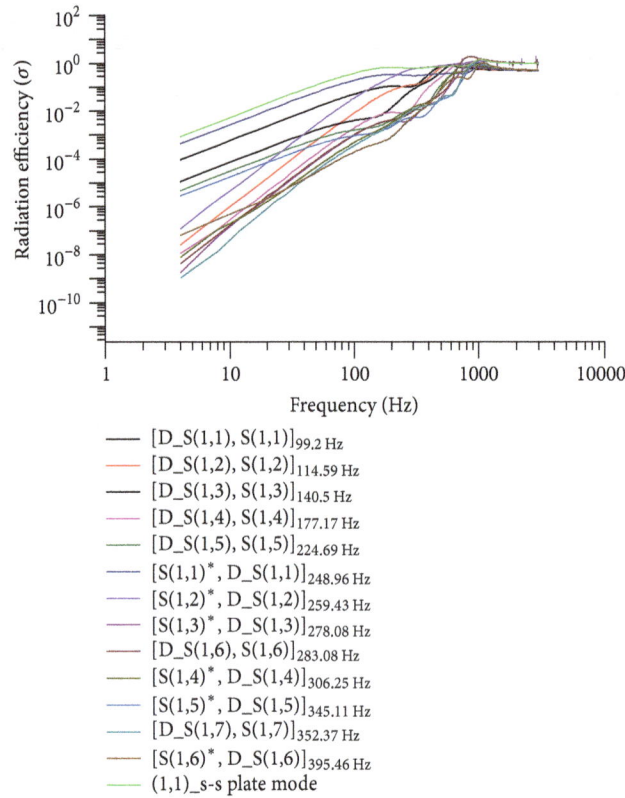

FIGURE 9: Calculation of duct modal radiation efficiency with respect to frequency for [(D_S, S)] group and plate mode (1, 1).

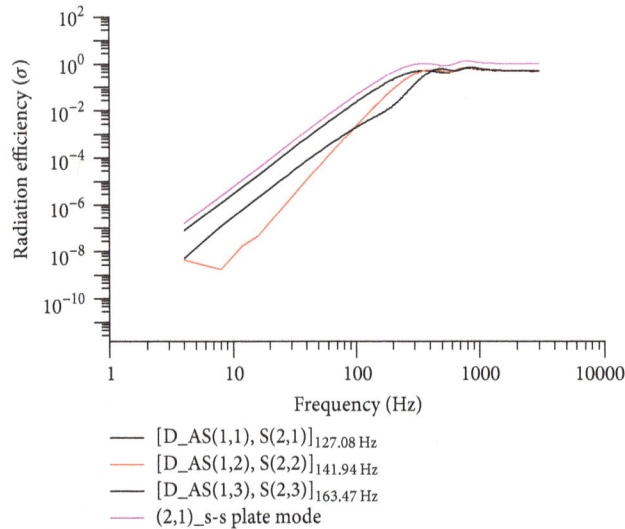

FIGURE 10: Calculation of duct modal radiation efficiency with respect to frequency for [D_AS, S] group and plate mode (2, 1).

displacement of (odd, odd) modes when compared to others. An analytical model of the total sound power radiated from duct walls has been validated with numerical results. Duct radiation pattern is similar to standard sound sources such as monopole and dipole. The effect of acoustic and structural mode coupling on radiation efficiency has been studied and compared to a simply supported plate behaviour. Slope of radiation efficiency curves for four duct mode groups have

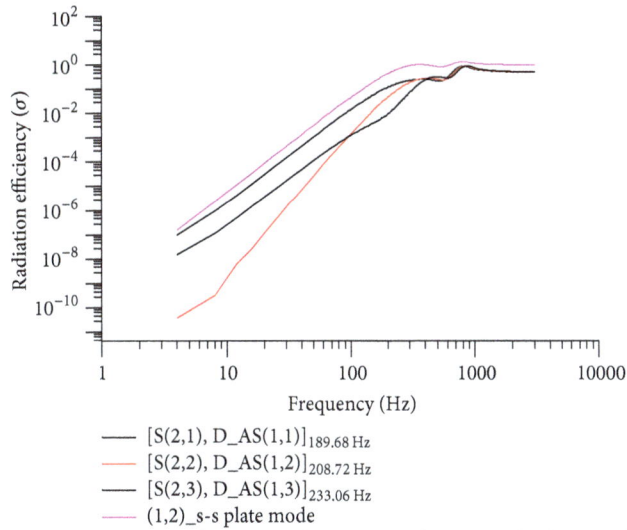

FIGURE 11: Calculation of duct modal radiation efficiency with respect to frequency for [S, D_AS] group and (1, 2) plate mode.

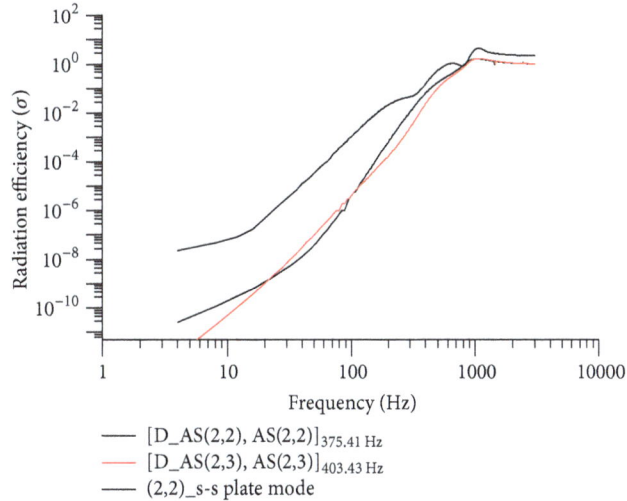

FIGURE 12: Calculation of duct modal radiation efficiency with respect to frequency for ([D_AS, AS]) and (2, 2) plate mode.

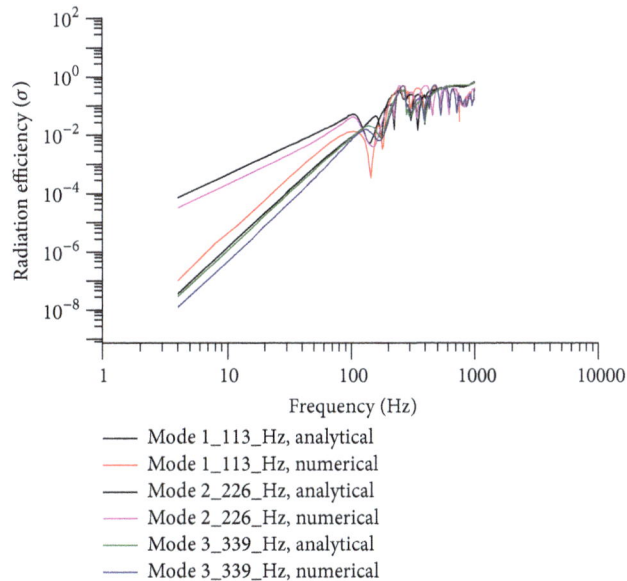

FIGURE 13: Radiation efficiency of the first three acoustic modes coupled with multiple structural modes (numerical versus analytical).

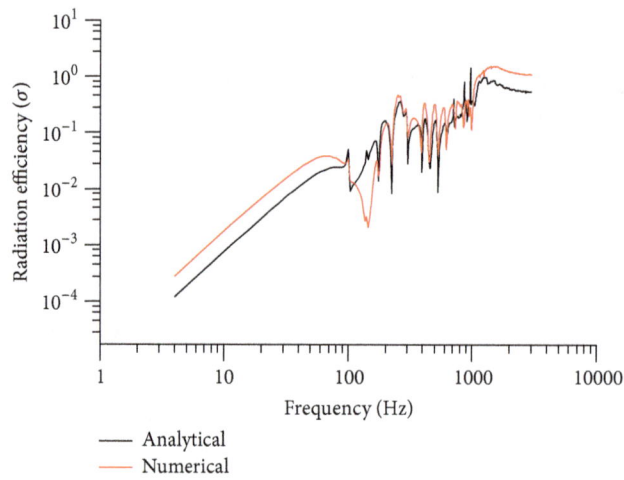

FIGURE 14: Total radiation efficiency of the rectangular duct (numerical versus analytical).

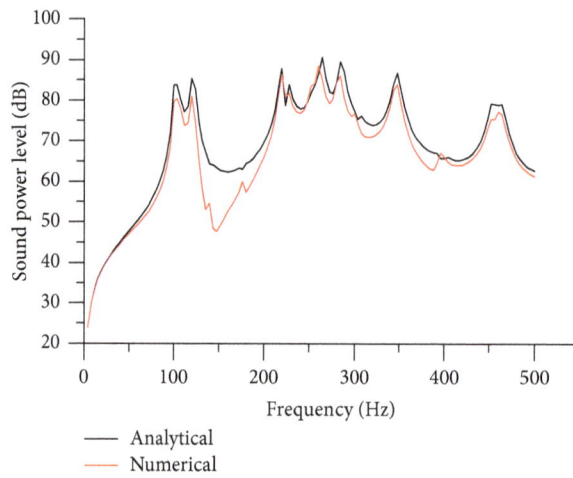

FIGURE 15: Total sound power radiated from the rectangular duct.

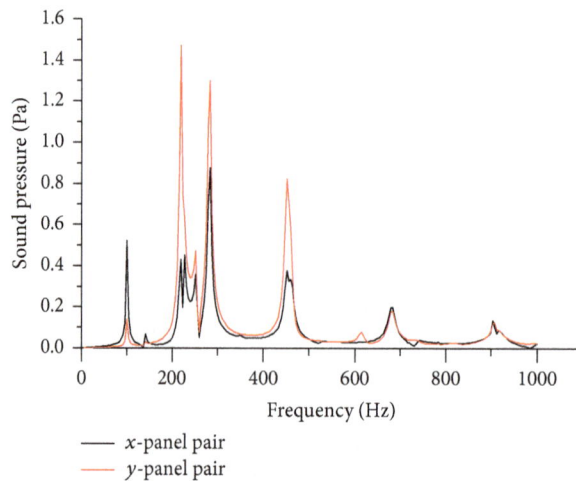

FIGURE 16: Sound pressure at field point on x-panel pair (0.35 m, 0.75 m, and 0.25 m) and z-panel pair (0.15 m, 0.75 m, and 0.45 m).

Pressure average Iso amplitude on deformed mesh dB(RMS).3
Occurrence 129
Max: 83.27
Min: 1.43988

Pressure average Iso amplitude on deformed mesh dB(RMS).3
Occurrence 151
Max: 87.2666
Min: 15.104

Pressure average Iso amplitude on deformed mesh dB(RMS).3
Occurrence 174
Max: 92.2676
Min: 19.1232

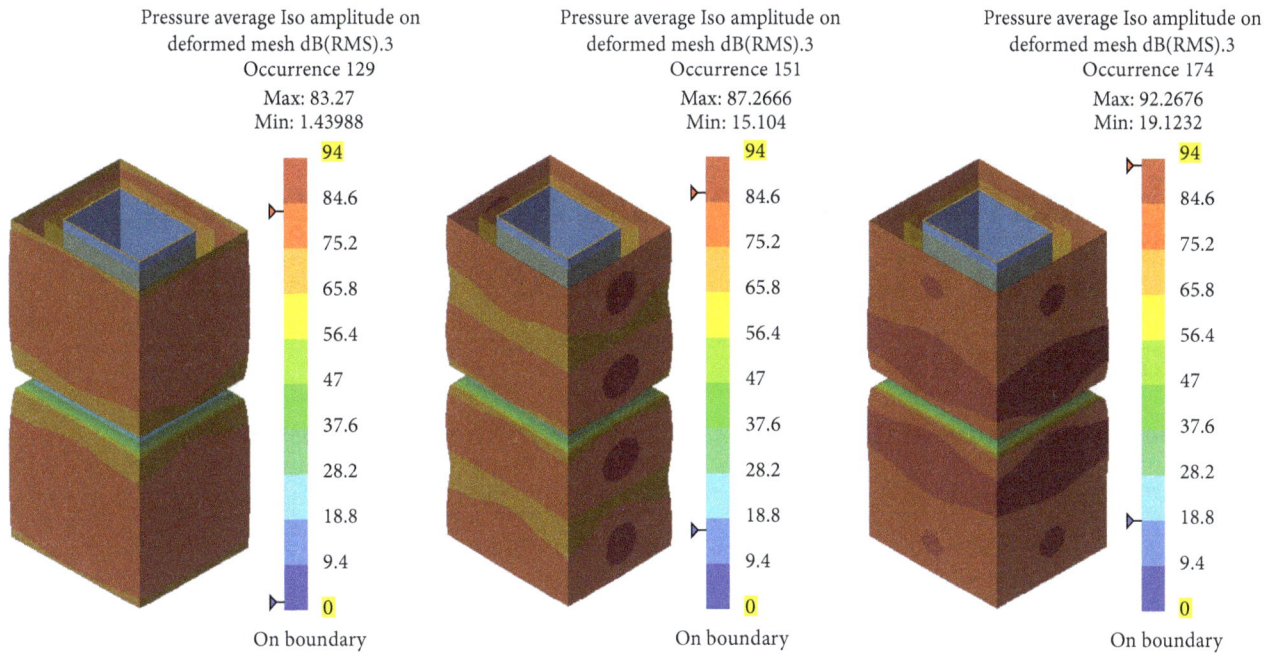

FIGURE 17: Sound radiation pattern of the first three coupled acoustic modes at 113 Hz, 226 Hz, and 340 Hz.

been compared to simply supported plate curves. It has been shown that sound radiation efficiency of the rectangular duct is similar to a plate.

Competing Interests

The authors declare that they have no competing interests.

Acknowledgments

The authors would like to thank Indian Institute of Technology Hyderabad for providing the required resources to conduct the current research work.

References

[1] A. Cummings, "Sound transmission through duct walls," *Journal of Sound and Vibration*, vol. 239, no. 4, pp. 731–765, 2001.

[2] W. H. Louisell, *Coupled Mode and Parametric Electronics*, John Wiley & Sons, New York, NY, USA, 1960.

[3] I. L. Ver and L. L. Beranek, *Noise and Vibration Control Engineering: Principles and Applications*, John Wiley & Sons, New York, NY, USA, 2006.

[4] B. Venkatesham, M. Tiwari, and M. L. Munjal, "Prediction of breakout noise from a rectangular duct with compliant walls," *International Journal of Acoustics and Vibrations*, vol. 16, no. 4, pp. 180–190, 2011.

[5] B. Venkatesham, M. Tiwari, and M. L. Munjal, "Analytical prediction of break-out noise from a reactive rectangular plenum with four flexible walls," *The Journal of the Acoustical Society of America*, vol. 128, no. 4, pp. 1789–1799, 2010.

[6] C. E. Wallace, "Radiation resistance of a rectangular panel," *Journal of the Acoustical Society of America*, vol. 51, no. 3, pp. 946–952, 1972.

[7] T. Ran Lin and J. Pan, "Sound radiation characteristics of a box-type structure," *Journal of Sound and Vibration*, vol. 325, no. 4-5, pp. 835–851, 2009.

[8] S. M. Kim and M. J. Brennan, "A compact matrix formulation using the impedance and mobility approach for the analysis of structural-acoustic systems," *Journal of Sound and Vibration*, vol. 223, no. 1, pp. 97–113, 1999.

[9] H. P. Lee, "Natural frequencies and modes of cylindrical polygonal ducts," *Journal of Sound and Vibration*, vol. 164, no. 1, pp. 182–187, 1993.

[10] ANSYS Inc, ANSYS 13, User guide, 2013.

[11] LMS International, LMS Virtual Lab Rev 13, User's Manual, 2013.

9

Mathematical Modelling and Acoustical Analysis of Classical Guitars and Their Soundboards

Meng Koon Lee, Mohammad Hosseini Fouladi, and Satesh Narayana Namasivayam

School of Engineering, Taylor's University, No. 1 Jalan Taylor's, 47500 Subang Jaya, Selangor, Malaysia

Correspondence should be addressed to Mohammad Hosseini Fouladi; mfoolady@gmail.com

Academic Editor: Kim M. Liew

Research has shown that the soundboard plays an increasingly important role compared to the sound hole, back plate, and the bridge at high frequencies. The frequency spectrum of investigation can be extended to 5 kHz. Design of bracings and their placements on the soundboard increase its structural stiffness as well as redistributing its deflection to nonbraced regions and affecting its loudness as well as its response at low and high frequencies. This paper attempts to present a review of the current state of the art in guitar research and to propose viable alternatives that will ultimately result in a louder and better sounding instrument. Current research is an attempt to increase the sound level with bracing designs and their placements, control of natural frequencies using scalloped braces, as well as improve the acoustic radiation of this instrument at higher frequencies by deliberately inducing asymmetric modes in the soundboard using the concept of "splitting board." Various mathematical methods are available for analysing the soundboard based on the theory of thin plates. Discrete models of the instrument up to 4 degrees of freedom are also presented. Results from finite element analysis can be utilized for the evaluation of acoustic radiation.

1. Introduction

Classical guitars are unique musical instruments as the acoustic response of each piece of a particular model is different from another one although they are dimensionally identical and are all made by the same luthier according to French [1]. Two reasons given for this lack of acoustic consistency are firstly the variations in the natural properties of wood and secondly the manual tuning process of the soundboard by experienced luthiers which is not well understood analytically. Borland [2] has determined that humidity of air and moisture content in the wood are important factors affecting how wood responds when it vibrates.

Technically, classical guitars have been modelled and analysed by using several mathematical models. These models were used for determining modal frequencies and frequency response function. Using these results, classical guitars can be objectively assessed by evaluating their acoustic radiation.

Throughout the evolution of the classical guitar since 1500 AD, it is generally agreed among luthiers that the type of wood, the design and placement of bracings on the

soundboard, and the reinforcement of the back plate play important roles in the production of a good acoustically radiating instrument. This consensus among luthiers creates an aura of mysticism that surrounds the construction of the guitars and translates into an enormous respect for top quality concert instruments. This intuitive analysis of the luthier can be reinforced by scientific knowledge through collaboration with research scientists in the fields of Mechanics of Solids and Continuum Mechanics. Such collaboration creates an interdisciplinary research in the field of musical acoustics such as that existing at the research centre of Universitat Politècnica de Catalunya (BarcelonaTech). The research here centres on a combination of experimental and numerical research and the experience of a well-known luthier [3].

Studies by Richardson et al. [4], Siminoff [5], and Bader [6] show the relatively greater importance of acoustic radiation from the soundboard as compared to those from the back plate and the bridge at high frequencies. Based on these findings, research on increasing the loudness of the guitar by focusing on the soundboard alone is a potential in future research.

The objectives of this paper are categorically summarised under the following sections:

Section 2: Mathematical Models

Section 3: Acoustical Analysis

The above two categories define the scope of review in this paper.

2. Mathematical Models

Richardson et al. [4] and Siminoff [5] have shown that the soundboard is the single most important component affecting the sound pressure level of the classical guitar. Factors that affect the performance of vibrating soundboards in terms of acoustic radiation are design, placement and arrangement of bracings, and thickness. These factors contribute to the musical acoustics of the classical guitar.

An in-depth understanding of the dynamic characteristics of the classical guitar can be obtained by considering some mathematical models. In particular, the simplest two-mass model of Christensen-Vistisen [7] provides a simple understanding of the interaction between a vibrating air mass and the soundboard. This model and the three-mass and four-mass models are three classical examples of discrete mathematical models of this instrument. These models are based on the mass-spring-damper mechanism. As Richardson et al. [4] and Siminoff [5] have shown that the soundboard is the single most important component affecting the sound pressure level of the classical guitar, it can therefore be modelled separately as a vibrating thin plate and the theory of thin plates can be applied to study its dynamic behaviour with the aim of improving its contribution to the sound pressure level of this instrument. This component, complete with design and arrangement of fan strutting, can then be assembled with the ribs and back plate to study the effects on modes and natural frequencies due to noncoupling between the soundboard and back plate versus coupling between these two components via the air mass inside the guitar body as was carried out by Elejabarrieta et al. [8].

2.1. Discrete and Continuous Systems

2.1.1. Discrete Systems. Christensen and Vistisen [7] proposed a simple 2-degree-of-freedom model, also known as the "Christensen-Vistisen" lumped parameter model. This model consisted of an air piston and the soundboard. Christensen [9] proposed a 3rd degree of freedom in the form of a back plate while Popp [10] proposed yet a 4th degree of freedom in the form of ribs. These are the 3- and 4-degree-of-freedom models, respectively. These additional degrees of freedom provided more realistic representations of the guitar. The range of frequencies investigated was 80–250 Hz.

A hybrid mechanical-acoustic model proposed by Sali and Hindryckx [11] and Sali [12] was used to investigate the changes in loudness relative to the first peak (the first resonance) of the complete instrument. This model consisted of a mass, spring, damper, and a massless membrane rigidly attached to the mass. The membrane had a constant area equivalent to that of the radiating surface.

2.1.2. Continuous Systems. As the number of degrees of freedom increases, modelling using discrete masses becomes cumbersome. To circumvent this problem, the complete instrument can be considered as a continuous system and its vibration characteristics can be effectively analysed using the finite element method as shown by Derveaux et al. [13] and Gorrostieta-Hurtado et al. [14]. The soundboard can also be shown to satisfy the criterion of a thin plate in flexure and the application of the theory of thin plates results in a fourth-order partial differential equation of the vibrating system. Attempts to solve this model analytically can be researched using current mathematical methods.

2.2. Vibration of the Complete Instrument.

Analytical models with 2-, 3-, and 4-degree-of-freedom have been formulated by Christensen and Vistisen [7], Christensen [9], and Popp [10], respectively, and are applicable to the complete instrument. Modal analysis of the 2- and 3-degree-of-freedom models by Caldersmith [15] and Richardson et al. [4], respectively, predicted two and three eigenvalues in the frequency range from 80 to 250 Hz. Similarly, modal analysis of the 4-degree-of-freedom model also predicted three eigenvalues in the frequency range of 80 to 250 Hz but a fourth eigenvalue was missing. It was concluded by Popp [10] that assigning a fourth-degree-of-freedom model in the form of a finite mass to the ribs does not introduce any new elastic restoring force and hence there is no fourth eigenvalue. Hence adding extra degrees of freedom beyond the fourth in this method of modelling the classical guitar would add unnecessary complications as even with the 4-degree-of-freedom model; the mechanics of the complete guitar body cannot be adequately represented [4].

Hess [16] conducted a parametric study with the two-mass model to identify a unique combination of physical parameters in an attempt to increase the sound level over a frequency range of 70 Hz to 250 Hz. Results showed that, by decreasing the stiffness and effective mass of the soundboard by 50% and decreasing the soundboard area by 28%, there was an increase of 3.2 dB in the sound pressure level per unit force over the entire frequency range. Although this investigation was performed on an acoustic guitar, there is no indication that these parameters could not be used to examine their influence on classical guitar soundboards.

The range of frequencies of a classical guitar investigated by Czajkowska [17] varies from 70 Hz to just under 2 kHz. However, there are also harmonic notes that the classical guitar can produce. To account for these higher frequency notes, Richardson [18] suggested that the range of frequencies is extended to 20 kHz, which is the upper threshold of human hearing. However, from ISO 226:2003, the minimum sound pressure level for any arbitrary loudness occurs within a bandwidth of 3 to 4 kHz. For practical purposes, experiments could be conducted up to 5 kHz.

Sakurai [19], a luthier, made some interesting video recordings of the vibration of the soundboard. He experimented with the traditional bracing structure and with diagonal braces and discovered that the soundboard could be made thinner and could vibrate with larger amplitudes

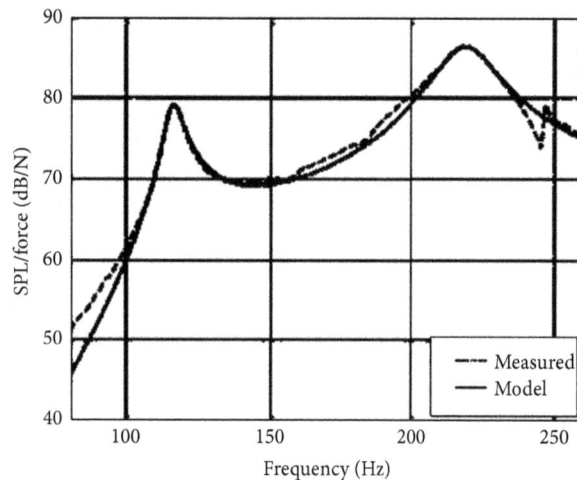

FIGURE 1: Comparison of theoretical and experimental results for sound pressure level at a distance of 2 m above the soundboard [16].

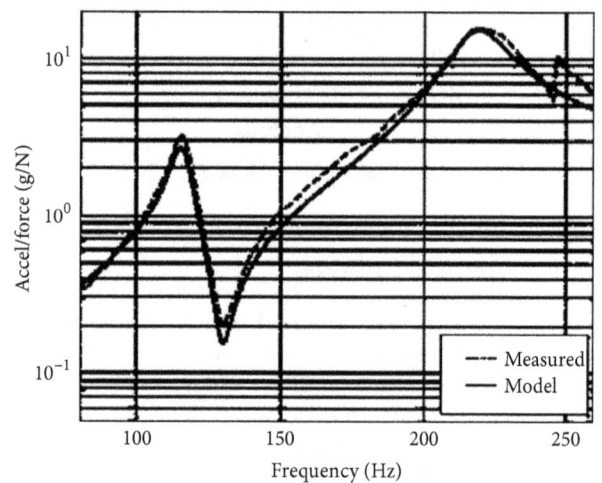

FIGURE 2: Comparison of theoretical and experimental results for acceleration per unit force [16].

without compromising on its structural integrity. However, there was no accompanying mathematical analysis.

Richardson et al. [4], in their revisitation of the 3-mass model, are of the opinion that the properties of the soundboard together with the design of bracings and the bridge play more important roles than results from the 3-mass model in relation to the fundamental top plate mode. They further concluded that low-order modes have significant controlling influence on the playing qualities of the guitar. This conclusion was based on informal listening tests. Studies by Richardson et al. [20] have shown that the noise components generated by these low-order modes are an important perceptual element in guitar sounds as perception is regarded as an important element in music.

2.2.1. Two-Degree-of-Freedom Model. The simplest model consists of two masses representing the soundboard and an air piston as proposed by Christensen and Vistisen [7]. Hess [16] has shown that this model gives good agreement between theoretical and experimental results for sound pressure and acceleration frequency response at low frequencies (80 to 250 Hz) as shown in Figures 1 and 2.

The first resonance typically occurs within a frequency range of 90–120 Hz while the second can be found in the range of 170–250 Hz. The model provided excellent quantitative fit for both sound pressure versus frequency and acceleration versus frequency responses. Hologram interferometry by Richardson and Walker [21] has shown that the second mode is the lowest (fundamental) mode of the soundboard alone. The first resonance is found only in the complete instrument made up of the soundboard, back plate, ribs, and neck. This implies that there is coupling between the soundboard and the air mass inside the cavity of the guitar (the Helmholtz resonator).

2.2.2. Three-Degree-of-Freedom Model. Christensen [9] proposed the addition of a third mass, the back plate of the guitar. The addition of a third degree of freedom depicts a more realistic guitar when it is played. Three resonant frequencies

were obtained and the phase relationships between the soundboard, the back plate, and the air piston were obtained. Results from the first three resonances were obtained by Richardson and Walker [21] using holographic interferometry. These results showed "strong" coupling between each resonance and the strings via the bridge as the latter lies on an antinodal area. That is, the bridge lies on an antinodal area since this is the location where the strings transfer vibration to the soundboard. "Strong" coupling refers to the large changes in volume of the air-cavity and this produces a large monopole contribution to the sound radiated from the guitar. However, according to Richardson [18], Wright [22], and McIntyre and Woodhouse [23], strong coupling produces undesirable "wolf-notes" due to overcoupling of the body to the strings.

Results for sound pressure versus frequency at the high frequency (above 400 Hz) spectrum showed that radiation from the soundboard dominates radiations from the back plate and the sound hole when the instrument is driven directly using an impact hammer with no strings attached as shown in Figure 3.

Richardson et al. [4] suggested a ratio A_t/m_t to indicate the "acoustical merit" of the instrument where A_t and m_t are the effective area and effective mass of the soundboard, respectively, since this ratio is directly proportional to the total sound radiation above 400 Hz. Therefore, if the soundboard can be made as thin as possible, then total sound radiation would increase. This is an important consideration for the classical guitar if the sound level from this instrument is to be increased. This is an attempt to quantify quality of the classical guitar. Thus, this property of the soundboard can be considered to contribute significantly to the "global" playing qualities of the instrument. "Global" refers to the perceptible changes in the "treble" and "base" playing ranges especially of the first and second body modes by changing the effective mass of the soundboard. Global properties could be measured in terms of the Q-value of resonances. The Q-values could also be a parameter associated with quality of the instrument according to Richardson et al. [4].

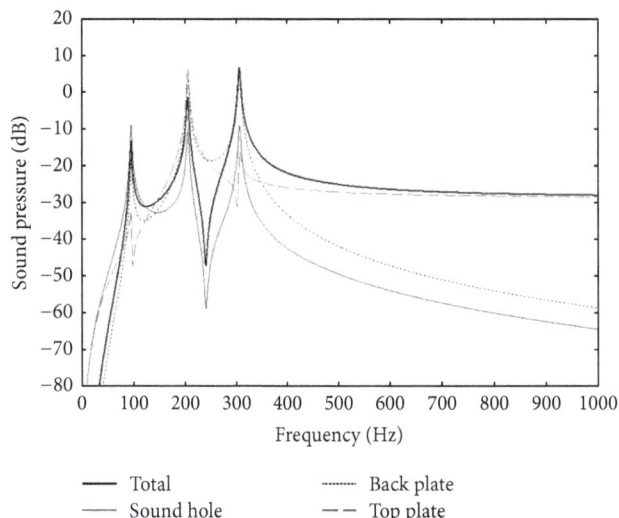

FIGURE 3: Monopole sound radiation from 3-degree-of-freedom model [4].

- — Total
- — Sound hole
- ······· Back plate
- — — Top plate

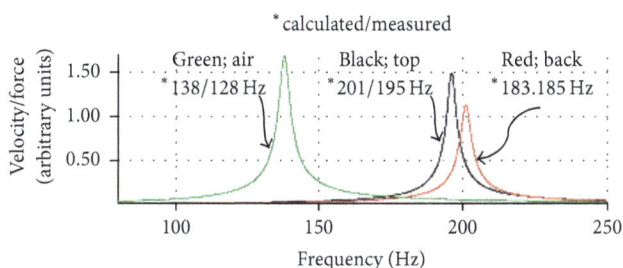

FIGURE 5: Phase relationship for Kohno classical guitar [10].

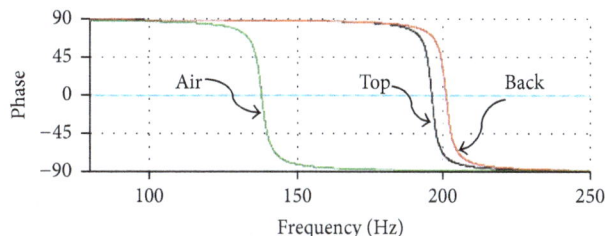

FIGURE 4: Isolated oscillator resonances for Kohno classical guitar [10].

2.2.3. Four-Degree-of-Freedom Model.

A 4-degree-of-freedom model by Popp [10] was used to gauge the relative importance of low-order modes in relation to midfrequency response. This model improved over the previous models by introducing a fourth oscillator known as the "ribs." This increased the number of degrees of freedom to four. The stiffness of the soundboard and back plate were measured directly and their effective areas and masses were used to calculate the resonances and phases. Vibrations of the neck were shown to significantly affect the frequency response in some guitars. The calculated and measured resonances agree reasonably well as shown in Figure 4 and the relative phases between the air piston, back plate, and top plate are as shown in Figure 5.

The addition of an extra degree of freedom in the form of "ribs" did not produce any significant phase difference between the results of this model compared to those of the 2- and 3-degree-of-freedom models.

2.2.4. Hybrid Mechanical-Acoustic Model.

This model was used to investigate the effects of brace positioning on the acoustics of the classical guitar in terms of loudness of tones based on the first resonant peak. This first resonant peak is a result of the coupling between the soundboard and the back

plate via the air mass inside the guitar box. It was found that brace positioning had an effect on the peak amplitudes of the frequency response function.

This model consists of a combination of mass, spring, damper, and a massless membrane as proposed by Sali and Hindryckx [11] and Sali [12]. This model was used to investigate the importance of the first mode on the tonal quality of the instrument. A comparison between good and bad quality guitars indicated that good quality guitars have lower frequency of the first mode and correspondingly higher amplitude in the frequency response function and lower or equal damping. This first mode corresponds to the first peak in the frequency response function of the instrument. It was found that the intensity or amplitude of the first mode was inversely proportional to the damping of the soundboard. The objective of this model was to optimize the placing of bracing for a better-quality instrument.

2.3. Vibration of the Soundboard.

Investigation of soundboard vibration up to 10 kHz is best performed using finite element analysis. Sumi and Ono [24] conducted experiments with three different quality guitars and modal analysis using ANSYS showed that the best quality guitar had a thickness of 3.0 mm at the centre part and 2.0 mm at the end whereas the more inferior ones had constant thicknesses of 2.8 mm and 2.6 mm. However, this is only an experimental work and no analytical model was available.

Dumond and Baddour [25, 26] studied the effects of scalloped braces on mode shapes. A simple analytical model based on Kirchhoff plate theory was used to study the vibration of a rectangular board with and without braces. The effects of rectangular braces on the resonant frequencies were compared with those from scalloped braces. Mathematically, it was shown that the shape of a scalloped brace can be modelled as a 2nd order piecewise polynomial function with peaks at positions 1/4 and 3/4 of the brace length. It was concluded that reducing the thickness of the brace reduced the lowest resonant frequency as this reduces the stiffness of the plate. It was also concluded that, by using scalloped braces, it was possible to control the 1st and the 4th natural frequencies of the brace-plate system simultaneously but control of two natural frequencies simultaneously is not possible using rectangular braces. Thus, scalloped braces will further assist the luthiers in controlling the type of soundboards they prefer their instruments to have. This simple model of the soundboard was modelled as a rectangular plate. Though this model is far from reality as the shape of the soundboard is more complex, consisting of a series of curves, it nevertheless suffices to explain the effects of plate thickness on modal frequencies. Davies [27] has shown that the boundary of the guitar soundboard could be successfully modelled using Chebyshev polynomials. Attempts were also made to model the boundary using Fourier and polynomial series but these resulted in large errors in their derivatives at the extremes of the fitted domain. The use of Chebyshev series minimized these errors. This mathematical concept could be used to modify the results of the rectangular plate in future research.

Besnainou et al. [28] conducted research into increasing the far-field radiation of the instrument using the concept of "splitting board." This was a deliberate attempt to create asymmetric modes of vibration which maximises acoustic radiation versus symmetric modes which minimizes the radiation due to destructive interference. The soundboard was split longitudinally along its axis of symmetry. One-half of the board below the bass strings had a thickness of 2 mm while the other half below the treble strings had a thickness of 3 mm. Accelerometers were placed in front of the 2nd and 5th strings. The frequency band of investigation was from 0 to approximately 22 kHz. Results showed an average increase of approximately 3 dB in sound pressure level of the instrument thus indicating that the concept of "splitting board" could be new concept of future soundboard design.

Caldersmith [15] discovered that the displacements of the back plate of an acoustic guitar are only a very small proportion of that of the top plate at the fundamental resonance. This observation was obtained from measurements with piezoelectric transducers attached to both the top and bottom plates. Based on this study and that of Richardson et al. [4], further research on improving the loudness of this instrument can be focusing on the top plate (soundboard) alone.

O'Donnell and McRobbie [29] experimented with a new material for the soundboard of an acoustic guitar. Instead of wood, carbon fibre reinforced polymer (CFRP) was used as a material for the soundboard. The soundboard was modelled as a rectangular plate in 3D with a thickness of 3 mm and COMSOL Multiphysics was used to obtain eigenvectors (mode shapes) corresponding to eigenfrequencies up to 100 Hz. These were compared to empirical results obtained from the soundboard of an acoustic guitar. It was found that there was a striking similarity in the mode shapes though the frequencies showed some variations as the shape of an actual guitar soundboard is different from that of a rectangular plate. Davies [27] had also arrived at a similar conclusion with regard to the similarity of mode shapes. Wegst [30, 31] has shown that wood is still the material of choice for soundboards of musical instruments due to its mechanical and acoustical properties.

2.3.1. Finite Element Method.
A finite element model is a discretization of a continuum into a large but finite number of nonoverlapping elements connected at their nodes. The response of the continuum is then approximated by the response of the finite element model. The finite element method is an appropriate approach for analysing the vibration of a continuum such as the soundboard over a wide range of frequencies. This range of frequencies is found in the work of Czajkowska [17], who attempted to differentiate higher quality instruments from lower quality ones. Experimental tests showed that higher quality instruments had larger top-back correlation coefficients compared with lower quality ones in the frequency range 1-2 kHz. It was also observed that higher quality instruments are characterized by stronger structural resonances of the soundboard in the range 4-5 kHz. These observations suggest that future research using finite element models is conducted at frequencies ranging from 70 Hz to 5 kHz with the objective of manufacturing better-quality instruments.

Modal analysis of soundboards made from a composite of polyurethane foam reinforced with carbon fibre was analysed by Okuda and Ono [32] using the finite element method. Results showed that the relationship between frequency and mode number could be freely controlled by adjusting the physical properties of this material. This is an attempt to introduce soundboards with consistent tones as those from wooden materials tend to be affected by humidity and moisture content of the surrounding air as shown by Borland [2]. Research into the potential use of an industrially moulded plastic component such as the guitar soundboard is given by Pedgley et al. [33].

Stanciu et al. [34] used finite element method to investigate the dynamic characteristics of acoustic plates as one of the components of a guitar. Plates without bracing, with 3 bracings, and with 5 bracings were studied. Parameters considered in this article were density of material, Young's modulus, thickness of the plates, and the number of bracings. Their influence on the resonant frequencies was obtained for the first 10 modes. It was concluded that, for a given design, plates with higher density have lower resonant frequencies and that lower frequencies resulted in greater acoustic power. Curtu et al. [35] obtained further correlations between these acoustic plates resonant frequencies and the mechanical, physical, and elastic properties of the composite materials of the complete guitar. Vernet [36] also investigated the influence of bracing on the mode shapes and resonant frequencies of the soundboards of guitars using the finite element method but did not use scalloped braces. da Silva Ribeiro et al. [37] conducted similar investigations on two different fan bracings using the finite element method. Results showed that there were significant variations of some of the mode shapes and modal frequencies due to differences in soundboard stiffness.

The influence of the bridge on the response of the soundboard was investigated by Torres and Boullosa [38] using finite element method. It was shown that the assembly and specific design of the bridge had considerable influence on the mode shapes at frequencies above 300 Hz.

Gorrostieta-Hurtado et al. [14] considered the soundboard as a thin plate whose motion is described by the Kirchhoff-Love equation. Its characteristics in various stages of development of the instrument were evaluated using modal analysis results from finite element method. The vibroacoustic characteristics of the complete instrument can also be investigated by finite element method as shown by Paiva and Dos Santos [39].

2.3.2. Boundary Element Method. The boundary element method belongs to the group of boundary type formulations. In this group, only the surface (boundary) of an acoustical fluid needs to be discretized. The number of degrees of freedom is considerably reduced as there is no need to discretize the entire volume of the fluid. This is especially applicable to our system comprising the soundboard (structure) surrounded by air (fluid) as only the sound pressure and sound velocity need to be defined at the boundary. The radiation of sound waves from the soundboard to "infinity" is implicitly included in the formulation by the inclusion of a perfectly matched layer.

Xu and Huang [40] showed that acoustic radiation of a three-dimensional structure could be computed using the finite element method as well with the boundary element method and that the latter method required less computation as only the surface needs to be meshed. In the case of the finite element method the volume of the object needs to be meshed and the boundary conditions of the exterior need to be specified as well. The boundary element method is further enhanced with the advent of a fast multipole algorithm which further reduces solution time and uses less computer memory. Future research into soundboard acoustic radiation could proceed along this concept.

A new approach to studying acoustic radiation of thin structures is to model them as surfaces without thickness and using the boundary element method as in Venkatesh et al. [41]. It was shown that the errors in their numerical solutions were better than those obtained by treating them as thin plates. This is also a possible alternative to investigate acoustic radiation from soundboards.

Investigations into the vibroacoustic behaviour of thin structures such as the soundboard could also proceed by modelling the soundboard using finite elements and the surrounding air by boundary elements. This results in coupling of both subsystems. The solution leads to the structural behaviour of the soundboard (structure) under the influence of air (fluid) as well as the propagation of acoustic waves within the air. Vibroacoustic applications such as this are given in Von Estorff [42] and in conjunction with LMS User's Manual [43].

2.3.3. Analytical Method. Exact solutions for an irregular-shaped plate such as the soundboard of a guitar which is subjected to various boundary conditions are difficult to obtain. This challenge prompts further research using current mathematical methods such as Variational Iteration, Adomian Decomposition, Perturbation, Least Squares, Collocation, and Rayleigh-Ritz. These have been used to solve various engineering problems involving fourth-order parabolic partial differential equations with constant coefficients as well as with variable coefficients.

Current trend in research indicates a return from numerical methods to analytical methods in attempts to seek exact solution for vibrating plates. This is evident from research on free vibration of irregular-shaped plates as well as rectangular plates with variable thickness by Sakiyama and Huang [44], [45], respectively, and on rectangular plates with central circular holes by Torabi and Azadi [46]. Cho et al. [47, 48]

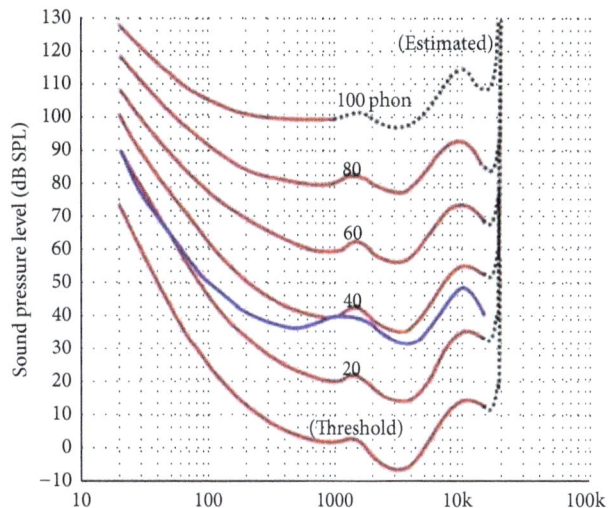

FIGURE 6: Equal-loudness contours [red: ISO 226:2003 revision, blue: ISO 226:1987 for 40 phons].

investigated vibration of rectangular plates with openings of different shapes as well as rectangular plates with holes and stiffeners. The soundboard with bracings is considered as a stiffened plate. The concept of equivalent rectangular plates, Davies [27], could be used to study the vibration of irregular-shaped plates. Mass remnant ratio as proposed by Mali and Singru [49] could be used to study the effect of holes on the natural frequencies of plates.

3. Acoustical Analysis

3.1. Range of Frequencies. Czajkowska [17] recommended that the bandwidth of investigations can be extended to 3 octaves above the E-note of the 1st string at the 12th fret. This frequency is 5.274 kHz. Based on ISO 226:1987, the minimum sound pressure level for any arbitrary loudness is between 3 kHz and 4 kHz as shown on equal-loudness curves provided by Moller and Lydoff [50]. This minimum is still valid based on revised ISO 226:2003 as shown in Figure 6. Thus, an attempt to increase the loudness of the classical guitar at frequencies up to 5 kHz could consider increasing its monopole radiation as dipole radiation tends to dominate at these frequencies as shown in Figure 7. It has also been demonstrated experimentally that frequency components above 5 kHz have little consequence on human perception of guitar tones. These factors suggest that further research on this instrument focuses on investigating its frequency response up to 5 kHz.

3.2. Radiated Power and Radiation Efficiency. Investigation of the acoustics of soundboards is concerned with the maximum acoustic power that it can radiate. Wood for guitars needs to be treated to provide minimum acoustic absorption. Special attention must be paid to the method of treatment as a study by Mamtaz et al. [52] has shown that treatment with natural fibre composites increases acoustic absorption instead of reducing it.

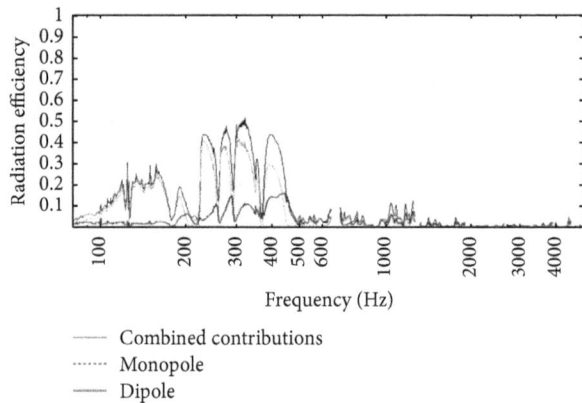

FIGURE 7: Effect of monopole and dipole components on radiation efficiency [51].

Expressions for the numerical evaluation of radiation efficiencies and radiation power of simply supported baffled plates such as those proposed by Lemmen and Panuszka [53] could be used as postprocessing tools in finite element analysis to evaluate the performance of soundboards for various boundary conditions. The boundary condition in the case of the guitar soundboard varies between simply supported and fixed support. An expression for the radiated power from forced vibration due to a harmonic point force of lightly damped simply supported plates is also available in [53]. At frequencies above 1 kHz, an expression for the frequency averaged radiation efficiency of a ribbed panel by Maidanik [54] gives good agreement with the numerical results of [53].

Van Engelen [55] has also proposed expressions for evaluating radiation power and radiation efficiency from velocities and pressures obtained from finite element analysis. The radiation efficiency of the soundboard can also be computed as shown by Perry and Richardson [51].

3.2.1. Effects of Monopole and Dipole Radiation on Radiation Efficiency. Perry and Richardson [51] have shown that, below 600 Hz, monopole radiation is the most important contributor to the total radiation efficiency of the classical guitar. Graphs of radiation efficiency versus frequency also show that, from 200 Hz to 550 Hz, there is an increase in dipole radiation and a reduction of monopole radiation resulting in a net reduction in the total radiation efficiency. The nature of the two lowest frequency modes was shown to be predominantly monopole. Sound radiation fields for a BR2 guitar at 350 Hz, 360 Hz, and 363 Hz show a change in radiation pattern from monopole to dipole as frequency increased as reproduced in Figure 8. Admittance (defined as velocity per unit force) versus frequency in [51] indicated two large and clearly defined peaks between 200 Hz and 300 Hz and as frequency increases to 4 kHz, the number of smaller peaks gets closer and closer together, making identifying individual modes difficult.

The radiation efficiency dropped from 0.37 at 350 Hz to 0.2 at 360 Hz and finally to 0.11 at 363 Hz. However, radiation,

defined as the sound pressure per unit force, at the front of the instrument, remained relatively constant and peaked at 0.7 Pa/N. Perry and Richardson [56] reinvestigated this instrument and found that the radiation efficiency dropped from 0.67 at 345 Hz to 0.10 at 458 Hz.

3.2.2. Relative Importance of Soundboard, Sound Hole, and Bridge in Acoustic Radiation. The relative importance of the radiation strengths of the sound hole, the bridge, and the soundboard of a classical guitar was investigated by Bader [6]. Radiation strength was measured in terms of the percentage of the whole radiation area. It was found that radiation from the sound hole dominates up to about 200 Hz. At frequencies above 200 Hz, radiation from the soundboard dominates. These relationships are as shown in Figure 9.

4. Factors to Consider in Experimental Modal Analysis

The aim of modal analysis is to obtain mode shapes and natural frequencies [57]. In mathematical terms, these are eigenvectors and eigenvalues of the instrument, respectively. The traditional method is based on total contact using mechanical impact hammers to cause an excitation and to record vibration using accelerometers. In this method, instrumentation consists of a signal generator, an exciter, a force transducer, an accelerometer, a signal conditioner, a data acquisition device, and a computer such as the experimental setup of Stanciu et al. [58].

Special consideration should be given to the choice of exciters for classical guitars as investigation is conducted over a wide range of frequencies. The fundamental frequency of a classical guitar is 82.4 Hz corresponding to the E-note of the 6th string. In some cases, this may be lowered to the D-note whose frequency is 73.4 Hz. This range of frequency can be obtained by using a modal hammer. However, the maximum frequency attainable with this hammer is between 1.8 and 2 kHz. Czajkowska [17] investigated the complete instrument over a range of frequencies varying from around 70 Hz to 5 kHz. This investigation was approximately 3 octaves above the E-note of the 1st string at the 12th fret. This range of frequencies was achievable using a bone vibrator attached to the bridge at a position closest to the 1st string. Vibration of the instrument was recorded using a noncontact method with a scanning laser vibrometer. Further advances in technology has led to totally noncontact excitations and measurements using ultrasound radiation force such as those used by Huber et al. [59].

5. Conclusion

The objective of this review is to assess the state of the art in guitar design and to explore research pathways for this instrument. It is evident that there is much scope for research into improving the performance of this musical instrument in terms of its loudness, the ability to control more frequencies simultaneously and to improve its acoustic radiation at higher frequencies. This process involved literature reviews

(a) 350 Hz

(b) 360 Hz

(c) 363 Hz

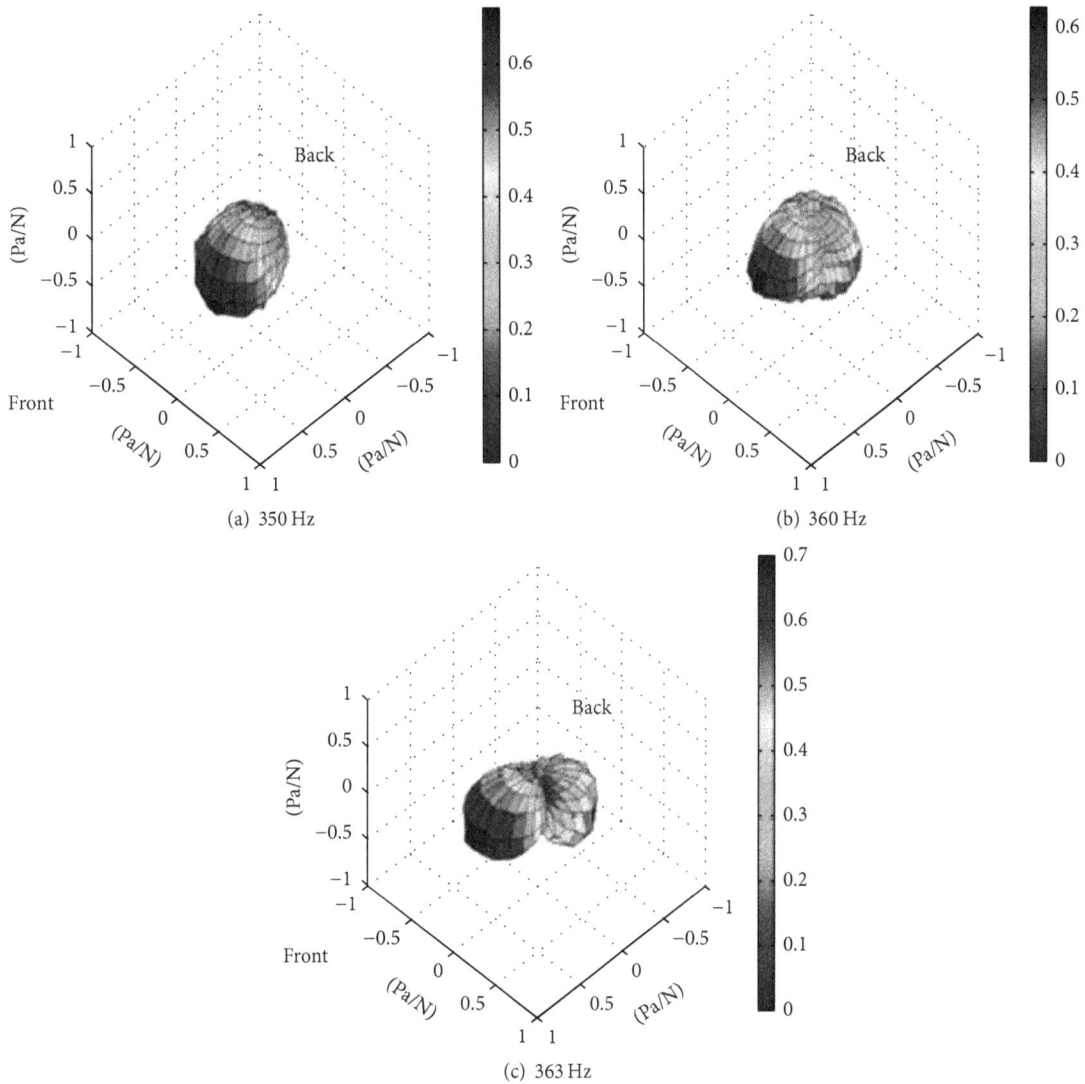

FIGURE 8: Change in radiation pattern from monopole to dipole for a BR2 guitar [51]. "Front" refers to the front of the guitar. The same is for "back."

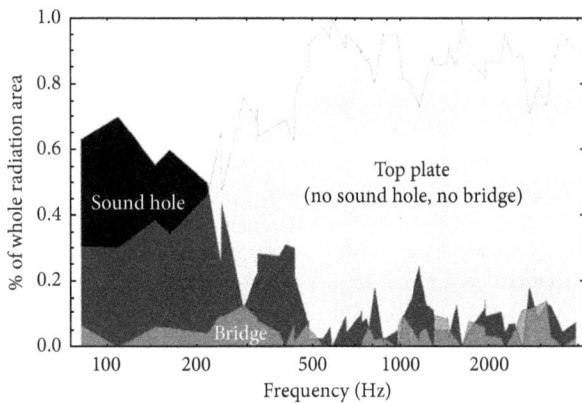

FIGURE 9: Frequency dependent areas for an LK model classical guitar [6].

on the complete instrument, its soundboard as a standalone component, and its acoustic radiation. Dipole radiation at

higher frequencies should be minimized to obtain higher radiation efficiency. Radiation power and radiation efficiency could also be computed analytically.

Research trends tend to focus on the resurgence of analytical methods of investigating the vibration of irregular-shaped plates and the use of equivalent rectangular plates. However, exact solutions for the transverse displacement of the soundboards of classical guitars as a special case of irregular-shaped plate are not yet forthcoming due to the difficulty caused by modelling its irregular shape and finding appropriate shape functions. However, popular numerical tools such as the finite element or boundary element methods will continue to be used to compare with results from analytical methods.

New bracing patterns, brace designs, and the concept of "splitting board" for redistributing modal patterns and their associated natural frequencies are possible pathways aimed at improving the design of the soundboard. A relook at the coupling mechanism between the soundboard and the back plate via the vibrating air mass inside the guitar body might also be a way forward in reinventing the guitar as a better musical instrument. Scalloped braces provide a method of controlling two natural frequencies simultaneously.

Traditional instrumentation involves total contact excitations and measurements while totally noncontact excitations and measurements involve the use of ultrasound radiation force.

Competing Interests

The authors declare that there is no conflict of interests regarding the publication of this paper.

References

[1] M. French, "Response variation in a group of acoustic guitars," *Sound & Vibration*, vol. 42, pp. 18–22, 2008.

[2] M. J. Borland, *The effect of humidity and moisture content on the tone of musical instruments [Ph.D. dissertation]*, University of Waterloo, Waterloo, Canada, 2014.

[3] M. A. Perez and A. Manjon, "Reigniting Guitar Evolution," Waves, 2014, https://www.bksv.com/en/about/waves/WavesArticles/2014/reigniting-quitar-evolution.

[4] B. E. Richardson, H. R. Johnson, A. D. Joslin, and I. A. Perry, "The three-mass model for the classical guitar revisited," in *Proceedings of the Acoustics 2012 Nantes Conference*, pp. 2777–2782, April 2012.

[5] R. Siminoff, *The Luthier's Handbook: A Guide to Building Great Tone in Acoustic Stringed Instruments*, Hal Leonard, 2002.

[6] R. Bader, "Radiation characteristics of multiple and single sound hole vihuelas and a classical guitar," *Journal of the Acoustical Society of America*, vol. 131, no. 1, pp. 819–828, 2012.

[7] O. Christensen and B. Vistisen, "Simple model for low-frequency guitar function," *The Journal of the Acoustical Society of America*, vol. 68, no. 3, p. 758, 1980.

[8] M. J. Elejabarrieta, A. Ezcurra, and C. Santamaría, "Coupled modes of the resonance box of the guitar," *Journal of the Acoustical Society of America*, vol. 111, no. 5, part 1, pp. 2283–2292, 2002.

[9] O. Christensen, "Quantitative models for low-frequency guitar function," *Journal of Guitar Acoustics*, vol. 6, pp. 10–25, 1982.

[10] J. E. Popp, "Four mass coupled oscillator guitar model," *Journal of the Acoustical Society of America*, vol. 131, no. 1, pp. 829–836, 2012.

[11] S. Sali and F. Hindryckx, "Modeling and optimizing of the first guitar mode," *Savart Journal*, vol. 1, no. 1, pp. 1–13, 2011.

[12] S. Sali, "Frequency response function of a guitar—a significant peak," *Catgut Acoustical Society Journal*, vol. 4, no. 6, pp. 1–13, 2002.

[13] G. Derveaux, A. Chaigne, P. Joly, and E. Bécache, "Time-domain simulation of a guitar: model and method," *Journal of the Acoustical Society of America*, vol. 114, no. 6, pp. 3368–3383, 2003.

[14] E. Gorrostieta-Hurtado, J.-C. Pedraza-Ortega, J.-M. Ramos-Arreguin, A. Sotomayor-Olmedo, and J. Perez-Meneses, "Vibration analysis in the design and construction of an acoustic guitar," *International Journal of the Physical Sciences*, vol. 7, no. 13, pp. 1986–1997, 2012.

[15] G. Caldersmith, "Guitar as a reflex enclosure," *Journal of the Acoustical Society of America*, vol. 63, no. 5, pp. 1566–1575, 1978.

[16] D. P. Hess, "Frequency response evaluation of acoustic guitar modifications," *Savart Journal*, vol. 1, no. 3, pp. 1–8, 2013.

[17] M. Czajkowska, "Analysis of classical guitars' vibrational behavior based on scanning laser vibrometer measurements," in *Proceedings of the 10th International Conference on Vibration Measurements by Laser and Noncontact Techniques (AIVEA '12)*, vol. 1457 of *AIP Conference Proceedings*, pp. 336–343, Ancona, Italy, June 2012.

[18] B. E. Richardson, "The acoustical development of the guitar," *Journal of the Catgut Acoustical Society*, vol. 2, no. 5, (Series II), pp. 1–10, 1994.

[19] M. Sakurai, *A Documentary by GSI guitars*, Japanese Television Show-Virtuosos, Tokyo, Japan, 2013.

[20] B. E. Richardson, S. A. H. Bryant, J. Rolph, and M. Weston, "Plucked string sound analysis and perception," in *Proceedings of the Institute of Acoustics*, vol. 30, pp. 437–444, January 2008.

[21] B. E. Richardson and G. P. Walker, "Mode coupling in the guitar," in *Proceedings of 12th International Congress of Acoustics*, vol. 3, K3-2, Toronto, Canada, 1986.

[22] H. Wright, *The acoustics and psychoacoustics of the guitar [Ph.D. thesis]*, University of Wales, College of Cardiff, Wales, UK, 1996.

[23] M. E. McIntyre and J. Woodhouse, "The acoustics of stringed musical instruments," *Interdisciplinary Science Reviews*, vol. 3, no. 2, pp. 157–173, 2013.

[24] T. Sumi and T. Ono, "Classical guitar top board design by finite element method modal analysis based on acoustic measurements of guitars of different quality," *Acoustical Science and Technology*, vol. 29, no. 6, pp. 381–383, 2008.

[25] P. Dumond and N. Baddour, "Effects of a scalloped and rectangular brace on the modeshapes of a brace-plate system," *International Journal of Mechanical Engineering and Mechatronics*, vol. 1, no. 1, pp. 1–8, 2012.

[26] P. Dumond and N. Baddour, "Can a brace be used to control the frequencies of a plate?" *SpringerPlus*, vol. 2, article no. 558, 2013.

[27] E. B. Davies, *On the structural and acoustics design of guitar soundboards [Ph.D. thesis]*, University of Washington, 1990.

[28] C. Besnainou, J. Frelat, and K. Buys, "A new concept for string-instrument sound board : the splitting board," in *Proceedings of the International Symposium on Music Acoustics*, pp. 25–27, Sydney, Australia, August 2010.

[29] J. O'Donnell and G. McRobbie, "A study of the acoustic response of carbon fiber reinforced plastic plates," in *Proceedings of the COMSOL Users Conference*, Stuttgart, Germany, 2011.

[30] U. G. K. Wegst, *The mechanical performance of natural materials [Ph.D. thesis]*, University of Cambridge, Cambridge, UK, 1996.

[31] U. G. K. Wegst, 'Natural Materials Selector, created using CES Constructor Software, Granta Design, Cambridge, UK' Cambridge, UK, 2004.

[32] A. Okuda and T. Ono, "Bracing effect in a guitar top board by vibration experiment and modal analysis," *Acoustical Science and Technology*, vol. 29, no. 1, pp. 103–105, 2008.

[33] O. Pedgley, E. Norman, and R. Armstrong, *Materials-Inspired Innovation for Acoustic Guitar Design*, METU Journal of the Faculty of Architecture—Middle East Technical University Faculty of Architecture, Ankara, Turkey, 2009.

[34] M. D. Stanciu, I. Curtu, C. Itu, and R. Grimberg, "Dynamical analysis with finite element method of the acoustic plates as constituents of the guitar," *ProLigno*, vol. 4, no. 1, pp. 41–52, 2008.

[35] I. Curtu, M. Stanciu, and R. Grimberg, "Correlations between the plates' vibrations from the guitar's structure and the physical, mechanical and elastically characteristics of the composite materials," in *Proceedings of the 9th WSEAS International Conference on Acoustics & Music: Theory & Applications*, pp. 55–60, Bucharest, Romania, June 2008.

[36] D. Vernet, "Influence of the guitar bracing using finite element method," Tech. Rep., The University of New South Wales and Ecole Normale Superiore, New South Wales, Australia, 2001.

[37] R. F. da Silva Ribeiro, A. J. da Silva, J. F. Silveira Feiteira, and N. de Medeiros, "Numerical analysis of acoustic guitars soundboards with different fan bracings," in *Proceedings of the 22nd International Congress of Mechanical Engineering (COBEM 2013)*, pp. 2506–2512, Ribeirão Preto, Brazil, November 2013.

[38] J. A. Torres and R. R. Boullosa, "Influence of the bridge on the vibrations of the top plate of a classical guitar," *Applied Acoustics*, vol. 70, no. 11-12, pp. 1371–1377, 2009.

[39] G. Paiva and J. M. C. Dos Santos, "Modal analysis of a Brazilian guitar body," in *Proceedings of the ISMA International Symposium on Music Acoustics*, pp. 233–239, Le Mans, France, 2014.

[40] Z. Xu and Q. Huang, "The study of three-dimensional structural acoustic radiation using FEM and BEM," *Advances in Theoretical and Applied Mechanics*, vol. 3, no. 4, pp. 189–194, 2010.

[41] P. R. Venkatesh, B. Chandrasekhar, and M. M. Benal, "A new approach to model acoustic radiation and scattering from thin bodies using boundary element method," *International Journal of Applied Engineering Research*, vol. 6, no. 6, pp. 745–761, 2011.

[42] O. Von Estorff, *Boundary Elements in Acoustics, Advances & Applications*, WIT Press, Southhampton, UK, 2008.

[43] *SYSNOISE Rev 5.6, User's Manual*, LMS International, Leuven, Belgium, 2005.

[44] T. Sakiyama and M. Huang, "Elastic bending analysis of irregular-shaped plates," *Structural Engineering and Mechanics*, vol. 7, no. 3, pp. 289–302, 1999.

[45] T. Sakiyama and M. Huang, *Free Vibration Analysis of Rectangular Plates with Variable Thickness*, Reports of the Faculty of Engineering, Nagasaki University, Nagasaki, Japan, 1998.

[46] K. Torabi and A. R. Azadi, "Vibration analysis for rectangular plate having a circular central hole with point support by Rayleigh-Ritz method," *Journal of Solid Mechanics*, vol. 6, no. 1, pp. 28–42, 2014.

[47] D. S. Cho, N. Vladimir, and T. M. Choi, "Approximate natural vibration analysis of rectangular plates with openings using assumed mode method," *International Journal of Naval Architecture and Ocean Engineering*, vol. 5, no. 3, pp. 478–491, 2013.

[48] D. S. Cho, N. Vladimir, and T. M. Choi, "Simplified procedure for the free vibration analysis of rectangular plate structures with holes and stiffeners," *Polish Maritime Research*, vol. 22, no. 2, pp. 71–78, 2015.

[49] K. D. Mali and P. M. Singru, "Determination of modal constant for fundamental frequency of perforated plate by Rayleigh's method using experimental values of natural frequency," *International Journal of Acoustics and Vibrations*, vol. 20, no. 3, pp. 177–184, 2015.

[50] H. Moller and M. Lydoff, "Background for revising equal-loudness contours," in *Proceedings of the 29th International Congress and Exhibition on Noise Control Engineering*, pp. 1–6, August 2000.

[51] I. A. Perry and B. E. Richardson, "Radiation efficiency and sound radiation fields of classical guitars," *Proceedings of the Institute of Acoustics*, vol. 35, part 1, pp. 365–372, 2013.

[52] H. Mamtaz, M. H. Fouladi, M. Al-Atabi, and S. Narayana Namasivayam, "Acoustic absorption of natural fiber composites," *Journal of Engineering*, vol. 2016, Article ID 5836107, 11 pages, 2016.

[53] R. L. C. Lemmen and R. J. Panuszka, "Numerical evaluation of acoustic power radiation and radiation efficiencies of baffled plates," in *Proceedings of the V School of Energy Methods in Vibro-Acoustics*, Technical University of Mining and Metallurgy, 1996.

[54] G. Maidanik, "Response of ribbed panels to reverberant acoustic fields," *The Journal of the Acoustical Society of America*, vol. 34, no. 6, pp. 809–826, 1962.

[55] A. J. Van Engelen, "Sound radiation from a baffled plate; theoretical and numerical approach," Research Report DCT 2009.005, University of Eindhoven, Department of Mechanical Engineering, Dynamics and Control Group, Eindhoven, The Netherlands, 2009.

[56] I. Perry and B. Richardson, "Radiation efficiency and sound field measurements on stringed musical instruments," in *Proceedings of the International Symposium on Music Acoustics (ISMA '14)*, pp. 77–82, Le Mans, France, 2014.

[57] P. Avitabile, "Experimental modal analysis," *Sound and Vibration*, vol. 35, no. 1, pp. 20–31, 2001.

[58] M. D. Stanciu, I. Curtu, I. C. Rosca, and N. C. Cretu, "Diagnosis of dynamic behaviour of ligno-cellulose composite plates in the construction of the classical guitar," *Bulletin of the Transilvania University of Brasov*, vol. 1, no. 50, 2008.

[59] T. M. Huber, N. M. Beaver, and J. R. Helps, "Noncontact modal excitation of a classical guitar using ultrasound radiation force," *Experimental Techniques*, vol. 37, no. 4, pp. 38–46, 2013.

10

Overlapping Signal Separation Method Using Superresolution Technique Based on Experimental Echo Shape

Jihad Al-Oudatallah,[1] **Fariz Abboud,**[1] **Mazen Khoury,**[2] **and Hassan Ibrahim**[2]

[1]*Department of Electronics and Communications, Damascus University, Damascus, Syria*
[2]*Higher Institute for Applied Science and Technology, Damascus, Syria*

Correspondence should be addressed to Jihad Al-Oudatallah; j_aloudh@hotmail.com

Academic Editor: Jorge P. Arenas

Overlapping signals separation is a difficult problem, where time windowing is unable to separate signals overlapping in time and frequency domain filtering is unable to separate signals with overlapping spectra. In this work, a simulation under MATLAB is implemented to illustrate the concept of overlapping signals. We propose an approach for resolving overlapping signals based on Fourier transform and inverse Fourier transform. The proposed approach is tested under MATLAB, and the simulation results validate the effectiveness and the accuracy of the proposed approach. The approach is developed using Gerchberg superresolution technique to cope with signals with low signal-to-noise ratio. For practical work, an echo shape determination is required to apply the proposed technique. The experimental results show accurate localization of multiple targets.

1. Introduction

In measurement systems based on pulse echo techniques, some problems may occur when received signal contains many overlapping reflected echoes due to many factors: multiple targets, structure of the propagation media that may be consisting of several layers; and the complex physical properties of the propagation path. Therefore, it is important to find an approach for overlapping echoes separation encountered in a wide range of applications such as military and biomedical applications, [1–5].

Overlapping problem occurs in both time and frequency domains. Time overlapping signals are encountered when time delay between echoes is shorter than the emitted signal's duration. Overlapping is due to a combination of layer thickness or distance between reflectors [1, 5].

A difficulty arises in separating the composite signal, when two or more of the individual component signals have spectra that are the same or similar. This is also a well-known difficulty in musical sound separation when the harmonics of two or more pitched instruments have similar frequencies [6].

The frequency-phase relationship of the individual signals is incoherent causing constructive and destructive interference, resulting in irregular characteristics of the spectrum of the composite signal rather than the individual signals [1]. So the relative phase of overlapping signals is found to influence the precision of frequency determination and plays a critical role in the observed amplitude of the combined signal; thus the amplitude and the phase of individual signals become unobservable [6, 7].

So it is a challenging problem in signal processing to obtain the spectrum and the time localization of the individual component signals, as they overlap in both time and frequency domains, because it is difficult for traditional methods like spectrum method and time-frequency method to separate overlapping signals. Time domain windowing is unable to separate signals overlapping in time. Frequency domain filtering is unable to separate spectrally overlapped signals due to the similarity between the spectra of the signals.

Previous studies related to signal separation have emerged in recent years. Several different approaches such as short-time Fourier transform, Wigner-Ville distribution, discrete wavelet transform, discrete cosine transform, and chirplet transform, and Fractional Fourier transform have been proposed to cope with the overlapping problem [1, 8].

In this paper, we propose a method for resolving overlapping signals based on Fourier transform and inverse Fourier transform. Proposed method performance for noisy signals is improved using Gerchberg superresolution technique.

To apply the proposed method in practical application (directed acoustic transmission-reception system in our case), it is necessary to obtain the echo shape derived from experiments, so Hilbert transform is used to extract the received signal envelop, and the extracted envelop is used to modulate the transmitted signal.

This paper is organized as follows: Section 2 introduces the concept of overlapping signals and illustrates the problem encountered by a simulation under MATLAB. In Section 3, the proposed technique is introduced and illustrated by simulation results. Section 4 presents experimental echo shape determination by extracting the received signal envelop using Hilbert transform, where the transmitted signal is modulated by the extracted envelop. Section 5 shows experimental examples results of applying the proposed technique to received signal, using acoustic transmission-reception system, and multiple targets. Finally, conclusion is drawn in Section 6.

2. Concept of Overlapping Signals

The concept of overlapping signals is illustrated by a simulation implemented in MATLAB. The composite signal is a summation of a number of individual component signals (three overlapping echoes); the individual component signals have the same frequency (2 kHz), different magnitudes, and different time delays as shown in Figure 1. The summation of overlapping signals is shown in Figure 2. Figure 3 shows the spectra of the three individual component signals after Fourier transformation, and Figure 4 shows the spectrum of the composite signal after Fourier transformation.

In all simulations, the following parameters will be used: the sampling frequency f_s = 100 kHz and the acquisition time T = 1 sec, so the number of samples N = 100000, discrete Fourier transform (DFT) computed with a 100000-point FFT, and the frequency increment df = 1 Hz.

As illustrated in Figure 4, frequency domain analysis of time overlapping signals using Fourier transformation does not produce the spectrum of the individual component signals. The individual component signals have regular spectra and main lobe centred at (2 kHz), but the spectrum of the composite signal has distortions and irregular characteristics.

3. Proposed Method

3.1. Theoretical Background. Assuming that the transmitted signal is $x(t)$,

$$x(t) = w(t) \cdot \sin 2\pi f_0 t, \qquad (1)$$

where $w(t)$ is the function shape of the transmitted pulse and f_0 is the transmission frequency. The received signal $y(t)$

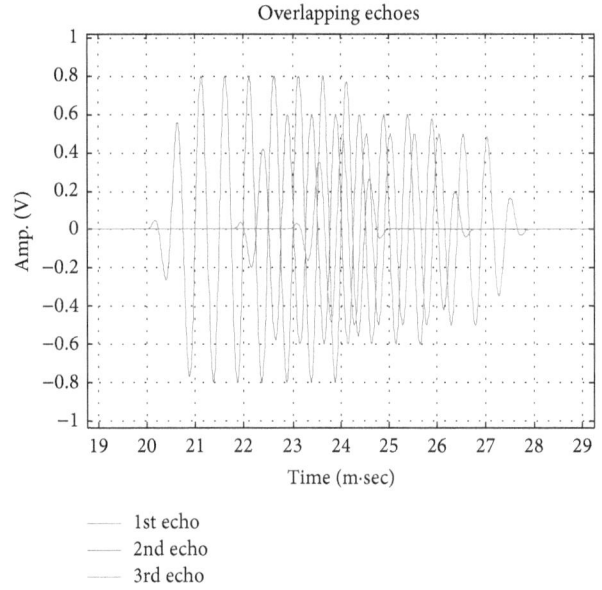

FIGURE 1: Three overlapping echoes.

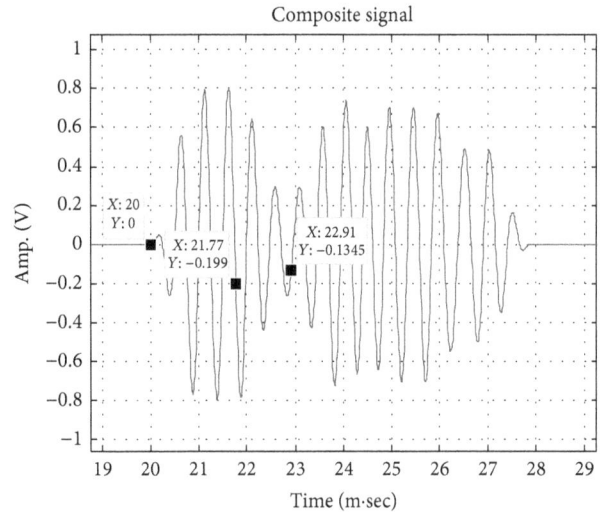

FIGURE 2: Composite signal.

could, for instance, be several (n) scaled and time delayed copies of a known signal $x(t)$:

$$y(t) = \sum_{k=1}^{n} a_k x(t - \tau_k), \qquad (2)$$

where a_k is the amplitude of the shape of the kth echo pulse and τ_k is its time delay. By taking the Fourier Transform of (2), we get

$$Y(f) = \sum_{k=1}^{n} a_k X(f) e^{-j2\pi f_0 \tau_k}, \qquad (3)$$

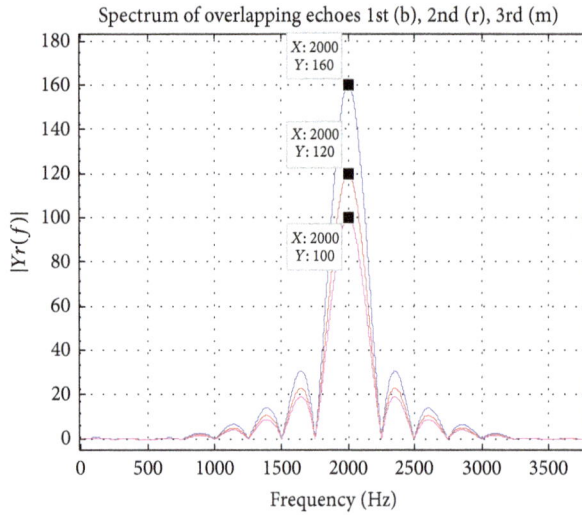

FIGURE 3: Spectra of the individual signals.

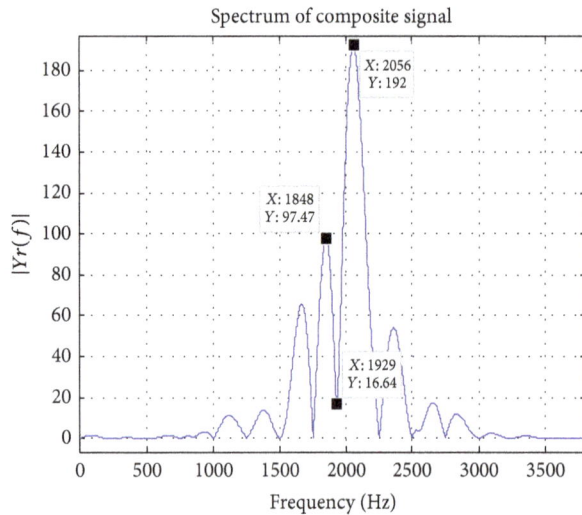

FIGURE 4: Spectrum of the composite signal.

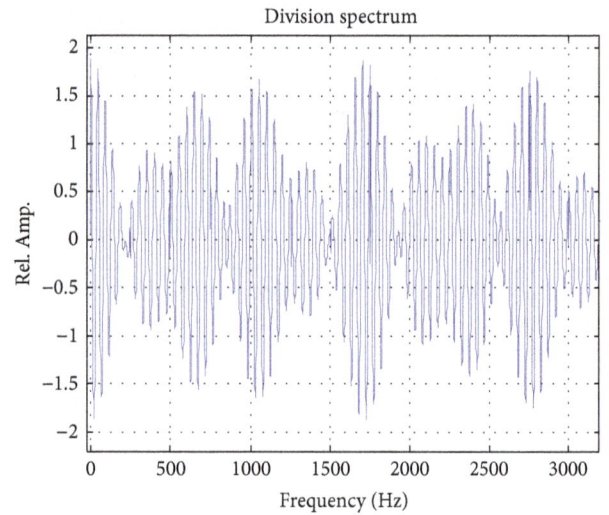

$$— H(f)$$

FIGURE 5: Transform function $H(f)$ in frequency domain.

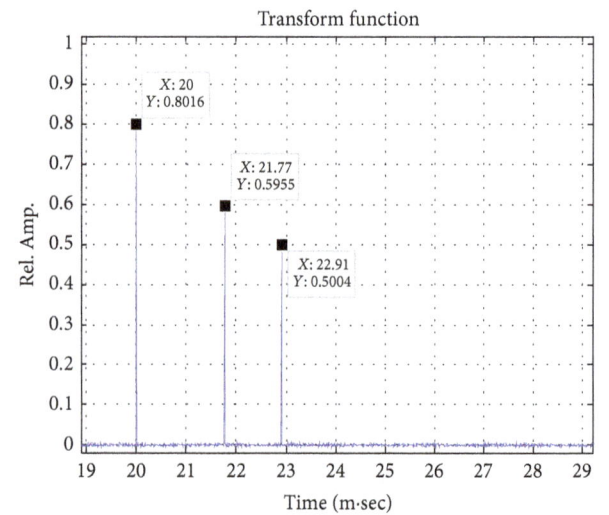

$$— h(t)$$

FIGURE 6: Transform function $h(t)$ in the time domain.

where $X(f)$ is the Fourier Transform of (1). We define the transform function $H(f)$ in the frequency domain by

$$H(f) = \frac{Y(f)}{X(f)} = \sum_{k=1}^{n} a_k e^{-j2\pi f_0 \tau_k}. \qquad (4)$$

By taking the Inverse Fourier Transform of (4), we get

$$h(t) = \sum_{k=1}^{n} a_k \delta(t - \tau_k). \qquad (5)$$

Equation (5) shows that the resulting transform function in the time domain is several (n) pulses $\delta(t)$ with amplitude a_k and time delay τ_k.

By applying this method on signal illustrated in Figure 2, where $y(t)$ is the composite signal, we get transform functions in frequency domain $H(f)$ and time domain $h(t)$ as shown in Figures 5 and 6.

Note that time delays and amplitudes of echoes, obtained in Figure 6, are in agreement with those given in Figure 1.

3.2. Effect of Signal-to-Noise Ratio (SNR).
In practical application, signals could not be as the simulated transmitted and received signals due to the noise, frequency shift, and system response.

Unfortunately, the proposed method is very sensitive to noise, which may cause a catastrophic effect on results; to illustrate this point, a white noise with signal-to-noise ratio (SNR = 20) will be added to the composite signal shown in Figure 2, so the noisy composite signal becomes as shown in Figure 7. Transform functions $H(f)$ and $h(t)$ in frequency and time domain are shown in Figures 8 and 9, respectively.

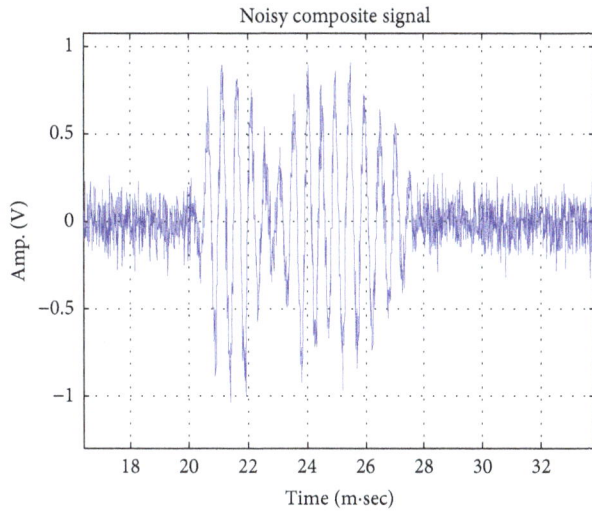

FIGURE 7: The composite signal corrupted by additive noise.

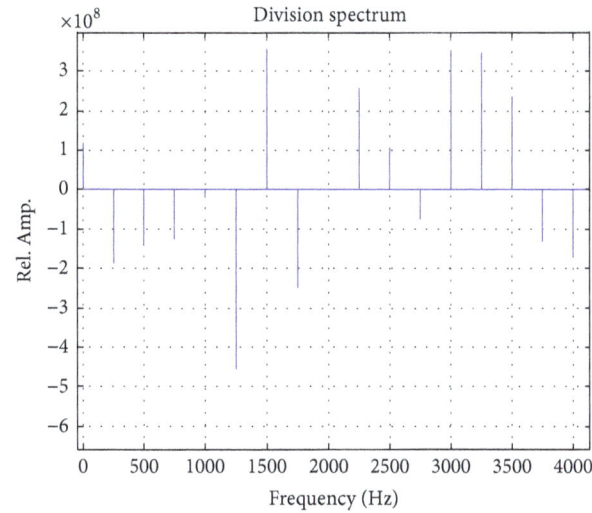

FIGURE 9: Transform function $h(t)$ of composite signal corrupted by additive noise in time domain.

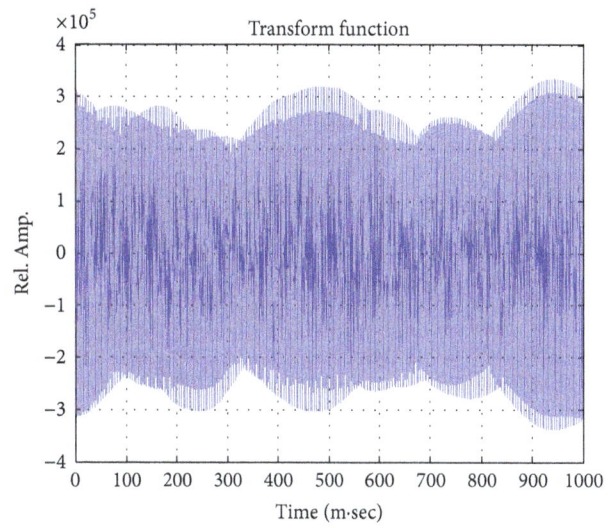

FIGURE 8: Transform function $H(f)$ of composite signal corrupted by additive noise in frequency domain.

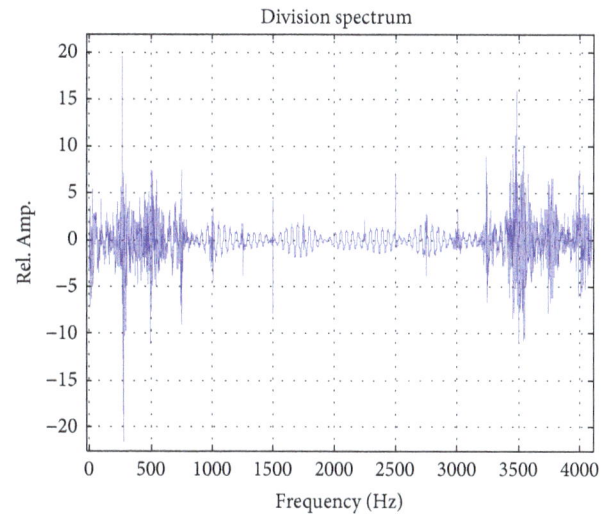

FIGURE 10: Transform function $H(f)$ according to (6) with $\gamma = 0.05$.

Narrow pulses with high amplitude shown in Figure 8 are caused by dividing by zeros (of $X(f)$), resulting in periodic sinusoidal signals in all the time domain, as shown in Figure 9; these signals cover pulses at time delays τ_k of our interest.

To cope with problem, (4) could be developed as the following form:

$$H(f) = \frac{Y(f) \cdot X^*(f)}{X(f) \cdot X^*(f) + \gamma} = \frac{F(R_{yx}(\tau))}{F(R_{xx}(\tau)) + \gamma}, \quad (6)$$

where γ is a small real value related to SNR, representing the ratio of the power spectral density of the noise and that of the signal [9], $X^*(f)$ is the complex conjugate of $X(f)$, $F(R_{yx}(\tau))$, and $F(R_{xx}(\tau))$ are Fourier transforms of cross-correlation and autocorrelation functions, respectively [10]. This technique used in (6) is similar to the Wiener Filter used in image processing [9], and it is powerful in measurement

conditions with low S/N ratio, where the autocorrelation function contains the same information about the signal, and, by using the cross-correlation function, it is possible to obtain an increase in the S/N ratio (the Wiener-Khintchine theorem) [11, 12].

Figure 10 shows the transform function $H(f)$ of previous example according to (6) with $\gamma = 0.05$.

As shown in Figure 10, the technique in used in (6) has reduced the noise effect (the effect of dividing by zero), but this is not enough to recognize pulses of interest in transform function $h(t)$, so an improvement based on super resolution technique is required.

3.3. Superresolution Technique. The proposed method performance is improved using Gerchberg superresolution technique [13], as illustrated in the following steps:

(1) The received signal is windowed by a time window [t1 t2], with zero padding outside the window, where t1 and t2 are limits of time window of interest; this is done by detection algorithm or by dividing time domain of the received signal into predefined time-length frames.

(2) Taking the Fourier transform of the signal obtained in step 1, (6) is used with a suitable predefined value for γ; then we get $H_p(f) = H(f)$.

(3) Transform function $H(f)$ obtained in step 2 is filtered by time band filter [t1 t2], where t1 and t2 are limits of the same time window in step 1, $H_c(f)$ is the filtered transform function of $H(f)$, and $H_w(f)$ is defined as the following:

$$H_w(f) = \begin{cases} H_p(f); & \text{for } f_1 < f < f_2 \\ H_c(f); & \text{otherwise}, \end{cases} \quad (7)$$

where f_1 and f_2 are frequency band limits of interest (around the transmission frequency $f_1 < f_0 < f_2$).

(4) Taking the inverse Fourier transform of $H_w(f)$, obtained transform function $h(t)$ is windowed by the same time window [t1 t2], but, to compensate the filter delay used in step 3, time window is shifted by the same filter delay and then $h_w(t)$ is obtained.

(5) Taking Fourier transform of $h_w(t)$ to get $H_c(f)$ and $H_p(f) = H_w(f)$, (7) is used to get a new $H_w(f)$.

(6) $H_m(f)$ is filtered $H_w(f)$ by time band filter [t1 t2], and $h_m(t)$ is the inverse Fourier transform of $H_m(f)$.

(7) A predefined number of iterations are used to get the best resolution for $h_m(t)$ by successive steps: taking the Fourier transform of the windowed $h_m(t)$ with filter delay compensation to get $H_c(f)$ and $H_p(f) = H_m(f)$, (7) is used to get a new $H_w(f)$, filtering $H_w(f)$ by time band filter [t1 t2] to get new $H_m(f)$, taking inverse Fourier transform of $H_m(f)$ to get new $h_m(t)$.

Modified transform functions $Hm(f)$, $hm(t)$ (red curve), and its absolute value (blue curve) of the previous example, in frequency and time domain, of the third-order iteration are shown in Figures 11 and 12, respectively.

Time delays obtained in Figure 12 are in agreement with those obtained in Figure 6 and given in Figure 1.

In practical application, the problem encountered is to get the reference transmitted signal that really propagates in the medium, where the received signal is considered to be several scaled and time delayed copies of it. So we have to construct heuristically the reference transmitted signal that has the same transmission frequency f_0 but modulated by shape function (pulse envelope); it is supposed that echoes have the same shape but with scaled amplitude.

In the following paragraph, we propose an approach to extracting the pulse envelope from the received signal using Hilbert transform.

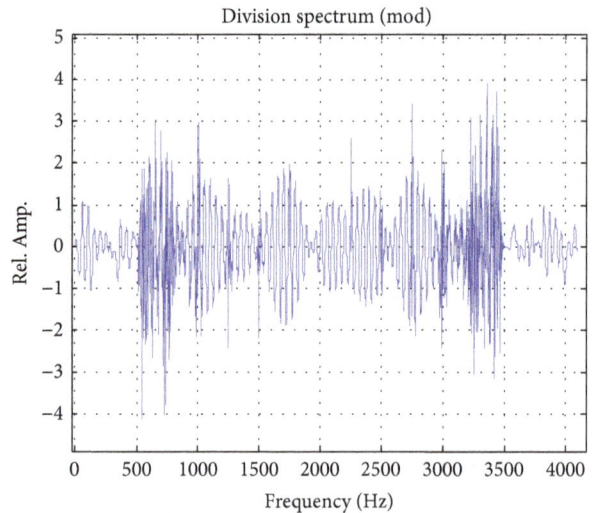

FIGURE 11: Modified transform function $Hm(f)$ in frequency domain.

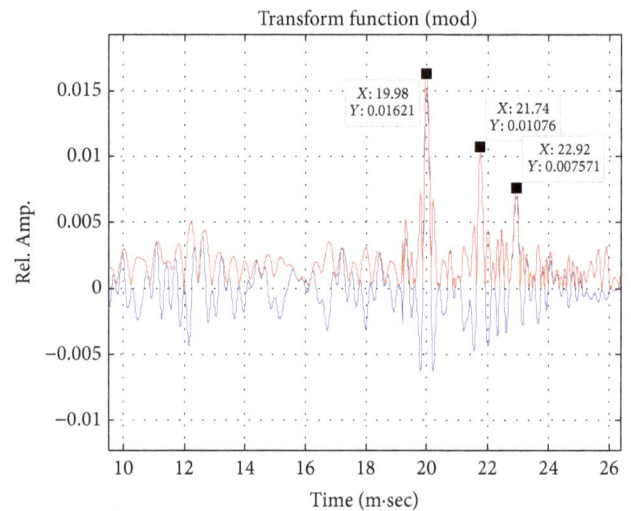

FIGURE 12: Modified transform function $hm(t)$ in time domain.

4. Experimental Envelope Detection

The Hilbert transform facilitates the formation of the analytic signal $x_a(t)$. The analytic signal is useful in the area of communications, particularly in pass band signal processing.

Figure 13 shows the schematic representation of experimental setup used for envelope detection.

This experimental setup is used in a closed and empty Basketball Hall that is considered a nearly anechoic environment, where microphone and loudspeaker are located about 2 m above the ground, and the distance from the microphone or the loudspeaker to the closest object is greater than 5 m.

First, we generate 10 periods of mathematical sine wave at (f_0 = 2000 Hz) frequency with a rectangular envelope as a transmission pulse (pulse width = 5 m·sec). Using a laptop, this pulse is transmitted through a sound card, an acoustic power amplifier, and a loudspeaker. Using a microphone, the pulse is received directly and acquired through the sound

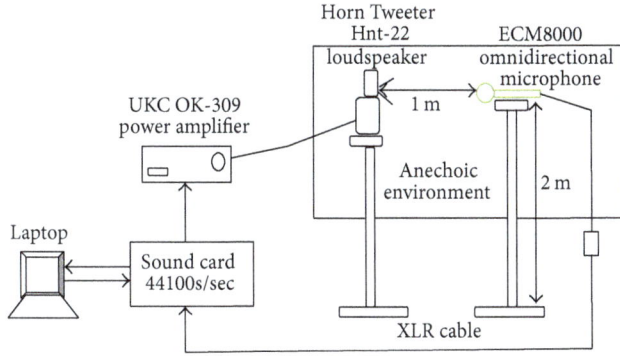

FIGURE 13: Schematic representation of experimental setup used for envelope detection.

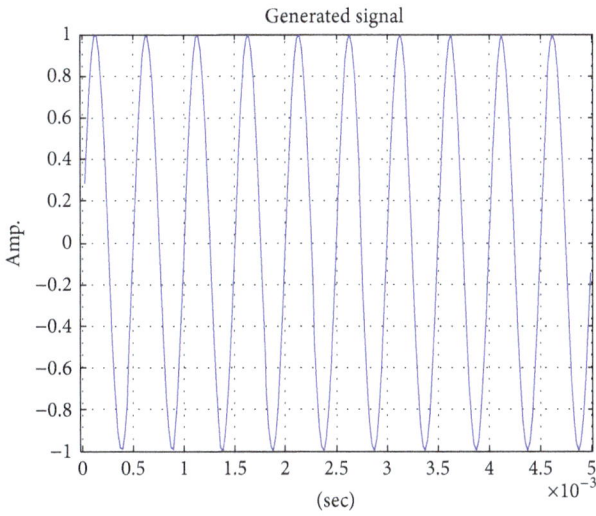

FIGURE 14: Mathematical generated signal.

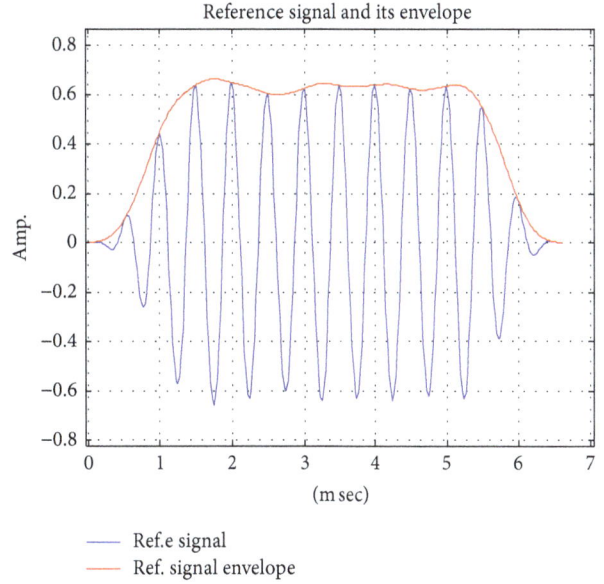

FIGURE 15: Reference signal and its envelope using Hilbert transform.

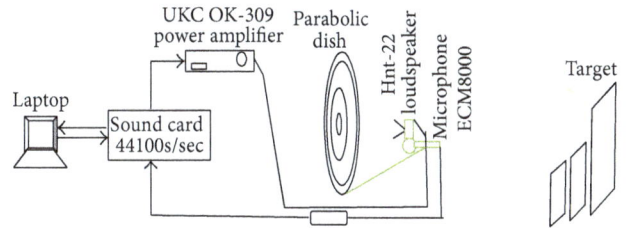

FIGURE 16: Schematic representation of experimental setup used for proposed method verification.

card; the received signal $x(t)$ is processed to obtain its analytic signal $x_a(t)$, that is,

$$x_a(t) = A(t) e^{j\Phi(t)} = x(t) + jH(x(t)), \quad (8)$$

where $A(t)$ is the instantaneous amplitude and $H(x(t))$ is the Hilbert Transform of $x(t)$, which is given by [14, 15]

$$H(x(t)) = \frac{1}{\pi} \int_{-\infty}^{\infty} \frac{x(\tau)}{t - \tau} d\tau. \quad (9)$$

Figure 14 shows the mathematical generated signal. Figure 15 shows the experimental reference signal and its extracted envelope using Hilbert transform, implemented under MATLAB. This envelope derived from experiments (received signal shifted to time 0), supposed to be similar to echo shape, is used to modulate mathematical sine wave at f_0 frequency to be the reference transmitted signal. This model of transmitted signal and reflected echoes takes into account the response of the transmitter and receiver sensors, the number of cycles of the excitation pulse, and the propagation medium characteristics.

5. Experimental Examples

The proposed method is experimentally tested using single, double, and triples targets and a directed acoustic transmission-reception system with a parabolic dish with a focal length of 35 cm. The parabolic dish has an elliptical aperture with a major axis of 65 cm and a minor axis of 60 cm. Figure 16 shows a schematic representation of experimental setup used for proposed method verification.

In all experiments, signals are sampled at the sampling frequency $f_s = 44.1$ kHz over an acquisition time $T = 1$ sec, so the number of samples acquired is $N = 44100$ samples, discrete Fourier transform (DFT) is computed with a 100000-point FFT, and the frequency increment is $df = 1$ Hz.

5.1. Single Target. At first, a single board of wood with dimensions 90×60 cm is used as a target, which is located at about 330 cm from the directed acoustic system, as shown in Figure 16. The modified transform function $hm(t)$ (red curve) and its absolute value (blue curve) are shown in Figure 17.

From Figure 17, target distance is $d = ((19.27 \text{ m·sec} \times 340 \text{ m}))/2 = 327.6$ cm; the result is satisfactorily close to the real value.

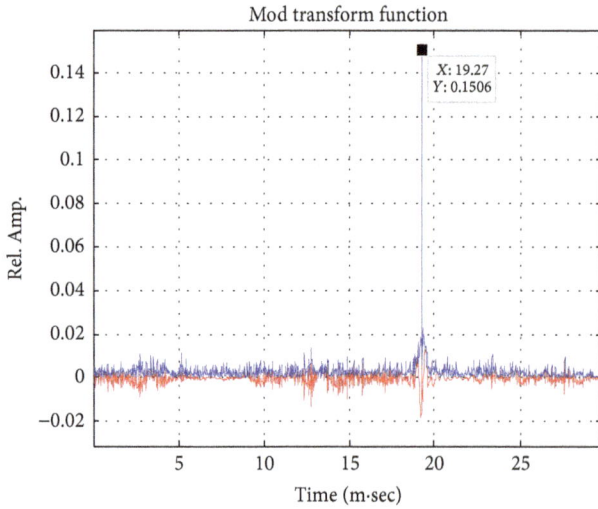

FIGURE 17: Modified transform function $hm(t)$ of single target.

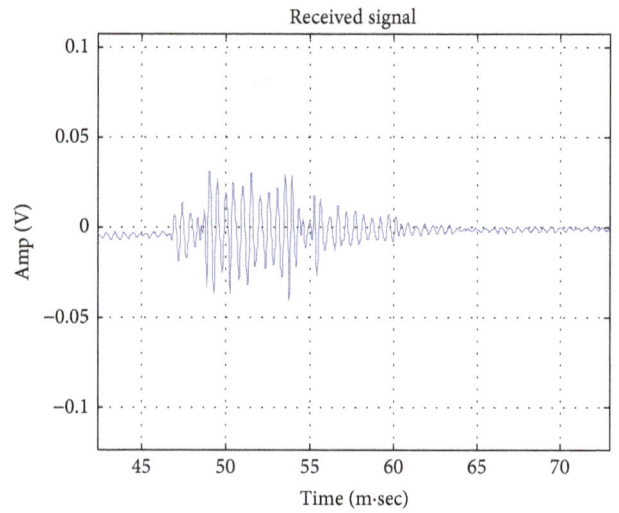

FIGURE 18: Received signal of double target at 8 m distance.

5.2. Double Target. Two boards of wood with dimensions 60 × 40 cm and 90 × 60 cm are used as a target (the smaller is in front, and the distance between them is about 30 cm); the target is located at about 8 m from the directed acoustic system, as shown in Figure 16. Figures 18, 19, and 20 show the received signal, the transform function $h(t)$ with $\gamma = 0.001$, and the modified transform function $hm(t)$ of third-order iteration, respectively.

From Figure 20, target distance is $d = ((46.69 \text{ m·sec} \times 340 \text{ m}))/2 = 7.94$ m, and distance between two targets is $d_{1,2} = (((48.5 - 46.69) \text{ m·sec} \times 340 \text{ m}))/2 = 30.77$ cm; results are satisfactorily close to real values.

5.3. Triple Target. Three boards of wood with dimensions 60 × 40 cm, 90 × 60 cm and 160 × 140 cm are used as a target (the smallest board is the closest to the system and the biggest one is the farthest); the target is located at about 9.25 m from the directed acoustic system, as shown in Figure 16, the distance between the first board and the second one is about 45 cm, and the distance between the second board and the third one is about 80 cm; Figure 21 shows the received signal, Figure 22 shows the transform $h(t)$ with $\gamma = 0.006$, and the modified transform function $hm(t)$ of third-order iteration is shown in Figure 23.

From Figure 23, target distance is $d = ((54.44 \text{ m·sec} \times 340 \text{ m}))/2 = 9.255$ m, the distance between first and second target is $d_{1,2} = (((57.19 - 54.44) \text{ m·sec} \times 340 \text{ m}))/2 = 46.75$ cm, and the distance between second and third target is $d_{2,3} = (((62.13 - 57.19) \text{ m·sec} \times 340 \text{ m}))/2 = 83.98$ cm; results are satisfactorily close to real values.

6. Conclusion

This paper presents a method to separate overlapping echoes using a superresolution algorithm based on experimental echo shape modelling.

The simulation results implemented under MATLAB show a good performance of the method at low SNR.

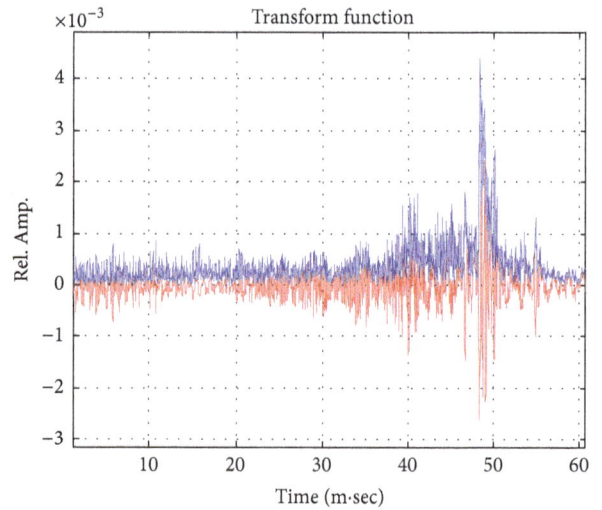

FIGURE 19: Transform function $h(t)$ with $\gamma = 0.001$.

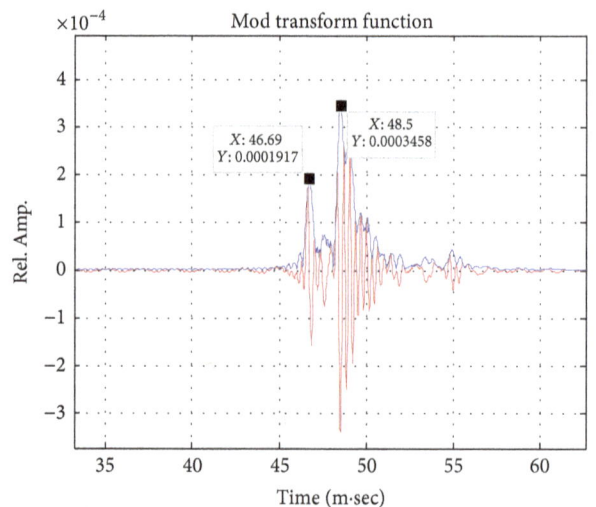

FIGURE 20: Modified transform function $hm(t)$ of third-order iteration with $\gamma = 0.001$.

Received signal

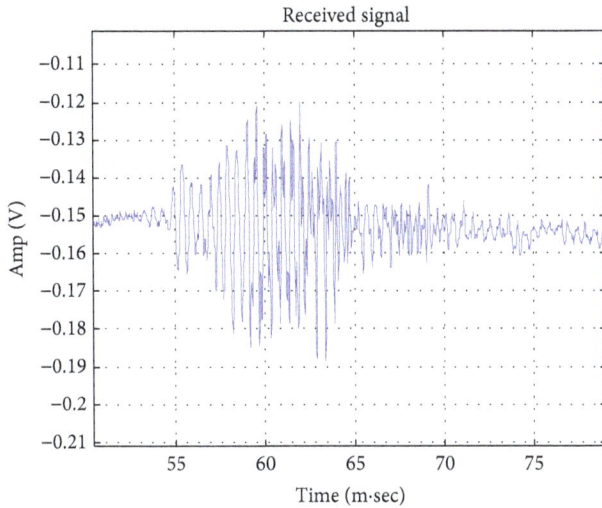

FIGURE 21: Received signal of triple target at 9.25 m distance.

Transform function

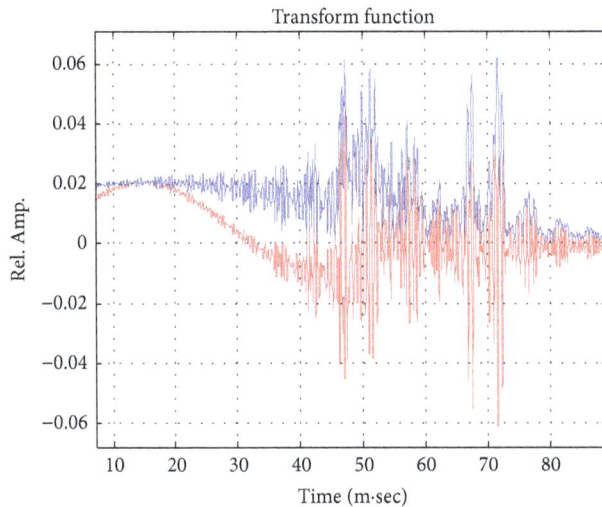

FIGURE 22: Transform function $h(t)$ with $\gamma = 0.006$.

Mod transform function

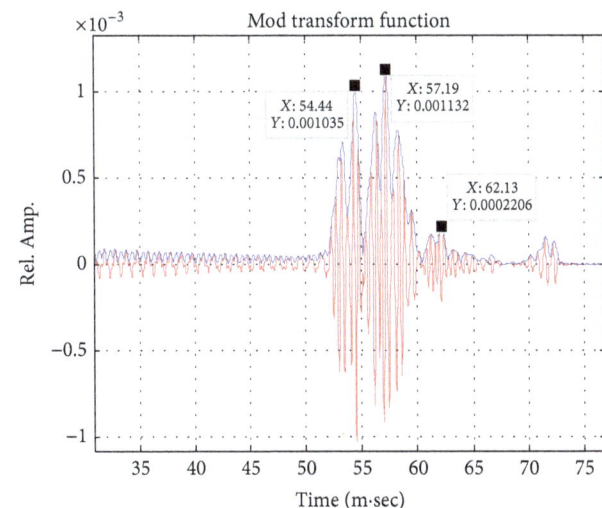

X: 54.44
Y: 0.001035

X: 57.19
Y: 0.001132

X: 62.13
Y: 0.0002206

FIGURE 23: Modified transform function $hm(t)$ of third-order iteration with $\gamma = 0.006$.

The method is experimentally tested using acoustic transmission-reception system and multiple targets; the method provides accurate localization of multiple targets due to a good accuracy in determining the time delays of individual echoes reflected from each target. Although amplitudes of individual echoes, obtained by this method, are different from the real amplitudes and smaller than them, each echo's amplitude relative to others is respected.

This study adds a new possibility of overlapping echoes separation and can be widely used in a range of applications such as military and biomedical applications.

Developments can be added to improve SNR and determine the amplitude of individual echoes that allow reconstructing the composite signal.

References

[1] D. M. J. Cowell and S. Freear, "Separation of overlapping linear frequency modulated (LFM) signals using the fractional fourier transform," *IEEE Transactions on Ultrasonics, Ferroelectrics, and Frequency Control*, vol. 57, no. 10, pp. 2324–2333, 2010.

[2] Y. Lu, A. Kasaeifard, E. Oruklu, and J. Saniie, "Fractional fourier transform for ultrasonic chirplet signal decomposition," *Advances in Acoustics and Vibration*, vol. 2012, Article ID 480473, 13 pages, 2012.

[3] E. G. Sarabia, J. R. Llata, S. Robla, C. Torre, and J. P. Oria, "Accurate estimation of airborne ultrasonic time-of-flight for overlapping echoes," *Sensors (Switzerland)*, vol. 13, no. 11, pp. 15465–15488, 2013.

[4] R. S., S. C. X., and M. S., "Estimation of signal parameters for multiple target localization," *ICTACT Journal on Communication Technology*, vol. 5, no. 4, pp. 1039–1044, 2014.

[5] J. Martinsson, F. Hägglund, and J. E. Carlson, "Complete post-separation of overlapping ultrasonic signals by combining hard and soft modeling," *Ultrasonics*, vol. 48, no. 5, pp. 427–443, 2008.

[6] J. Woodruff, Y. Li, and D. Wang, "Resolving overlapping harmonics for monaural musical sound separation using pitch and common amplitude modulation," in *Proceedings of the 9th International Conference on Music Information Retrieval (ISMIR'08)*, pp. 538–543, USA, September 2008.

[7] K. Reinhold, "Comparison of frequency estimation methods for reflected signals in mobile platforms," *World Academy of Science, Engineering and Technology*, vol. 57, pp. 147–150, 2009.

[8] S. Zhang, M. Xing, R. Guo, L. Zhang, and Z. Bao, "Interference suppression algorithm for SAR based on time-frequency transform," *IEEE Transactions on Geoscience and Remote Sensing*, vol. 49, no. 10, pp. 3765–3779, 2011.

[9] R. C. Gonzalez and R. E. Woods, *Digital Image Processing*, Prentice Hall, 3rd edition, 2008.

[10] J. DiBiase Hector, *A high-accuracy, low-latency technique for talker localization in reverberant environments using microphone arrays*, Diss. Brown University, 2000.

[11] G. John and G. Proakis Dimitris, *Digital Signal Processing. Principles, Algorithms and Applications*, Prentice Hall International, 3rd edition, 1996.

[12] S. Adrián-Martínez, M. Bou-Cabo, I. Felis et al., "Acoustic signal detection through the cross-correlation method in experiments with different signal to noise ratio and reverberation conditions," in *Proceedings of the International Conference on Ad-Hoc Networks and Wireless*, Springer, Berlin, Germany, 2014.

[13] R. W. Gerchberg, "Super-resolution through error energy reduction," *Optica Acta*, vol. 21, no. 9, pp. 709–720, 1974.

[14] M. A. Azpúrua, M. Pous, and F. Silva, "Decomposition of Electromagnetic Interferences in the Time-Domain," *IEEE Transactions on Electromagnetic Compatibility*, vol. 58, no. 2, pp. 385–392, 2016.

[15] K. Dragomiretskiy and D. Zosso, "Variational mode decomposition," *IEEE Transactions on Signal Processing*, vol. 62, no. 3, pp. 531–544, 2014.

11

Generation of Hydroacoustic Waves by an Oscillating Ice Block in Arctic Zones

Usama Kadri

Department of Mathematics, Massachusetts Institute of Technology, Cambridge, MA 02139, USA

Correspondence should be addressed to Usama Kadri; ukadri@mit.edu

Academic Editor: Luc Gaudiller

The time harmonic problem of propagating hydroacoustic waves generated in the ocean by a vertically oscillating ice block in arctic zones is discussed. The generated acoustic modes can result in orbital displacements of fluid parcels sufficiently high that may contribute to deep ocean currents and circulation. This mechanism adds to current efforts for explaining ocean circulation from a snowball earth Neoproterozoic Era to greenhouse earth arctic conditions and raises a challenge as the extent of ice blocks shrinks towards an ice-free sea. Surprisingly, unlike the free-surface setting, here it is found that the higher acoustic modes exhibit a larger contribution.

1. Introduction

The majority of research on ocean circulation neglects the compressibility of water, which is mostly justified. However, it is well known that considering the slight compressibility of water may give rise to compression-type waves [1], known as hydroacoustic waves, or acoustic-gravity waves (under free-surface conditions); see also [2]. It has been shown recently that these waves may contribute to deep ocean currents and circulation [3], as they are continuously generated in the ocean by wind-wave interactions [4, 5], interaction of nearly opposing waves [1, 6–10], and submarine earthquakes [2].

To our knowledge, the vast majority of literature related to this type of waves considers a free-surface boundary condition. As a result, the first mode (often referred to as the zeroth mode) is a surface gravity mode, and its evanescent compartments turn into progressive acoustic modes above a certain cut-off frequency [2]. However, in the case of a rigid boundary replacing the free-surface, propagating and trapped acoustic modes might behave differently as elegantly presented by [11]. In this paper we limit our study to the generation of progressive acoustic modes, by an oscillating block of horizontal ice sheet, and their propagation is confined by horizontal rigid boundaries: the sea floor at the bottom and ice sheets at the top. Obviously, under these conditions and neglecting the elasticity of the boundaries, there can be no

gravity surface mode, and all energy is transferred to the acoustic modes. Thus, studying such conditions sheds light on wave energy transfer, as well as water transportation and circulation in confined seas. Here, we present results relevant to arctic zones, from a Neoproterozoic Era [12] to current arctic zones conditions. Oscillations of large ice blocks may be triggered by atmospheric and ocean currents [13], localised wind storms, ice quakes, or other violent geophysical processes.

2. Theoretical Analysis

2.1. Assumptions and Governing Equation. The system considered here is two-dimensional Cartesian coordinates (x, z) with the origin in the midpoint of the ice block; the x-axis is horizontally parallel to the ice sheet (and sea floor), and the z-axis is vertically upwards as shown in Figure 1. The ocean has a constant depth h. The ice block, assumed infinitely long with a total width of $2b$, is oscillating vertically at the angular frequency ω and a small amplitude ζ_0 such that the displacement of the ice block is given by

$$\zeta(x, t) = \zeta_0 \mathcal{H}\left(b^2 - x^2\right) \exp\left(i\omega t\right), \tag{1}$$

where t denotes time, $\mathcal{H}$ is the Heaviside step function, and $i \equiv \sqrt{-1}$.

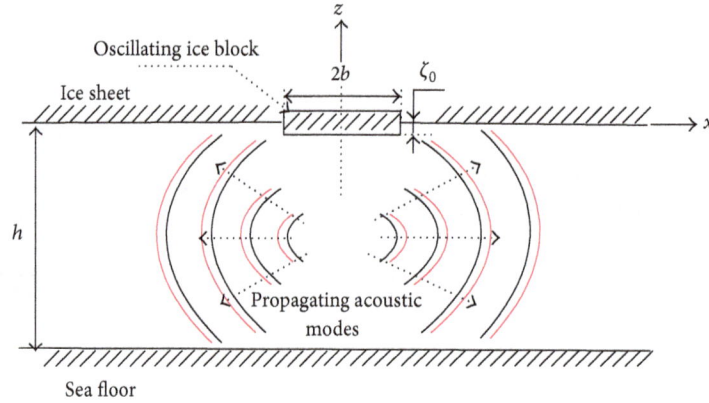

FIGURE 1: Schematic representation of the flow domain.

The governing equation is the standard two-dimensional compressible wave equation [14, 15]. Applying the appropriate boundary conditions and using standard techniques the following horizontal and vertical fluid velocities, u_n and w_n, induced by the nth acoustic mode can then be derived (see the Appendix):

$$u_n(x,z,t) = \begin{cases} \mp 2i\zeta_0 \dfrac{\mu_n \cos[\mu_n(z+h)]\sin(k_n b)\exp[-i(k_n x - \omega t - \pi/2)]}{k_n \cos(\mu_n h)}, & |x| > b, \\[2ex] -2i\zeta_0 \dfrac{\mu_n \cos[\mu_n(z+h)]\cos(k_n b)\exp[-i(k_n|x| - \omega t)]}{k_n \cos(\mu_n h)}, & |x| < b, \end{cases} \quad (2)$$

$$w_n(x,z,t) = \begin{cases} \mp 2\zeta_0 \dfrac{\mu_n^2 \sin[\mu_n(z+h)]\sin(k_n b)\exp[-i(k_n x - \omega t - \pi/2)]}{k_n^2 \cos(\mu_n h)}, & |x| > b, \\[2ex] -2\zeta_0 \dfrac{\mu_n^2 \cos[\mu_n(z+h)]\sin(k_n b)\exp[-i(k_n|x| - \omega t)]}{k_n^2 \cos(\mu_n h)}, & |x| < b, \end{cases} \quad (3)$$

where μ_n is the eigenvalue and k_n is the horizontal wavenumber.

3. Results and Discussion

3.1. Displacements. In order to obtain the horizontal and vertical particle displacement components $(X_n; Z_n)$ we integrate (2) with respect to time, so that, for $|x| > b$, we can write $X_n = A\sin(k_n x - \omega t)$ and $Z_n = B\cos(k_n x - \omega t)$, where $(X_n/A)^2 + (Z_n/B)^2 = 1$ is the equation of an ellipse with a horizontal semiaxis A and a vertical semiaxis B given by

$$A = \pm \frac{2\zeta_0}{\omega}\frac{\mu_n \cos[\mu_n(z+h)]\sin(k_n b)}{k_n \cos(\mu_n h)};$$

$$B = \mp \frac{2\zeta_0}{\omega}\frac{\mu_n^2 \sin[\mu_n(z+h)]\sin(k_n b)}{k_n^2 \cos(\mu_n h)}. \quad (4)$$

The orbital behaviours of the acoustic modes are periodic with depth as shown in Figure 2, with $n/2$ wavelengths fitting the depth h, in opposition to the free-surface setting in which $(2n-1)/4$ wavelengths fit the depth [3, 16]. For the numerical calculations we considered a speed of sound $c = 1500\,\text{m/s}$, $h = 4500\,\text{m}$, $b = 10^4\,\text{m}$, $\zeta_0 = 1\,\text{m}$, and $\omega = 5\,\text{rad/s}$. With

the current setting, the obtained displacements correspond to horizontal and vertical velocities reaching up to tens of centimetres per second for the first mode, which is comparable to those obtained by an equivalent underwater earthquake with a free surface [2, 3]. However, a distinguishable result here is that the leading mode is not necessarily the first acoustic, but the highest acoustic mode (which can be obtained from (A.9)) given by

$$N_{\max} = \left\lfloor \frac{\omega h}{\pi c} \right\rfloor, \quad (5)$$

where the special brackets represent the floor function. For the case of $n = N_{\max}$, the displacements and velocities may become one order of magnitude larger than those of the first acoustic mode, $n = 1$, and thus, unlike the free-surface problem, all progressive acoustic modes have to be considered.

Note that the choice of $\zeta_0 = 1\,\text{m}$ far exceeds observed amplitudes of oscillating sea-ice blocks in the late Common Era [17]. Nevertheless this choice allows a proper comparison with the free-surface underwater earthquake problem [2] emphasising the importance of the higher modes and directly reflects on two actual scenarios. The first scenario describes an underwater earthquake in the arctic ocean, in particular

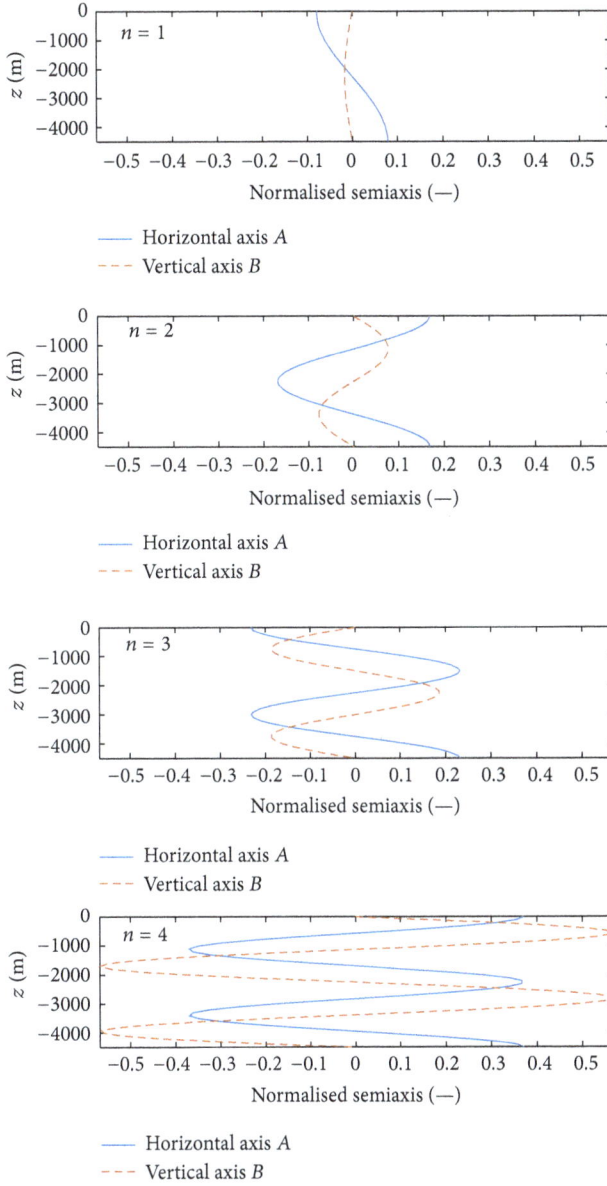

FIGURE 2: Semiaxis of water parcels for acoustic modes, from top to bottom: $n = 1, 2, 3, 4$; $h = 4500$ m, $c = 1500$ m/s, $b = 10^4$ m, $\zeta_0 = 1$ m, and $\omega = 5$ rad/s.

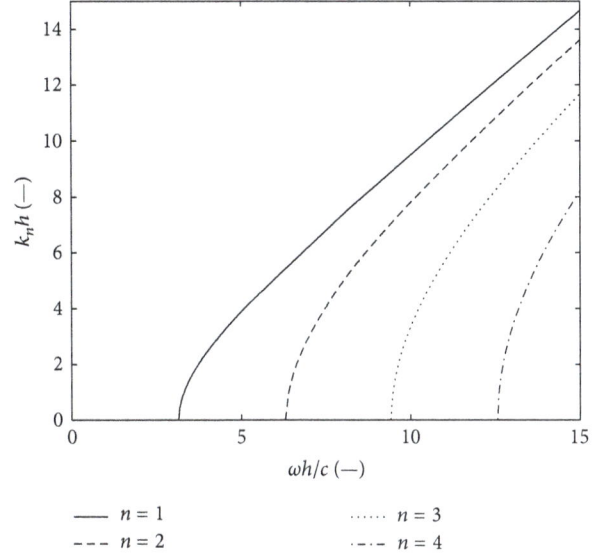

FIGURE 3: Dispersion of modes; $h = 4500$ m and $c = 1500$ m/s.

in winter if the sea surface is completely frozen. Recall that in the absence of gravity and elasticity, while the mathematical solution in hand is valid for a rigid sea floor and an oscillating ice block at the surface (oscillating ice block problem), it is equally valid for the setting with a rigid frozen surface and an oscillating sea floor block at the bottom (e.g., earthquake in the arctic ocean). The second scenario considers the snowball earth Neoproterozoic Era, where sea-ice thickness might have exceeded a kilometre. Under such conditions, ice quakes and other violent geophysical phenomena could have caused oscillations with amplitudes that are comparable with the chosen $\zeta_0 = 1$ m. Further qualitative analysis through different climate eras is given in the last section.

3.2. *Dispersion Relation.* The dispersion relation is given in nondimensional form by

$$\mu_n k_n h^2 \sinh(\mu_n h) = 0. \tag{6}$$

For the incompressible case $\omega h/c \to 0$, and therefore $\mu_n \to k_n$. In this case there is no nontrivial solution, and no waves are formed. For the compressible case, the wavenumber of the acoustic mode is given by

$$k_n h = \left(\frac{\omega^2 h^2}{c^2} - n^2 \pi^2 \right)^{1/2}, \quad n = 1, \dots, N_{\max}, \tag{7}$$

where $N_{\max}$ is the maximum number of progressive acoustic modes. Thus, the cut-off frequency for each mode is given by $\omega_c = n\pi c/h$, so that, at a given depth and relatively low frequencies $\omega < \omega_c$, only evanescent modes can exist, whereas when $\omega > \omega_c$ there should be at least one nonevanescent acoustic mode. For the case of $h = 4500$ m the cut-off frequency is $\omega_c = \pi/3$. The dispersion relations of $k_n h$ versus $\omega h/c$ are shown in Figure 3. It is easy to show that the group velocities, defined as $c_g = (\mathrm{d}\omega/\mathrm{d}k_n)$, differ from one mode to another though all tend to the sound speed c if the normalised frequency $(\omega h/c)$ is large enough. These can be obtained from the slopes of the dispersion relations presented in Figure 3.

3.3. *Energy Aspects.* The total wave energy is composed of two parts: kinetic and elastic potential; with the absence of a free surface there is no gravity potential energy. Following similar steps as given by [2] we obtain an expression for the kinetic energy per unit volume:

$$E_{\kappa,n} = \frac{1}{2} \rho \zeta_0^2 \frac{\omega^4 \mu_n \left[\sin(2\mu_n h) + 2\mu_n h \right] \sin^2(k_n b)}{hc^2 k_n^4 \cos^2(\mu_n h)} \tag{8}$$

and the elastic energy is $E_{\epsilon,n} = E_{\kappa,n}/4$. Above the cut-off frequency the oscillation of the ice block results in energy "transfer" to the corresponding acoustic mode/modes (Figure 4).

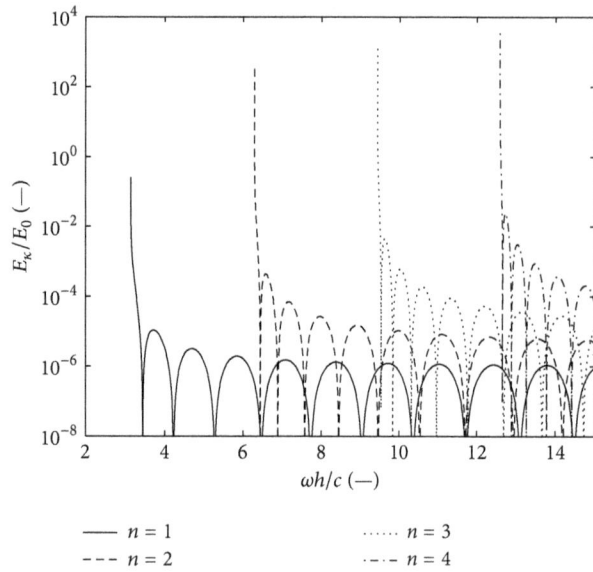

FIGURE 4: Kinetic energy versus frequency; $h = 4500$ m, $c = 1500$ m/s, $b = 10^4$ m, and $\zeta_0 = 1$ m.

However, below the cut-off frequency, the fluid becomes incompressible and no energy transfer occurs. With regard to this point it is worth noting that below the cut-off frequency the stratification of the ocean or the elasticity of the sea floor and ice sheets, which was neglected in the presented analysis, might become important [18]. Considering stratification may give rise to internal gravity waves that are believed to be a major driver of deep ocean mixing [19]. On the other hand, considering the elasticity of the boundaries, it is anticipated that oscillations below the cut-off frequency would give rise to a Scholte-type modes that propagate through both interfaces, with the sea floor and with the ice sheet [20]. Other types of trapped modes are discussed in detail by [11].

It is also notable that the fact that the higher acoustic modes have a larger contribution on water transport compared to the lower modes suggests that the flow field would become temporally and spatially nonuniform, if stratification is considered, as also found in observations [21–25]. Nevertheless, stratification or consequently variations in the speed of sound do not practically have a direct effect on the fluid motion induced by a hydroacoustic wave, whose group velocity remains almost unaltered. Hence, stratification is not considered in the current analysis, though a numeric comparison with the effects of internal gravity waves due to stratification is briefly presented in the following section.

3.4. From Snowball to Greenhouse Earth.

The water motion in the ocean plays a prominent role in distributing heat, originated from solar radiation or hydrothermal activities, around the globe. In free-surface open seas, ocean water results in humidification and increasing the temperature of the surrounding air, which in turn forms storms and rain. Without currents and underwater streams localised temperatures become more severe leading to extreme global weather. Hence, water motion in the ocean acts as a global climate regulator.

The flow generated by the motion of the ice block is equivalent to the superposed flow by two line sources located at the edges of the ice block $x = \pm b$. Such superposition can be either destructive or constructive, resulting in various peak velocities, as given qualitatively in Figure 5. A minimal flow field is created when the oscillating ice block has a width $2b$ multiples of a wavelength for each acoustic mode; that is, $2b = NL_n$, where $L_n = 2\pi/k_n$ is the wavelength, and $N = 1, 2, 3, \ldots$. On the other hand, a maximal flow field is induced when $2b = L_n(N - 1/2)$. Thus, the mutual existence of multiple acoustic modes is expected to create regions of minimal and maximal peak velocities as confirmed by Figure 6. It is worth mentioning here that deep ocean currents induced by internal gravity waves can travel at speeds varying from 0.02 to 0.1 m/s, whereas surface currents (that do not exist in snowball earth) can travel at speeds as high as 2.5 m/s. Thus, from Figure 6, it is clear that internal gravity waves become more important in a greenhouse earth, as opposed to hydroacoustic waves, keeping in mind that the latter, despite the lower induced peak velocities, would still span the entire depth.

It is believed that during the Neoproterozoic Era the earth was fully covered by snow, with ice blocks that extended over hundreds of metres thickness [12, 26]. It was suggested by [26] that during the Neoproterozoic Era ice could have extended to hundreds or thousands of kilometres, with sea ice that could have covered up to 14×10^6 km^2 of the Arctic Ocean during winter [27]. Thus, ice blocks could have been of the scale of tectonic plates [26], with similar vertical displacements, that probably exceeded our choice of $\zeta_0 = 1$ m, in particular with violent geophysical events that characterised that period. In this regard, it is also expected that multiple extremely long period oscillations could have formed high amplitude progressive waves causing circulations at velocity magnitudes of centimetres per second that are many orders of magnitudes higher than those from the suggested geothermal mechanism (i.e., micrometers per second) by [12] for that era. Here, it is also notable that, due to the differences in scales and distribution of events, the significance of two mechanisms could be decoupled in the space-time domain.

In the eighties of the last century the mean extent of the ice covering the arctic ocean had shrank to a value of less than 16 million squared kilometres, decreasing at a rate of 3% per decade [13]. Clearly, the ice pack thickness was dramatically reduced, and as the sea ice thins and shrinks further, it is expected that ice block oscillations will take place at relatively smaller areas and have less impact. In addition, recent observations suggest that, under current conditions, ice block oscillation amplitude may not exceed a few millimetres [17], producing a weak pressure signature on fluid parcels residing through the water column. Due to the different nature of waves radiated from ice-sheet movements, which is relatively a new scientific discipline, or even near sea glaciers, the available high quality data goes back to a few years only. It is well known that hundreds of ice quakes occur annually in localised regions, some of which are related to glacial-sea interface and have magnitudes that may exceed 5 [28]. At such discontinuous space-time interaction events, any impact is expected to be localised, and the contribution

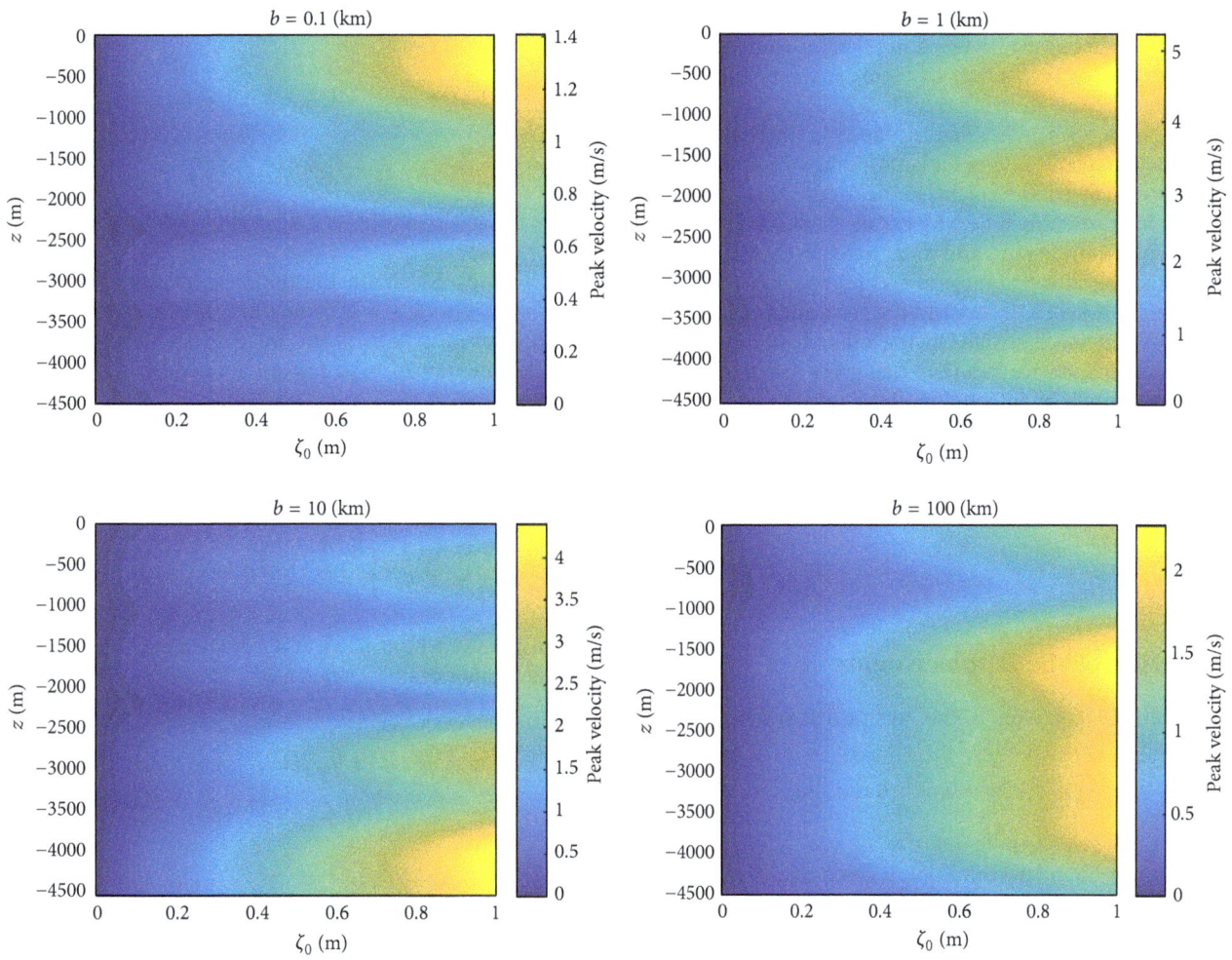

FIGURE 5: Peak velocities induced by four different oscillating ice blocks, $b = 0.1, 1, 10,$ and $100\,\mathrm{km}$, with various amplitudes $0 < \zeta_0 < 1$; $h = 4500\,\mathrm{m}$ and $c = 1500\,\mathrm{m/s}$.

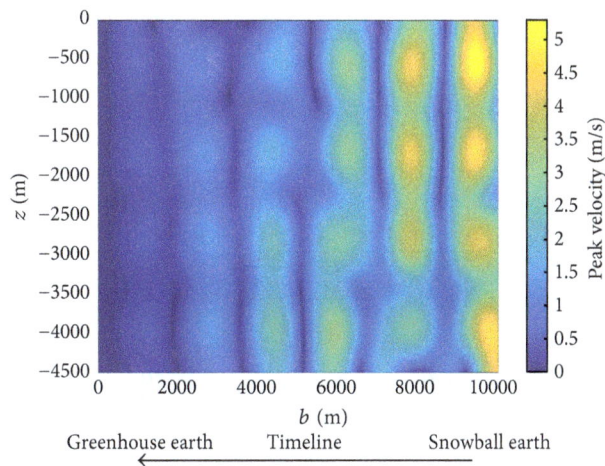

FIGURE 6: Peak velocities from various ice block widths representing timeline from snowball to greenhouse earth, from right to left. The amplitude $\zeta_0 = 0.0001b$; $h = 4500\,\mathrm{m}$ and $c = 1500\,\mathrm{m/s}$.

to water circulation becomes far less efficient, compared to snowball earth conditions. A qualitative demonstration of the change of impact of the ice block oscillations from snowball to greenhouse earth is given in Figure 6 where the x-axis represents the timeline, from right to left.

The Arctic Ocean is relatively poor in marine life, in general, and in plants, in particular. Although Phytoplankton is amongst the few marine plants in the Arctic Ocean, they are essential for sustaining a healthy ocean. Phytoplankton feed on nutrients from currents and water transport. As the greenhouse effect sustains and more areas of ice packs shrink and thin, one can expect a decrease in the generation of acoustic, by the present mechanism. On the other hand, if the sea becomes completely ice free [29], that gives rise to local formation of acoustic by interacting surface gravity waves [7, 9, 10]. The transition between the different mechanisms would possibly disturb global current patterns. Such disturbance is expected to affect nutrients transport and Phytoplankton population and potentially cause severe climate changes [30, 31], which is left for future work.

Appendix

Derivation of the Displacements

The governing equation is the standard two-dimensional compressible wave equation [14, 15]:

$$\varphi_{tt} - c^2 \nabla^2 \varphi = 0, \quad -h \leq z \leq 0, \quad (A.1)$$

where φ is the velocity potential.

The upper boundary condition is given by

$$\varphi_z = \zeta_t, \quad z = 0, \quad (A.2)$$

and the bottom boundary condition is simply

$$\varphi_z = 0, \quad z = -h. \quad (A.3)$$

Since here we only consider time harmonic problems we can write

$$\varphi(x, z, t) = f(x, z) \exp(i\omega t). \quad (A.4)$$

Substituting (A.4) into (A.1), (A.2), and (A.3) and making use of (1) give

$$\nabla^2 f + \frac{\omega^2}{c^2} f = 0, \quad -h \leq z \leq 0,$$

$$f_z = \zeta_0 \mathcal{H}(b^2 - x^2), \quad z = 0, \quad (A.5)$$

$$f_z = 0, \quad z = -h.$$

In the space domain the upper boundary condition is inhomogeneous. To overcome this difficulty we apply a Fourier transformation to the wavenumber domain, where $F(w, z)$ is the transform of $f(x, z)$, defined by

$$F(w, z) = \int_{-\infty}^{\infty} \exp(-iwx) f(x, z) \, dx \quad (A.6)$$

and the inverse transform, defined by

$$f(x, z) = \frac{1}{2\pi} \int_{-\infty}^{\infty} \exp(iwx) F(w, z) \, dx. \quad (A.7)$$

Defining $w = k + i\lambda$, where k and λ are real positive numbers, the eigenvalues can be written as

$$w = 0, \pm k_s, \pm k_n, \pm i\lambda_n, \quad n = 1, 2, \ldots, \quad (A.8)$$

where

$$k_s = \frac{\omega}{c};$$

$$k_n = \left(k_s^2 - \mu_n^2\right)^{1/2}, \quad \mu_n < k_s; \quad (A.9)$$

$$\lambda_n = \left(\mu_n^2 - k_s^2\right)^{1/2}, \quad \mu_n > k_s.$$

Using standard techniques as given by [2], in which a Fourier transform is applied over the horizontal axis, we obtain an expression for $f(x, z)$. For the range $|x| > b$ we write

$$f(x, z)$$

$$= \pm 2\zeta_0 \left\{ \sum_{n=1}^{N} \frac{\mu_n \cos[\mu_n(z+h)] \sin(k_n b) \exp[-i(k_n x - \pi/2)]}{k_n^2 \cos(\mu_n h)} \mp \sum_{n=N+1}^{\infty} \frac{\mu_n \cos[\mu_n(z+h)] \sinh(\lambda_n b) \exp(\mp i\lambda_n x)}{\lambda_n^2 \cos(\mu_n h)} \right\}, \quad (A.10)$$

where the upper signs of ($\pm$) and ($\mp$) in (A.10) are for $x > b$ and the lower ones are for $x < b$. On the other hand, for the range $|x| < b$, we write

$$f(x, z)$$

$$= 2\zeta_0 \left\{ \sum_{n=1}^{N} \frac{\mu_n \cos[\mu_n(z+h)] \cos(k_n b) \exp(-ik_n|x|)}{k_n^2 \cos(\mu_n h)} \right.$$

$$- \sum_{n=N+1}^{\infty} \frac{\mu_n \cos[\mu_n(z+h)] \cosh(\lambda_n x) \exp(-\lambda_n b)}{\lambda_n^2 \cos(\mu_n h)}$$

$$\left. + \frac{\cos[k_s(z+h)]}{k_s \sin(k_s h)} \right\}. \quad (A.11)$$

The first terms of right-hand side of (A.10) and (A.11) represent the acoustic modes, whereas the second terms represent the evanescent acoustic modes, which exponentially decay from the source and thus will not be considered in the sequel. The third term of (A.11) represents the surge mode which exists only below the oscillating ice block at $|x| < b$.

Ignoring the evanescent and surge terms in (A.10) and (A.11), substituting into the potential equation (A.4), and differentiating with respect to x and z give the horizontal and vertical velocities, respectively.

Competing Interests

The author declares that there are no competing interests.

Acknowledgments

The author thanks Carl Wunsch for raising the question that resulted in this paper and Raffaele Ferrari for useful discussions.

References

[1] M. S. Longuet-Higgins, "A theory of the origin of microseisms," *Philosophical Transactions of the Royal Society of London. Series A. Mathematical and Physical Sciences*, vol. 243, pp. 1–35, 1950.

[2] T. Yamamoto, "Gravity waves and acoustic waves generated by submarine earthquakes," *International Journal of Soil Dynamics and Earthquake Engineering*, vol. 1, no. 2, pp. 75–82, 1982.

[3] U. Kadri, "Deep ocean water transport by acoustic-gravity waves," *Journal of Geophysical Research: Oceans*, vol. 119, no. 11, pp. 7925–7930, 2014.

[4] F. Ardhuin, T. Lavanant, M. Obrebski et al., "A numerical model for ocean ultra-low frequency noise: wave-generated acoustic-gravity and Rayleigh modes," *The Journal of the Acoustical Society of America*, vol. 134, no. 4, pp. 3242–3259, 2013.

[5] S. Kedar, M. Longuet-Higgins, F. Webb, N. Graham, R. Clayton, and C. Jones, "The origin of deep ocean microseisms in the North Atlantic Ocean," *Proceedings of the Royal Society A: Mathematical, Physical and Engineering Sciences*, vol. 464, no. 2091, pp. 777–793, 2008.

[6] W. Farrell and W. Munk, "Booms and busts in the deep," *Journal of Physical Oceanography*, vol. 40, no. 9, 2010.

[7] U. Kadri, "Wave motion in a heavy compressible fluid: revisited," *European Journal of Mechanics B. Fluids*, vol. 49, pp. 50–57, 2015.

[8] U. Kadri, "Triad resonance between a surface-gravity wave and two high frequency hydro-acoustic waves," *European Journal of Mechanics B: Fluids*, vol. 55, no. 1, pp. 157–161, 2016.

[9] U. Kadri and T. R. Akylas, "On resonant triad interactions of acoustic—gravity waves," *Journal of Fluid Mechanics*, vol. 788, article R1, 12 pages, 2016.

[10] U. Kadri and M. Stiassnie, "Generation of an acoustic-gravity wave by two gravity waves, and their subsequent mutual interaction," *Journal of Fluid Mechanics*, vol. 735, article R6, 9 pages, 2013.

[11] C. M. Linton and P. McIver, "Embedded trapped modes in water waves and acoustics," *Wave Motion*, vol. 45, no. 1-2, pp. 16–29, 2007.

[12] Y. Ashkenazy, H. Gildor, M. Losch, and E. Tziperman, "Ocean circulation under globally glaciated snowball earth conditions: steady-state solutions," *Journal of Physical Oceanography*, vol. 44, no. 1, pp. 24–43, 2014.

[13] C. L. Parkinson, D. J. Cavalieri, P. Gloersen, H. J. Zwally, and J. C. Comiso, "Arctic sea ice extents, areas, and trends, 1978–1996," *Journal of Geophysical Research: Oceans*, vol. 104, no. 9, pp. 20837–20856, 1999.

[14] R. Stoneley, "The effect of the ocean on Rayleigh waves," *Geophysical Journal International*, vol. 1, pp. 349–356, 1926.

[15] H. Lamb, *Hydrodynamics*, Cambridge University Press, Cambridge, UK, 6th edition, 1932.

[16] F. B. Jensen, W. A. Kuperman, M. B. Porter, and H. Schmidt, *Computational Ocean Acoustics*, Springer, New York, NY, USA, 2011.

[17] D. R. MacAyeal, E. A. Okal, R. C. Aster, and J. N. Bassis, "Seismic and hydroacoustic tremor generated by colliding Icebergs," *Journal of Geophysical Research: Earth Surface*, vol. 113, no. 3, 2008.

[18] E. Eyov, A. Klar, U. Kadri, and M. Stiassnie, "Progressive waves in a compressible-ocean with an elastic bottom," *Wave Motion*, vol. 50, no. 5, pp. 929–939, 2013.

[19] R. Ferrari and C. Wunsch, "Ocean circulation kinetic energy: reservoirs, sources, and sinks," *Annual Review of Fluid Mechanics*, vol. 41, no. 1, pp. 253–282, 2009.

[20] P. D. Bromirski and R. A. Stephen, "Response of the ross ice shelf, antarctica, to ocean gravity-wave forcing," *Annals of Glaciology*, vol. 53, no. 60, pp. 163–172, 2012.

[21] E. A. D'Asaro and J. H. Morison, "Internal waves and mixing in the Arctic Ocean," *Deep Sea Research Part A: Oceanographic Research Papers*, vol. 39, no. 2, pp. S459–S484, 1992.

[22] C. Halle and R. Pinkel, "Internal wave variability in the Beaufort Sea during the winter of 1993/1994," *Journal of Geophysical Research C: Oceans*, vol. 108, no. 7, article 3210, 2003.

[23] M. D. Levine, C. A. Paulson, and J. H. Morison, "Internal waves in the arctic ocean: comparison with lower-latitude observations," *Journal of Physical Oceanography*, vol. 15, no. 6, pp. 800–809, 1985.

[24] R. Pinkel, "Near-inertial wave propagation in the western Arctic," *Journal of Physical Oceanography*, vol. 35, no. 5, pp. 645–665, 2005.

[25] A. J. Plueddemann, "Internal wave observations from the Arctic environmental drifting buoy," *Journal of Geophysical Research: Oceans*, vol. 97, no. 8, pp. 12619–12638, 1992.

[26] P. F. Hoffman, A. J. Kaufman, G. P. Halverson, and D. P. Schrag, "A neoproterozoic snowball earth," *Science*, vol. 281, no. 5381, pp. 1342–1346, 1998.

[27] J. Weiss, "Scaling of fracture and faulting of ice on earth," *Surveys in Geophysics*, vol. 24, no. 2, pp. 185–227, 2003.

[28] I. Joughin, "Greenland rumbles louder as glaciers accelerate," *Science*, vol. 311, no. 5768, pp. 1719–1720, 2006.

[29] J. E. Overland and M. Wang, "When will the summer Arctic be nearly sea ice free?" *Geophysical Research Letters*, vol. 40, no. 10, pp. 2097–2101, 2013.

[30] J. O. Sewall and L. C. Sloan, "Disappearing arctic sea ice reduces available water in the american west," *Geophysical Research Letters*, vol. 31, no. 6, Article ID L06209, 2004.

[31] J. A. Francis and S. J. Vavrus, "Evidence linking arctic amplification to extreme weather in mid-latitudes," *Geophysical Research Letters*, vol. 39, no. 6, Article ID L06801, 2012.

12

Sound Transmission in a Duct with Sudden Area Expansion, Extended Inlet, and Lined Walls in Overlapping Region

Ahmet Demir

Engineering Faculty, Department of Mechatronics, Karabuk University, 78100 Karabuk, Turkey

Correspondence should be addressed to Ahmet Demir; ademir@karabuk.edu.tr

Academic Editor: Toru Otsuru

The transmission of sound in a duct with sudden area expansion and extended inlet is investigated in the case where the walls of the duct lie in the finite overlapping region lined with acoustically absorbent materials. By using the series expansion in the overlap region and using the Fourier transform technique elsewhere we obtain a Wiener-Hopf equation whose solution involves a set of infinitely many unknown expansion coefficients satisfying a system of linear algebraic equations. Numerical solution of this system is obtained for various values of the problem parameters, whereby the effects of these parameters on the sound transmission are studied.

1. Introduction

One can reduce the unwanted noise propagating along a duct by using a reactive or a dissipative silencer. In reactive silencers sudden area changes in cross-sectional area help to reduce the energy in the transmitted wave via internal reflections. Having a sudden area expansion together with a sudden area contraction simple expansion chambers works in accordance with this principle and is widely investigated in literature [1–4]. In further investigations it has been shown that the extension of inlet and outlet tubes into the expansion chamber increased the acoustic attenuation performance [5–7].

On the other hand, it has been proved that the treatment of the duct walls with an acoustically absorbent lining is another effective method in reducing unwanted noise [8]. Application of locally reacting linings or expansion chambers in ducts are efficient methods for noise reduction. These two methods were combined in [9] to discover transmission properties of a combination silencer consist of an expansion chamber whose walls are treated by acoustic liners and have been analysed by the author previously.

In this paper, the transmission of sound in an extended tube resonator whose walls are in overlapping region, where extended inlet and expanding duct walls overlap, are treated by locally reacting lining is investigated. So the main objective

of this paper is to reveal the influence of the partial lining on the transmitted field and to present an alternative method of formulation. The method previously employed in [10, 11] consists of expanding the field in the overlap region into a series of complete set of orthogonal eigenfunctions and using the Fourier transform technique elsewhere. The problem is then reduced directly into a Wiener-Hopf equation whose solution involves a set of infinitely many unknown expansion coefficients satisfying an infinite systems of linear algebraic equations. Numerical solution to these systems is obtained for various values of the parameters of the problem such as the radii of the semi-infinite waveguides, the overlap length, and the impedance loading whereby the effects of these parameters on the transmitted field are presented graphically.

The time dependence is assumed to be $\exp(-i\omega t)$ with ω being the angular frequency and suppressed throughout.

2. Materials and Methods

Consider two opposite semi-infinite circular cylindrical waveguides of different radii with common longitudinal axis, say z, in a cylindrical polar coordinate system (ρ, ϕ, z). They occupy the regions $\rho = a$ and $z < l$ and $\rho = b > a$ and $z > 0$, respectively, where l represents the overlap length. These two waveguides are connected with a vertical wall at

$a < \rho < b$ and $z = 0$. The parts of the surfaces $r = a + 0$ and $\rho = b - 0$ lying in the overlap region $0 < z < l$ of the waveguides and the vertical wall are assumed to be treated by acoustically absorbing linings which are characterized by constant but different surface admittances, say η_1, η_2, and η_3, respectively, while the remaining parts are perfectly rigid (see Figure 1). The waveguides are immersed in an inviscid and compressible stationary fluid of density $\bar{\rho}_0$ and sound speed c. A plane sound wave is incident from the positive z-direction, through the waveguide of radius $\rho = a$. From the symmetry of the geometry of the problem and the incident field the acoustic field everywhere will be independent of the ϕ coordinate. We shall therefore introduce a scalar potential $u(\rho, z)$ which defines the acoustic pressure and velocity by $p = i\omega\bar{\rho}_0 u$ and $\mathbf{v} = \operatorname{grad} u$, respectively.

Let the incident field be given by

$$u^i = \exp(ikz), \tag{1}$$

where $k = \omega/c$ denotes the wave number. For the sake of analytical convenience we will assume that the surrounding medium is slightly lossy and k has a small positive imaginary part. The lossless case can be obtained by letting $\operatorname{Im} k \to 0$ at the end of the analysis.

The total field $u^T(\rho, z)$ can be written as

$$u^T(\rho, z) = \begin{cases} u_1(\rho, z) + u^i(\rho, z), & \rho \in (0, a), \ z \in (-\infty, \infty), \\ u_2^{(1)}(\rho, z) [\mathscr{H}(z) - \mathscr{H}(z - l)] + u_2^{(2)}(\rho, z) \mathscr{H}(z - l), & \rho \in (a, b), \ z \in (0, \infty). \end{cases} \tag{2}$$

$u_1(\rho, z)$ and $u_2^{(j)}(\rho, z)$ ($j = 1, 2$) denote the scattered fields which satisfy the Helmholtz equation

$$\left[\frac{1}{\rho} \frac{\partial}{\partial \rho} \left(\rho \frac{\partial}{\partial \rho} \right) + \frac{\partial^2}{\partial z^2} + k^2 \right] \begin{bmatrix} u_1(\rho, z) \\ u_2^{(j)}(\rho, z) \end{bmatrix} = 0, \tag{3}$$

$$j = 1, 2$$

and are to be determined with the help of the following boundary and continuity relations:

$$\frac{\partial}{\partial \rho} u_1(a, z) = 0, \quad z < l, \tag{4a}$$

$$\frac{\partial}{\partial z} u_2^{(1)}(\rho, 0) = 0, \tag{4b}$$
$$a < \rho < b,$$

$$\left[ik\eta_1 + \frac{\partial}{\partial \rho} \right] u_2^{(1)}(a, z) = 0, \tag{4c}$$
$$0 < z < l,$$

$$\left[ik\eta_2 - \frac{\partial}{\partial \rho} \right] u_2^{(1)}(b, z) = 0, \tag{4d}$$
$$0 < z < l,$$

$$\left[ik\eta_3 + \frac{\partial}{\partial z} \right] u_2^{(1)}(\rho, 0) = 0, \tag{4e}$$
$$a < \rho < b,$$

$$\frac{\partial}{\partial \rho} u_2^{(2)}(b, z) = 0, \quad z > l, \tag{4f}$$

$$u_2^{(1)}(\rho, l) - u_2^{(2)}(\rho, l) = 0, \tag{4g}$$
$$a < \rho < b,$$

$$\frac{\partial}{\partial z} u_2^{(1)}(\rho, l) - \frac{\partial}{\partial z} u_2^{(2)}(\rho, l) = 0, \tag{4h}$$
$$a < \rho < b,$$

$$u_1(a, z) + u_i(a, z) - u_2^{(2)}(a, z) = 0, \quad z > l, \tag{4i}$$

$$\frac{\partial}{\partial \rho} u_1(a, z) + \frac{\partial}{\partial \rho} u_i(a, z) - \frac{\partial}{\partial \rho} u_2^{(2)}(a, z) = 0, \quad z > l. \tag{4j}$$

In addition to these boundary and continuity relations one has to take into account the following radiation and edge conditions to ensure the uniqueness of the mixed boundary value problem stated by (3) and (4a)–(4j):

$$u \sim \frac{e^{ikr}}{r}, \quad r = \sqrt{\rho^2 + z^2} \longrightarrow \infty, \tag{5}$$

$$u^T(\rho, z) = O(1), \quad z \longrightarrow l, \tag{6a}$$

$$\frac{\partial}{\partial \rho} u^T(\rho, z) = O\left((z - l)^{-1/2}\right), \quad z \longrightarrow l. \tag{6b}$$

2.1. The Wiener-Hopf Equations. Consider the Fourier transform of the Helmholtz equation satisfied by the scattered field $u_1(\rho, z)$ in the region $\rho < a$ for $z \in (-\infty, \infty)$; namely,

$$\left[\frac{1}{\rho} \frac{\partial}{\partial \rho} \left(\rho \frac{\partial}{\partial \rho} \right) + K^2(\alpha) \right] F(\rho, \alpha) = 0, \tag{7}$$

where $F(\rho, \alpha)$ is the Fourier transform of the field $u_1(\rho, z)$ defined to be

$$F(\rho, \alpha) = \int_{-\infty}^{\infty} u_1(\rho, z) e^{i\alpha z} dz \tag{8a}$$

$$= e^{i\alpha l} [F_+(\rho, \alpha) + F_-(\rho, \alpha)]$$

with

$$F_- (\rho, \alpha) = \int_{-\infty}^{l} u_1 (\rho, z) e^{i\alpha(z-l)} dz, \qquad (8b)$$

$$F_+ (\rho, \alpha) = \int_{l}^{\infty} u_1 (\rho, z) e^{i\alpha(z-l)} dz. \qquad (8c)$$

Owing to the analytical properties of Fourier integrals, $F_+(\rho, \alpha)$ and $F_-(\rho, \alpha)$ are regular functions in the upper half-plane $\operatorname{Im} \alpha > \operatorname{Im}(-k)$ and in the lower half-plane $\operatorname{Im} \alpha < \operatorname{Im} k$, respectively. The solution of (7) reads

$$F (\rho, \alpha) = -A (\alpha) \frac{J_0 (K\rho)}{K (\alpha) J_1 (Ka)}, \qquad (9)$$

where $A(\alpha)$ is a spectral coefficient to be determined and $K(\alpha)$ is the square-root function

$$K (\alpha) = \sqrt{k^2 - \alpha^2} \qquad (10)$$

which is defined in the complex α-plane cut as shown in Figure 2 such that $K(0) = k$. Consider now the Fourier transform of (4a); namely,

$$\dot{F}_- (a, \alpha) = 0. \qquad (11)$$

The differentiation of (9) with respect to ρ and putting $\rho = a$ gives

$$e^{i\alpha l} \dot{F}_+ (a, \alpha) = A (\alpha). \qquad (12)$$

Substituting (12) into (9) yields

$$F_+ (\rho, \alpha) = -\dot{F}_+ (a, \alpha) \frac{J_0 (K\rho)}{K (\alpha) J_1 (Ka)} - F_- (\rho, \alpha). \qquad (13)$$

In the region $a < \rho < b$ the field $u_2^{(2)}(\rho, z)$ satisfies the Helmholtz equation for $z \in (l, \infty)$ as denoted in (3). The Fourier transform of this equation for the region in question is

$$\left[\frac{1}{\rho} \frac{\partial}{\partial \rho} \left(\rho \frac{\partial}{\partial \rho} \right) + K^2 (\alpha) \right] G_+ (\rho, \alpha)$$
$$= f (\rho) - i\alpha g (\rho), \qquad (14)$$

where

$$f (\rho) = \frac{\partial}{\partial z} u_2^{(2)} (\rho, l), \qquad (15a)$$

$$g (\rho) = u_2^{(2)} (\rho, l). \qquad (15b)$$

In (14), $G_+(\rho, \alpha)$ is a regular function in the upper half of the complex α-plane which is defined as

$$G_+ (\rho, \alpha) = \int_{l}^{\infty} u_2^{(2)} (\rho, z) e^{i\alpha(z-l)} dz. \qquad (16)$$

Particular solutions to (14) can be found easily by using Green's function which satisfies the Helmholtz equation

$$\left[\frac{1}{\rho} \frac{\partial}{\partial \rho} \left(\rho \frac{\partial}{\partial \rho} \right) + K^2 (\alpha) \right] \mathscr{G} (\rho, \alpha) = 0, \qquad (17)$$
$$\rho \neq t, \ \rho, t \in (a, b)$$

with the following conditions:

$$\mathscr{G} (t + 0, t, \alpha) = \mathscr{G} (t - 0, t, \alpha), \qquad (18a)$$

$$\frac{\partial}{\partial \rho} \mathscr{G} (t + 0, t, \alpha) - \frac{\partial}{\partial \rho} \mathscr{G} (t - 0, t, \alpha) = \frac{1}{t}, \qquad (18b)$$

$$\frac{\partial}{\partial \rho} \mathscr{G} (b, t, \alpha) = 0, \qquad (18c)$$

$$\frac{\partial}{\partial \rho} \mathscr{G} (a, t, \alpha) = 0. \qquad (18d)$$

The solution is

$$\mathscr{G} (\rho, t, \alpha) = \frac{1}{K^2 (\alpha) M (\alpha)} Q (\rho, t, \alpha) \qquad (19)$$

with

$$Q (\rho, t, \alpha) = \frac{\pi}{2} \begin{cases} [J_0 (K\rho) KY_1 (Ka) - KJ_1 (Ka) Y_0 (K\rho)] [J_0 (Kt) KY_1 (Kb) - KJ_1 (Kb) Y_0 (Kt)], & a \le \rho \le t, \\ [J_0 (K\rho) KY_1 (Kb) - KJ_1 (Kb) Y_0 (K\rho)] [J_0 (Kt) KY_1 (Ka) - KJ_1 (Ka) Y_0 (Kt)], & t \le \rho \le b, \end{cases} \qquad (20a)$$

$$M (\alpha) = [J_1 (Ka) Y_1 (Kb) - J_1 (Kb) Y_1 (Ka)]. \qquad (20b)$$

Together with the boundary condition (4f) the solution can be written as

$$G_+ (\rho, \alpha) = \frac{1}{K^2 (\alpha) M (\alpha)} \Bigg\{ B (\alpha)$$
$$\cdot [J_0 (K\rho) KY_1 (Kb) - Y_0 (K\rho) KJ_1 (Kb)]$$

$$+ \int_{a}^{b} [f (t) - i\alpha g (t)] Q (t, \rho, \alpha) t \, dt \Bigg\}. \qquad (21)$$

Here, $B(\alpha)$ is a spectral coefficient to be determined.

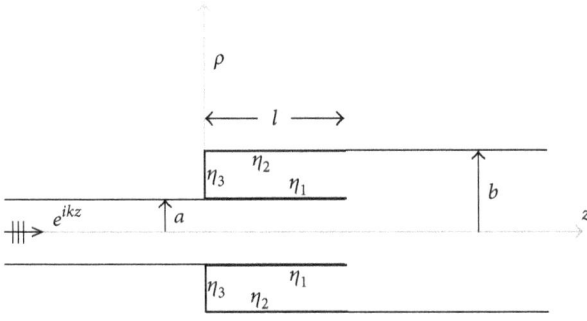

FIGURE 1: Geometry of the problem.

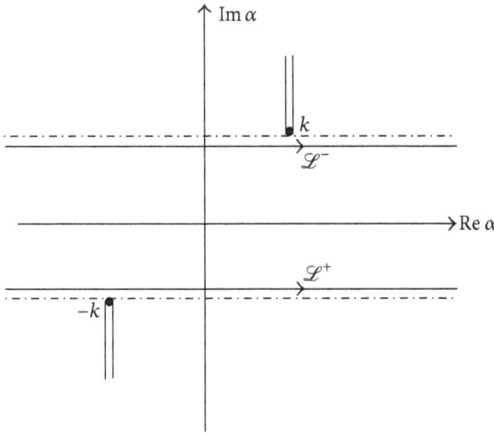

FIGURE 2: Complex α-plane.

The continuity relation in (4j) requires

$$\dot{G}_+(a,\alpha) = \dot{F}_+(a,\alpha).$$ (22)

$B(\alpha)$ can be solved uniquely from (22) as

$$B(\alpha) = -\dot{F}_+(a,\alpha).$$ (23)

The substitution of (23) into (21) gives

$$G_+(\rho,\alpha) = \frac{1}{K^2(\alpha)M(\alpha)} \left\{ -\dot{F}_+(a,\alpha) \right.$$

$$\cdot \left[J_0(K\rho)KY_1(Kb) - Y_0(K\rho)KJ_1(Kb) \right]$$ (24)

$$\left. + \int_a^b \left[f(t) - i\alpha g(t) \right] Q(t,\rho,\alpha)t\,dt \right\}.$$

Although the left-hand side of (24) is regular in the half-plane $\text{Im}(\alpha) > \text{Im}(-k)$, the regularity of the right hand side is violated by the presence of simple poles lying at the upper half-plane, namely, at $\alpha = \alpha_m$ ($\text{Im}(\alpha_m) > \text{Im}(k)$) with

$$K^2(\alpha_m) \left[J_1(K_m a)Y_1(K_m b) - J_1(K_m b)Y_1(K_m a) \right]$$

$$= 0, \quad K_m = K(\alpha_m), \ m = 0,1,2,\ldots.$$ (25)

In order to provide regularity of the right hand side of (24) in the upper half of the α-plane, these poles must be eliminated by imposing that their residues are zero. This gives

$$\dot{F}_+(a,k) = \frac{1}{\pi}\left[f_0 - ikg_0 \right]\frac{a^2 - b^2}{ab},$$ (26a)

$$\dot{F}_+(a,\alpha_m) = \frac{1}{\pi}\left[f_m - i\alpha_m g_m \right]\frac{J_1^2(K_m a) - J_1^2(K_m b)}{J_1(K_m a)J_1(K_m b)}$$ (26b)

for $m = 0,1,2,\ldots$. Here, f_m and g_m's are the expansion coefficients of the functions $f(\rho)$ and $g(\rho)$, respectively, which may be represented trough the following complete sets of orthogonal functions

$$\begin{bmatrix} f(\rho) \\ g(\rho) \end{bmatrix} = \frac{2}{\pi b}\begin{bmatrix} f_0 \\ g_0 \end{bmatrix}$$

$$+ \sum_{m=1}^{\infty}\begin{bmatrix} f_m \\ g_m \end{bmatrix}\left[J_0(K_m\rho)K_m Y_1(K_m b) \right.$$ (27)

$$\left. - K_m J_1(K_m b)Y_0(K_m\rho) \right].$$

Using the continuity relation (4i) together with (24) and taking into account (13) give

$$\frac{\dot{F}_+(a,\alpha)}{K^2(\alpha)}\frac{J_1(Kb)}{\pi M(\alpha)J_1(Ka)} + \frac{a}{2}F_-(a,\alpha) = \frac{1}{2}$$

$$\cdot \frac{1}{K^2(\alpha)M(\alpha)}\int_a^b \left[f(t) - i\alpha g(t) \right]$$

$$\cdot \left[J_0(Kt)KY_1(Kb) - KJ_1(Kb)Y_0(Kt) \right]t\,dt - \frac{a}{2}$$

$$\cdot \frac{e^{ikl}}{i(\alpha+k)},$$ (28)

$$\frac{\dot{F}_+(a,\alpha)}{K^2(\alpha)}N(\alpha) + \frac{a}{2}F_-(a,\alpha) = \frac{b}{\pi a}\frac{\left[f_0 - i\alpha g_0 \right]}{k^2 - \alpha^2} + \frac{1}{\pi}$$

$$\cdot \sum_{m=1}^{\infty}\frac{\left[f_m - i\alpha g_m \right]}{\alpha_m^2 - \alpha^2}\frac{J_1(K_m b)}{J_1(K_m a)} - \frac{a}{2}\frac{e^{ikl}}{i(\alpha+k)}$$

which is the Wiener-Hopf equation to be solved through classical procedures. Here, $N(\alpha)$ stands for

$$N(\alpha) = \frac{J_1(Kb)}{\pi M(\alpha)J_1(Ka)}.$$ (29)

The final solution of the W-H equation is determined to be

$$\frac{\dot{F}_+(a,\alpha)}{(k+\alpha)}N_+(\alpha) = -\frac{a}{\pi b}\frac{\left[f_0 + ikg_0 \right]}{(k+\alpha)N_+(k)} + \frac{1}{\pi}$$

$$\cdot \sum_{m=1}^{\infty}\frac{\left[f_m + i\alpha_m g_m \right]}{2\alpha_m(\alpha_m+\alpha)}\frac{J_1(K_m b)}{J_1(K_m a)}\frac{k+\alpha_m}{N_+(\alpha_m)}$$ (30)

$$- \frac{kae^{ikl}}{i(\alpha+k)N_+(k)},$$

where $N_{\pm}(\alpha)$ are the split functions resulting from the Wiener-Hopf factorization of $N(\alpha)$ as

$$N(\alpha) = N_+(\alpha) N_-(\alpha). \tag{31}$$

Their explicit expressions are given in [12] as

$$N_+(\alpha) = \left[\pi \right.$$
$$\left. \cdot \frac{J_1(ka)}{J_1(kb)} \left[J_1(ka) Y_1(kb) - J_1(kb) Y_1(ka) \right] \right]^{-1/2}$$
$$\cdot e^{-\alpha\chi} \prod_{n=0}^{\infty} \frac{\left(1 + \alpha/\sqrt{k^2 - (j_n/b)^2} \right)}{\left(1 + \alpha/\sqrt{k^2 - (j_n/a)^2} \right) \left(1 + \alpha/\alpha_n \right)}. \tag{32}$$

Here j_n's are the roots of the Bessel function of the first kind

$$J_1(j_n) = 0, \quad n = 0, 1, \ldots, \tag{33a}$$

$$\chi = \frac{i}{\pi} \left[b \ln b - a \ln a - (b - a) \ln (b - a) \right] \tag{33b}$$

with

$$N_-(\alpha) = N_+(-\alpha). \tag{33c}$$

2.2. Determination of the Unknown Coefficients. The field in the region $a < \rho < b$, $z \in (0, l)$ can be expressed is terms of the waveguide normal modes as

$$u_2^{(1)}(\rho, z)$$
$$= \sum_{n=0}^{\infty} a_n \left[e^{i\beta_n z} - P_n e^{-i\beta_n z} \right] \left[J_0(\xi_n \rho) - R_n Y_0(\xi_n \rho) \right] \tag{34}$$

with

$$P_n = \frac{ik\eta_3 + i\beta_n}{ik\eta_3 - i\beta_n}, \tag{35a}$$

$$R_n = \frac{ik\eta_1 J_0(\xi_n a) - \xi_n J_1(\xi_n a)}{ik\eta_1 Y_0(\xi_n a) - \xi_n Y_1(\xi_n a)}$$
$$= \frac{ik\eta_2 J_0(\xi_n b) + \xi_n J_1(\xi_n b)}{ik\eta_2 Y_0(\xi_n b) + \xi_n Y_1(\xi_n b)}. \tag{35b}$$

In (34), ξ_n's are the roots of the following equation:

$$\frac{ik\eta_1 J_0(\xi_n a) - \xi_n J_1(\xi_n a)}{ik\eta_1 Y_0(\xi_n a) - \xi_n Y_1(\xi_n a)}$$
$$- \frac{ik\eta_2 J_0(\xi_n b) + \xi_n J_1(\xi_n b)}{ik\eta_2 Y_0(\xi_n b) + \xi_n Y_1(\xi_n b)} = 0 \tag{36}$$

while β_n's are defined as

$$\beta_n = \sqrt{k^2 - \xi_n^2}. \tag{37}$$

Taking into account (27) and (34), the continuity relations (4g) and (4h) can be written in the following form:

$$\frac{2}{\pi b} \left[f_0 + i\alpha g_0 \right] + \sum_{m=1}^{\infty} \left[f_m + i\alpha g_m \right]$$
$$\cdot \left[J_0(K_m \rho) K_m Y_1(K_m b) \right.$$
$$\left. - K_m J_1(K_m b) Y_0(K_m \rho) \right] = i \sum_{n=0}^{\infty} a_n \left[(\alpha + \beta_n) e^{i\beta_n l} \right.$$
$$\left. - P_n (\alpha - \beta_n) e^{-i\beta_n l} \right] \left[J_0(\xi_n \rho) - R_n Y_0(\xi_n \rho) \right]. \tag{38}$$

Multiplying (38) by $2\rho/\pi b$ and $[J_0(K_m\rho)K_m Y_1(K_m b) - K_m J_1(K_m b) Y_0(K_m\rho)]\rho$, respectively, and then integrating over ρ from a to b read

$$\left[f_0 + i\alpha g_0 \right]$$
$$= \frac{i}{S_0} \sum_{n=0}^{\infty} a_n \left[(\alpha + \beta_n) e^{i\beta_n l} - P_n (\alpha - \beta_n) e^{-i\beta_n l} \right] \triangle_{0n}, \tag{39a}$$

$$\left[f_m + i\alpha g_m \right]$$
$$= \frac{i}{S_m} \sum_{n=0}^{\infty} a_n \left[(\alpha + \beta_n) e^{i\beta_n l} - P_n (\alpha - \beta_n) e^{-i\beta_n l} \right] \triangle_{mn} \tag{39b}$$

with

$$\triangle_{0n} = \frac{2}{\pi \xi_n} \left\{ \frac{a}{b} \left[J_1(\xi_n a) - R_n Y_1(\xi_n a) \right] - \left[J_1(\xi_n b) \right.\right.$$
$$\left.\left. - R_n Y_1(\xi_n b) \right] \right\}, \tag{40a}$$

$$\triangle_{mn} = \frac{2\xi_n}{\pi(\xi_n^2 - K_m^2)} \left\{ \frac{J_1(K_m b)}{J_1(K_m a)} \left[J_1(\xi_n a) \right.\right.$$
$$\left.\left. - R_n Y_1(\xi_n a) \right] - \left[J_1(\xi_n b) - R_n Y_1(\xi_n b) \right] \right\}, \tag{40b}$$

where S_0 and S_m stand for

$$S_0 = \frac{2}{\pi^2} \frac{a^2 - b^2}{b^2}, \tag{41a}$$

$$S_m = \frac{2}{\pi^2} \frac{J_1^2(K_m a) - J_1^2(K_m b)}{J_1^2(K_m a)}. \tag{41b}$$

Using the W-H solution (30) together with (26a) and (26b) we obtain a set of linear algebraic equations in terms of the unknown coefficients f_m and g_m.

$$\frac{\pi}{2}\frac{b}{a}\left[f_0 - ikg_0\right]S_0 N_+(k) = \frac{b}{\pi a}\frac{\left[f_0 + ikg_0\right]}{N_+(k)} + \frac{k}{\pi}$$

$$\cdot \sum_{m=1}^{\infty}\frac{\left[f_m + i\alpha_m g_m\right]}{\alpha_m N_+(\alpha_m)}\frac{J_1(K_m b)}{J_1(K_m a)} - \frac{kae^{ikl}}{iN_+(k)}, \tag{42a}$$

$$\frac{\pi}{2}\left[f_r - i\alpha_r g_r\right]\frac{J_1(K_r a)}{J_1(K_r b)}S_r N_+(\alpha_r) = \frac{b}{\pi a}\frac{\left[f_0 + ikg_0\right]}{N_+(k)}$$

$$+ \frac{1}{\pi}\left(k + \alpha_r\right)$$

$$\cdot \sum_{m=1}^{\infty}\frac{\left[f_m + i\alpha_m g_m\right]}{2\alpha_m\left(\alpha_m + \alpha_r\right)}\frac{J_1(K_m b)}{J_1(K_m a)}\frac{k + \alpha_m}{N_+(\alpha_m)} \tag{42b}$$

$$- \frac{kae^{ikl}}{iN_+(k)},$$

and taking into account (39a) and (39b) we obtain now a set of equations to determine the unknown expansion coefficient a_n as

$$-i\frac{\pi}{2}\frac{b}{a}N_+(k)\sum_{n=0}^{\infty}a_n\left[(k - \beta_n)e^{i\beta_n l} - P_n(k + \beta_n)e^{-i\beta_n l}\right]\triangle_{0n}$$

$$= \frac{b}{\pi a}\frac{1}{N_+(k)}\frac{i}{S_0}\sum_{n=0}^{\infty}a_n\left[(k + \beta_n)e^{i\beta_n l} - P_n(k - \beta_n)\right]$$

$$\cdot e^{-i\beta_n l}\right]\triangle_{0n} + i\frac{k}{\pi}\sum_{n=0}^{\infty}\sum_{m=1}^{\infty}a_n\left[(\alpha_m + \beta_n)e^{i\beta_n l}\right. \tag{43a}$$

$$\left. - P_n(\alpha_m - \beta_n)e^{-i\beta_n l}\right]\frac{\triangle_{mn}}{\alpha_m N_+(\alpha_m)}\frac{J_1(K_m b)}{S_m J_1(K_m a)}$$

$$- \frac{kae^{ikl}}{iN_+(k)},$$

$$-i\frac{\pi}{2}\frac{J_1(K_r a)}{J_1(K_r b)}N_+(\alpha_r)\sum_{n=0}^{\infty}a_n\left[(\alpha_r - \beta_n)e^{i\beta_n l} - P_n(\alpha_r\right.$$

$$\left. + \beta_n)e^{-i\beta_n l}\right]\triangle_{rn} = \frac{b}{\pi a}\frac{1}{N_+(k)}\frac{i}{S_0}\sum_{n=0}^{\infty}a_n\left[(k + \beta_n)e^{i\beta_n l}\right.$$

$$\left. - P_n(k - \beta_n)e^{-i\beta_n l}\right]\triangle_{0n} + \frac{i}{\pi}$$

$$\cdot \sum_{n=0}^{\infty}\sum_{m=1}^{\infty}a_n\left[(\alpha_m + \beta_n)e^{i\beta_n l} - P_n(\alpha_m - \beta_n)e^{-i\beta_n l}\right] \tag{43b}$$

$$\cdot \frac{(k + \alpha_r)\triangle_{mn}}{2\alpha_m(\alpha_m + \alpha_r)}\frac{J_1(K_m b)}{S_m J_1(K_m a)}\frac{k + \alpha_m}{N_+(\alpha_m)}$$

$$- \frac{kae^{ikl}}{iN_+(k)}.$$

2.3. Reflected and Transmitted Fields. According to (8a), the scattered field in the region $0 < \rho < a$, that is, $u_1(\rho, z)$, can be

obtained by taking the inverse Fourier transform of $F(\rho, \alpha)$. By considering (13) we write

$$u_1(\rho, z)$$

$$= -\frac{1}{2\pi}\int_{\mathscr{L}}\dot{F}_+(a, \alpha)\frac{J_0(K\rho)}{K(\alpha)J_1(Ka)}e^{-i\alpha(z-l)}d\alpha. \tag{44}$$

The evaluation of this integral for $z < l$ and $z > l$ will give us the reflected wave propagating backward in the inner cylinder and the transmitted wave, respectively.

For $z < l$, the integral is calculated by closing the contour in the upper half-plane and evaluating the residues contributions from the simple poles occurring at the zeros of $J_1(Ka)$ lying in the upper α-half-plane, namely, at $Ka = j_n$. The reflection coefficient $\mathscr{R}$ of the fundamental mode is defined as the complex coefficient multiplying the travelling wave term $\exp(-ikz)$ and is computed from the contribution of the first pole at $\alpha = k$. The result is

$$\mathscr{R} = -\frac{e^{i2kl}}{\left[N_+(k)\right]^2} - \frac{i}{\pi}\frac{\left[f_0 + ikg_0\right]}{kb\left[N_+(k)\right]^2}e^{ikl}$$

$$+ \frac{i}{\pi}\frac{e^{ikl}}{aN_+(k)}\sum_{m=1}^{\infty}\frac{\left[f_m + i\alpha_m g_m\right]}{\alpha_m N_+(\alpha_m)}\frac{J_1(K_m b)}{J_1(K_m a)}. \tag{45}$$

The first term is the reflection coefficient related to the case where a semi-infinite rigid duct is inserted axially into a larger rigid tube of infinite length [12] whereas the second one is the correction term involving the effect of the impedances of the annular region and the overlap length.

Similarly, the transmission coefficient $\mathscr{T}$ of the fundamental mode which is defined as the complex coefficient of $\exp(ikz)$ is obtained by evaluating the integral in (44) for $z > l$. This integral is now computed by closing the contour in the lower half of the complex α-plane. The pole of interest is at $\alpha = -k$ whose contribution gives

$$(-1 + \mathscr{T})e^{ikz} + O\left(e^{i\sqrt{k^2 - (j_n/b)^2}z}\right) \tag{46a}$$

with

$$\mathscr{T} = \frac{a^2}{b^2} + \frac{ie^{-ikl}}{\pi ka}\left(\frac{b}{a} - \frac{a}{b}\right)\left[f_0 + ikg_0\right]. \tag{46b}$$

The first term in (46a) cancels out the incident wave in the region $\rho < a$, $z > l$, while the second is the transmission coefficient of the fundamental mode.

3. Results and Discussion

In this section in order to show the effects of the parameters like the length of the extended inlet l and the surface admittance $\eta_{1,2,3}$ on the transmitted field, some numerical results showing variation of the transmission coefficient $\mathscr{T}$ with different parameters are presented. In all numerical calculations the solution of the infinite system of algebraic equations is obtained by truncating the infinite series at $N = 5$, since the transmission coefficient becomes insensitive for

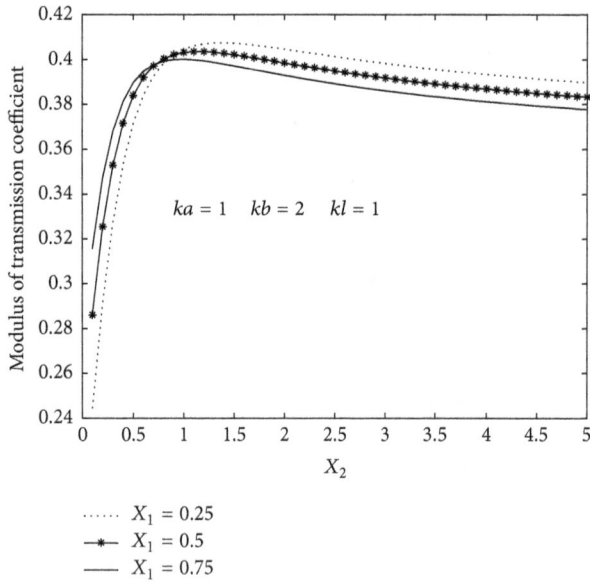

FIGURE 3: Transmission coefficient $\mathcal{T}$ versus the surface admittance $\eta_2 = iX_2$ $(X_2 > 0)$ for different values of $\eta_1 = iX_1$.

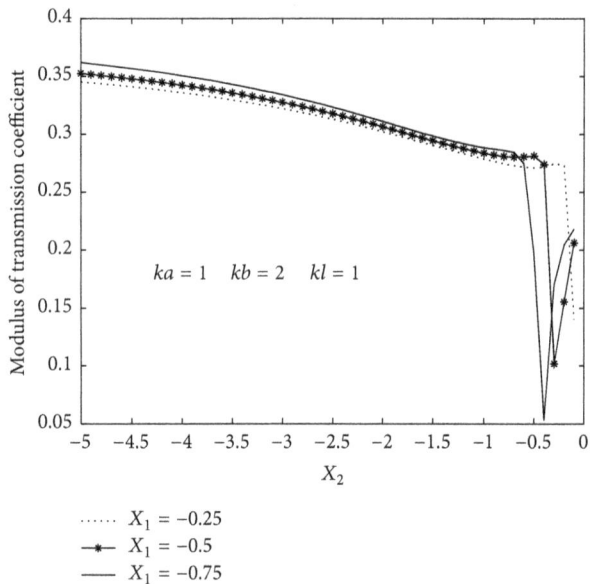

FIGURE 4: Transmission coefficient $\mathcal{T}$ versus the surface admittance $\eta_2 = iX_2$ $(X_2 < 0)$ for different values of $\eta_1 = iX_1$.

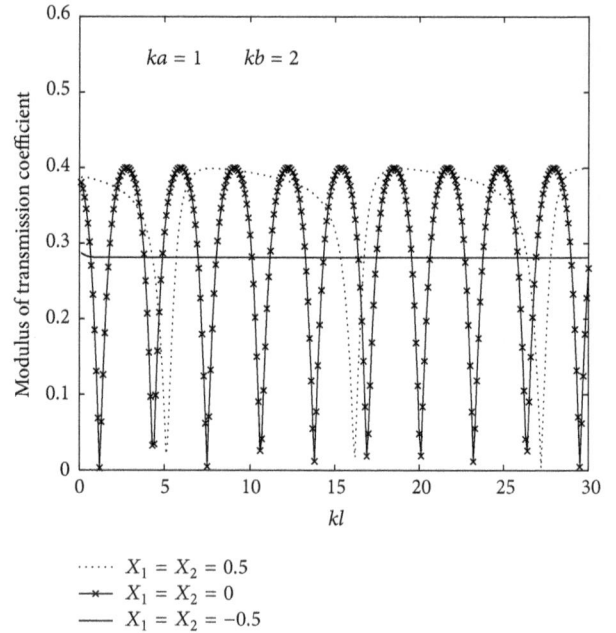

FIGURE 5: Transmission coefficient $\mathcal{T}$ versus the extended inlet length kl for different values of $\eta_1 = iX_1$ and $\eta_2 = iX_2$.

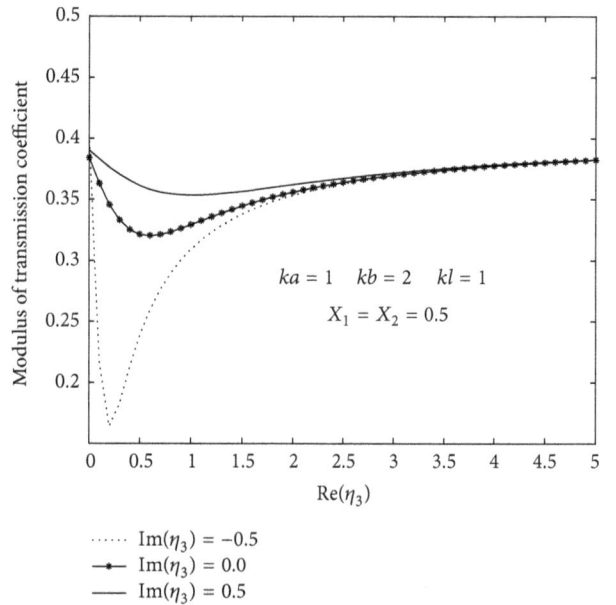

FIGURE 6: Transmission coefficient $\mathcal{T}$ versus the real part of η_3 for different values of $\text{Im}(\eta_3)$.

$N > 5$. We also limit ourselves with only imaginary values of surface admittance $\eta_{1,2}$ for simplicity.

In Figures 3 and 4, while the admittance $X_2 > 0$ of the lateral wall of the expanding duct increases the transmitted field is ascending until some value of X_2; then it starts to attenuate gradually. But for negative values of X_2 the attenuation is more visible especially around $-0.5 < X_2 < 0$. For different values of X_1 not much but some decrease in the transmitted field is observed.

In Figure 5, an oscillatory behaviour is seen for increasing values of the extended inlet length kl, but this behaviour is

broken for negative values of X_1 and X_2. From Figure 6 it is observed that the transmission does not alter as the real part of η_3 increased. But for small positive values of $\text{Re}(\eta_3)$ imaginary part $\text{Im}(\eta_3)$ becomes effective. The most reduction on the sound transmission is seen for the negative value of $\text{Im}(\eta_3)$.

Figure 7 shows an excellent agreement between the present paper (for the case of extended inlet length $kl \rightarrow 0$) and the previous study [9] of the author (for the case of

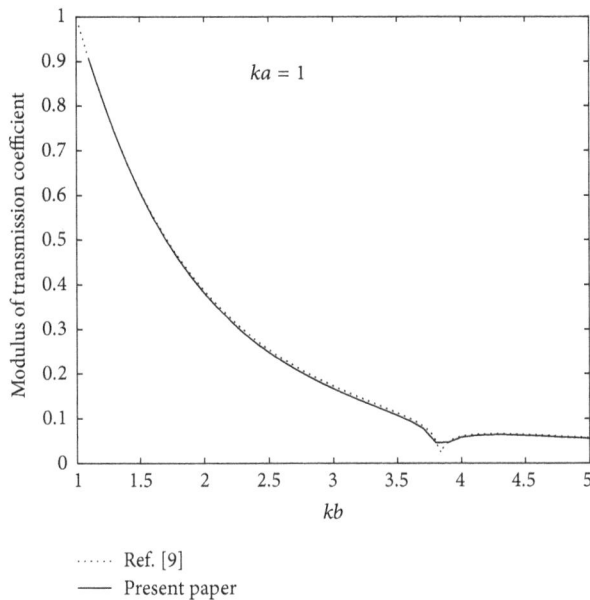

FIGURE 7: Transmission coefficient $\mathcal{T}$ versus the expansion chamber radius kb.

expansion chamber length $kl \rightarrow \infty$ and surface admittance is taken to be zero). In this comparison, transmission coefficient is calculated as though it is in a rigid-walled duct with sudden area expansion (without extended inlet).

4. Conclusions

This paper examines the transmission of sound waves in an extended tube resonator whose walls in overlapping region, where extended inlet and expanding duct walls overlap, are treated by acoustically absorbing materials of finite length. In the present work the lined region of the inner surface is assumed to be finite which makes the problem more complicated. To overcome the additional difficulty caused by the impedance discontinuity a hybrid method of formulation consisting of expressing the total field in terms of complete sets of orthogonal waveguide modes where available and using the Fourier transform elsewhere is adopted. The mixed boundary value problem is reduced to a Wiener-Hopf equation whose solution involves infinitely many expansion coefficients satisfying an infinite system of linear algebraic equations. These equations are solved numerically and the effects of various parameters on transmitted field such as the extended inlet length and the surface admittance of the lined section are displayed graphically. As a future work a similar problem now with an extended outlet will be studied following the same method used here.

Competing Interests

The author declares that there is no conflict of interests regarding the publication of this paper.

References

[1] J. W. Miles, "The analysis of plane discontinuities in cylindrical tubes. Part I," *The Journal of the Acoustical Society of America*, vol. 17, pp. 259–271, 1946.

[2] M. L. Munjal, *Acoustics of Ducts and Mufflers*, Wiley-Interscience, New York, NY, USA, 1987.

[3] J. Kergomard and A. Garcia, "Simple discontinuities in acoustic waveguides at low frequencies: critical analysis and formulae," *Journal of Sound and Vibration*, vol. 114, no. 3, pp. 465–479, 1987.

[4] A. Selamet and P. M. Radavich, "The effect of length on the acoustic attenuation performance of concentric expansion chambers: an analytical, computational and experimental investigation," *Journal of Sound and Vibration*, vol. 201, no. 4, pp. 407–426, 1997.

[5] M. Åbom, "Derivation of four-pole parameters including higher order mode effects for expansion chamber mufflers with extended inlet and outlet," *Journal of Sound and Vibration*, vol. 137, no. 3, pp. 403–418, 1990.

[6] K. S. Peat, "The acoustical impedance at the junction of an extended inlet or outlet duct," *Journal of Sound and Vibration*, vol. 9, pp. 101–110, 1991.

[7] A. Selamet and Z. L. Ji, "Acoustic attenuation performance of circular expansion chambers with extended inlet/outlet," *Journal of Sound and Vibration*, vol. 223, no. 2, pp. 197–212, 1999.

[8] A. D. Rawlins, "Radiation of sound from an unflanged rigid cylindrical duct with an acoustically absorbing internal surface," *Proceedings of the Royal Society. London. Series A. Mathematical, Physical and Engineering Sciences*, vol. 361, no. 1704, pp. 65–91, 1978.

[9] A. Demir and A. Büyükaksoy, "Transmission of sound waves in a cylindrical duct with an acoustically lined muffler," *International Journal of Engineering Science*, vol. 41, no. 20, pp. 2411–2427, 2003.

[10] A. Büyükaksoy and A. Demir, "Diffraction of sound waves by a rigid cylindrical cavity of finite length with an internal impedance surface," *Zeitschrift für Angewandte Mathematik und Physik*, vol. 56, no. 4, pp. 694–717, 2005.

[11] A. Büyükaksoy, G. Uzgören, and F. Birbir, "The scattering of a plane wave by two parallel semi-infinite overlapping screens with dielectric loading," *Wave Motion*, vol. 34, no. 4, pp. 375–389, 2001.

[12] A. D. Rawlins, "A bifurcated circular waveguide problem," *IMA Journal of Applied Mathematics*, vol. 54, no. 1, pp. 59–81, 1995.

13

Effects of the Cone and Edge on the Acoustic Characteristics of a Cone Loudspeaker

Yue Hu,[1,2] Xilu Zhao,[2] Takao Yamaguchi,[3] Manabu Sasajima,[1] Tatsushi Sasanuma,[1] and Akira Hara[1]

[1]Foster Electric Company, Limited, 1-1-109 Tsutsujigaoka, Akishima, Tokyo 196-8550, Japan
[2]College of Mechanical Engineering, Saitama Institute of Technology, 1690 Fusaiji, Fukaya, Saitama 369-0293, Japan
[3]Department of Mechanical Science and Technology, Faculty of Science and Technology, Gunma University, 1-5-1 Tenjin-cho, Kiryu, Gunma 376-8515, Japan

Correspondence should be addressed to Xilu Zhao; zhaoxilu@sit.ac.jp

Academic Editor: Kim M. Liew

Loudspeakers are designed for reproducing the original sound field as faithfully as possible. In order to faithfully reproduce sound, it is important to understand the relationships among the physical characteristics of the loudspeaker. This paper focuses on the cone, the edge, and the behavior of air around the voice coil, which are important elements in the design of cone loudspeakers and evaluates their effects on the acoustic characteristics of the loudspeaker.

1. Introduction

People hear a variety of sounds in their everyday lives. Loudspeakers are designed for faithfully reproducing the original sound field. Currently, the most commonly used loudspeaker design is the cone loudspeaker. For faithful sound reproduction, the relationships among the physical characteristics of the loudspeaker must be elucidated. One of the physical characteristics required for an ideal loudspeaker, assuming a constant input voltage, is a flat sound pressure in the frequency domain.

In pursuit of this goal, researchers have long studied vibration analysis of the loudspeaker. Investigations of sound radiation by Brown [1] and Bordoni [2] treated the cone loudspeaker as a rigid body. However, the rigid body model of the cone is valid only during the so-called piston motion, in which the vibration system components, including the cone, edge, center cap, voice coil, and spider (see Figure 1), vibrate together as a unit in the low-frequency range. Nimura et al. [3] carried out a theoretical analysis of the vibration of the cone and used a graphical method to calculate the eigenvalues of the vibration of a conical loudspeaker cone. Kagawa [4]

calculated the eigenvalues of the membrane vibrations of curved cones in addition to conical loudspeaker cones and found that curving the cone changed the eigenvalues of the membrane vibrations. Frankfort [5], also seeking solutions for the membrane vibrations of conical loudspeaker cones, formulated a differential equation that accounts for bending vibration and made detailed calculations of the sound pressure frequency response, vibration patterns, and driving-point admittance. However, Frankfort's analysis did not account for the effects of the edge and center cap, which play important roles in cone loudspeakers. Subsequently, the finite element method (FEM) has been used for vibration analysis of the cone [6–11]. Kyouno et al. analyzed the sound radiation from loudspeakers, accounting for coupling of the electrical, mechanical, and acoustic systems [12].

In these vibration analyses of cone loudspeakers, the cone maintains a piston motion in the low-frequency region. The biggest difficult point of these analyses is the split vibration of the vibration plate in the middle- and high-frequency regions. Control over the cone and the edge is very important in the design of a loudspeaker. However, there has been insufficient study of this topic in research up to now.

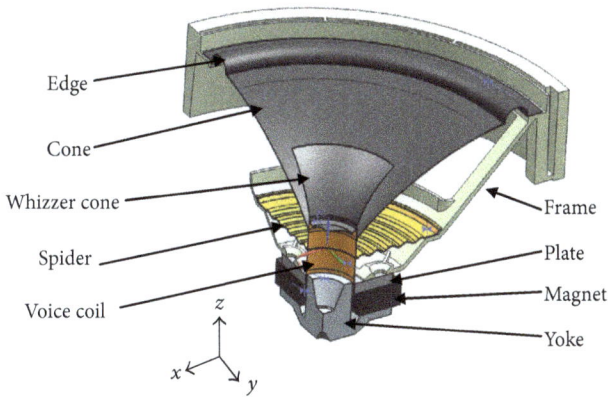

FIGURE 1: 1/4 model of a loudspeaker.

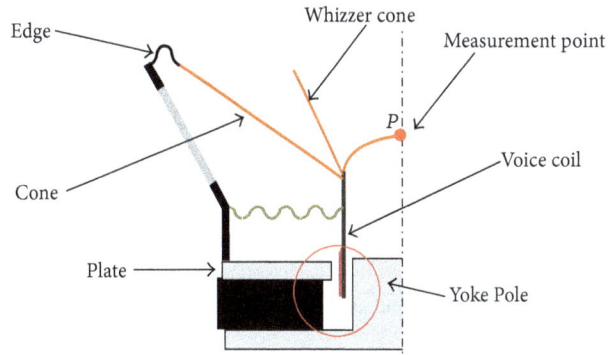

FIGURE 2: Important elements of the design of cone loudspeakers.

Furthermore, these acoustic analyses, particularly in the case of FEM, do not account for air viscosity, which may reduce the accuracy of their estimates of sound pressure frequency properties.

In particular, if the acoustic pathway in enclosure of small earphone is very narrow, air viscosity exerts a significant damping effect, reducing the estimation accuracy. Sasajima et al. succeeded at improving the estimation accuracy in the face of this problem by accounting for air viscosity using the finite element method [13].

Loudspeakers larger than earphones have a narrow space between the voice coil and the magnetic circuit, making it very likely that air viscosity has a significant impact and is a factor of estimation errors of the sound pressure frequency. Therefore, general acoustic analyses need to account for the damping effect of air viscosity to accurately analyze loudspeakers.

This paper examines cone loudspeakers, focusing intensively on the effects of the cone, the edge, and the narrow space between the voice coil and the magnetic circuit, which have a significant impact on the vibration characteristics and sound pressure frequency response. First, we perform an experimental study of the characteristics of a cone loudspeaker to determine its vibration characteristics and sound pressure frequency response. Next, we perform a vibration-acoustic analysis of the cone loudspeaker that accounts for the impact of air viscosity using our own independently developed acoustic analysis software. To verify the accuracy of our analysis, we compare the values from our analysis, which accounts for air viscosity, to the actual values of the vibration and sound pressure frequency properties. We also consider the behavior of the air around the voice coil. Finally, we calculate the optimal Young's modulus and density of the cone and edge and examine in detail their impact on the vibration characteristics.

2. Methods and Materials

Figure 1 illustrates the cone loudspeaker of diameter 0.16 m, which is the focus of this study. Because the cone exhibits

axial symmetry, we assume a model that is 1/4 symmetric with respect to the x-z and y-z planes.

The defining feature of cone loudspeakers is that the cone is approximately conical in shape, as illustrated in Figure 1. The cone is supported by an edge around the outer periphery and a spider that supports the center of the cone by holding the voice coil in position. The outer perimeter of the spider is anchored to the frame. The cover at the center of the cone is typically a center cap, but our model used a whizzer cone to support the high-frequency characteristics.

The voice coil is attached to the center of the cone, where it is supplied with the input signal current. These elements comprise the loudspeaker's vibration system. The outer periphery of the vibration system is supported by a frame.

The magnetic circuit for supplying magnetic flux to the voice coil consists of a plate, magnet, and yoke. Magnetic flux is supplied by the magnet through the yoke [14]. Sound waves are generated when the driving force generated according to Fleming's left hand rule drives the cone, which is attached to the voice coil.

A cross-sectional view of the cone loudspeaker is shown in Figure 2. The performance of a loudspeaker is largely governed by its cone, which radiates the sound directly. In this sense, the cone can be said to be the heart of the loudspeaker. In an idealized model, the cone is always a rigid body travelling with a back-and-forth piston motion; however, in the middle- and high-frequency regions, the shape and material property values of the cone generate split vibrations in axisymmetric and nonaxisymmetric modes, exerting a significant influence on the loudspeaker's vibration characteristics and sound pressure frequency response.

The edge, by undergoing large deformations and acting as a spring, not only facilitates movement of the cone in the z-axis direction but also plays a role in reducing movement in the x-y direction along the central axis. However, the edge does not always move back and forth together with the cone. The cone and the edge can vibrate in opposite phases, affecting the vibration characteristics.

By controlling the characteristics of the cone and edge, which are important design elements of a cone loudspeaker, the vibration characteristics of the cone loudspeaker can be improved. This study examines the vibration characteristics

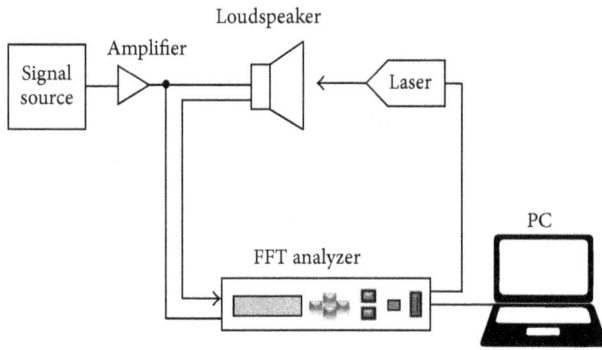

FIGURE 3: Cone vibration displacement measurement system.

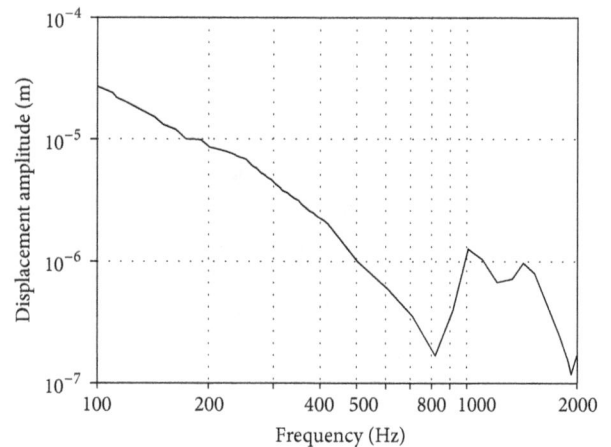

FIGURE 4: Comparison of the measurement result of cone vibration displacement amplitude.

of the loudspeaker by focusing attention on the materials of the cone and edge, which can be expected to have a particularly large impact.

In addition, the narrow space between the outer perimeter of the yoke pole and the inner perimeter of the plate, indicated in Figure 2, increases the air viscosity, and the flow of air in this gap can also be expected to have an important influence, which is very likely to be an error factor in the estimation of sound pressure frequency response. The damping effect of air viscosity, which is typically ignored in acoustic analysis, should therefore be taken into account for accurate analysis of loudspeaker performance.

3. Experimental

The performance of a cone loudspeaker is evaluated in terms of its vibration characteristics and sound pressure frequency properties; therefore, in our investigation, we took actual measurements of these characteristics.

3.1. Measurement of Cone Vibration Displacement. The cone is the core of a loudspeaker's vibration system, and it is therefore very important to understand the vibration characteristics of the cone when designing a loudspeaker's vibration system.

We measured the cone's vibration displacement using a laser displacement meter. Measurements were taken using a sweep signal of 10 Hz to 5 kHz, with the measurement point set to position P at the center of the whizzer cone, which is located at the center of the loudspeaker cone (see Figure 2). The measurement system is shown in Figure 3.

Figure 4 shows the actual displacement characteristics at the measurement point. From 100 Hz to 800 Hz, the displacement can be seen to decrease nearly linearly; however, beyond 800 Hz, peaks occur at approximately 1 and 1.5 kHz.

3.2. Measurement of Sound Pressure Frequency Properties. We tested the sound pressure frequency of the loudspeaker by taking measurements in a reverberation room and an anechoic room, in accordance with the standards established in JIS Z 8732. Figure 5(a) shows a schematic of the two-room measurement environment, with the reverberation room on the left and the anechoic room on the right. Figure 5(b)

shows a photo of the reverberation room and anechoic room as seen from the outside. The insides of the reverberation room and anechoic room are shown in Figures 5(c) and 5(d), respectively.

The measurement system included an audio analyzer (Etani Electronics ASA-10 Mark II) and a microphone (Aco type 4012). The tests were performed using a sweep signal in the frequency range of 20 Hz to 20 kHz. The observation point for measurement of the sound pressure in the anechoic room was a position 0.8 m in front of the loudspeaker, on its central axis. The measurement system is shown in Figure 6.

The measurement results are shown in Figure 7. An initial peak can be seen at approximately 80 Hz, and then the sound pressure frequency characteristic is generally flat from 200 Hz to 1 kHz. Above 1 kHz, a pattern of repeated peaks and dips can be observed.

We were able to determine the acoustic characteristics of the cone loudspeaker from the results of the measurements of the cone vibration displacement and the sound pressure frequency properties.

However, these results by themselves do not reveal the impact of the cone and edge on the acoustic characteristics of the loudspeaker. This test was also inadequate in terms of the limited types of loudspeaker we were able to measure.

Therefore, to elucidate the effect of the materials of the cone and edge on the acoustic properties of the loudspeaker, we continued our investigation as follows. First, we performed a coupled analysis of vibration and acoustics, taking air viscosity into account. Next, we compared the results of the analysis to the experimental values to verify the accuracy of the analysis. In lieu of an experiment, we used the acoustic analysis to reproduce the effects of the characteristics of the cone and edge on the acoustic characteristics of the loudspeaker. Finally, we applied the response surface methodology to determine the optimal cone and edge materials in order to design a loudspeaker with the desired acoustic characteristics.

(a) Schematic of measurement environment

(b) Reverberation room and anechoic room

(c) Reverberation room

(d) Anechoic room

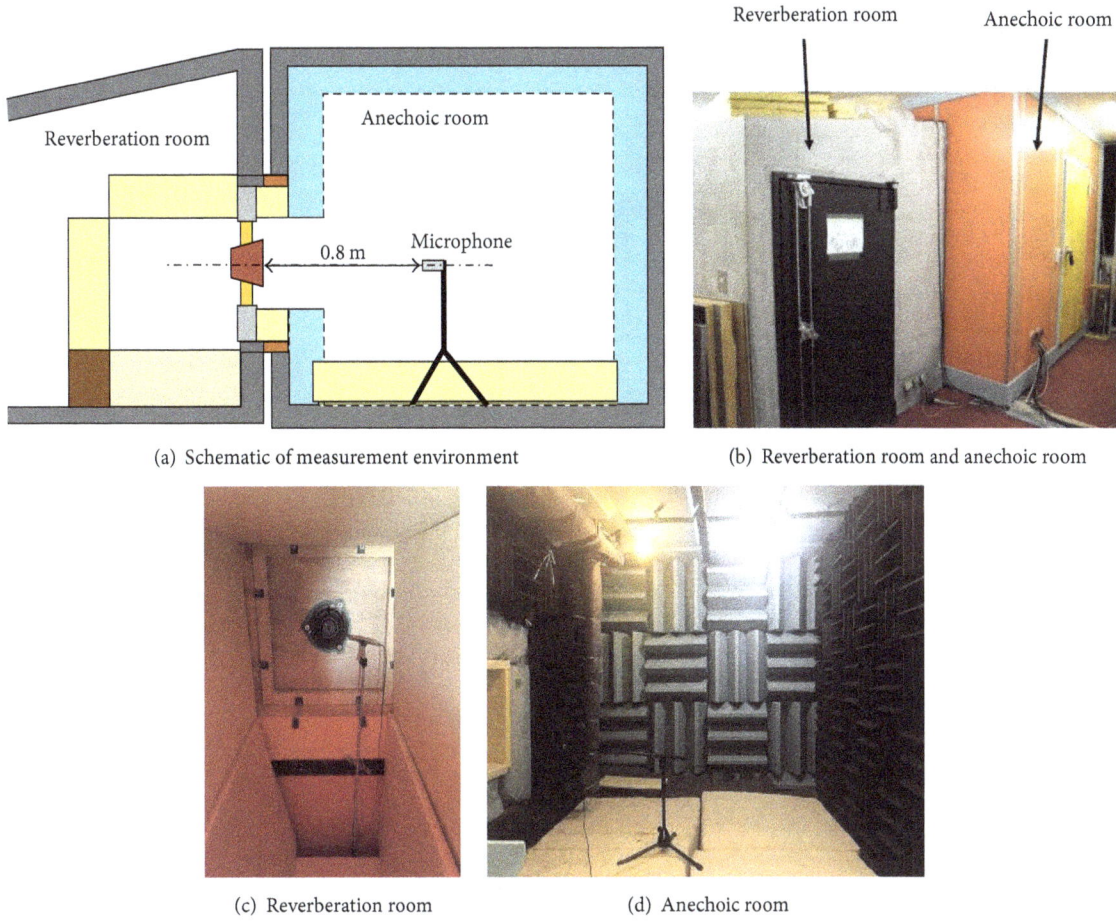

FIGURE 5: Measurement environment.

4. Analysis and Results

In order to account for air viscosity, which is typically ignored in acoustic analyses of cone loudspeakers, we propose an application of the finite element method to perform a coupled analysis of vibration and acoustics that takes into account air viscosity.

4.1. Method of Analysis. As the acoustic element, consider a three-dimensional tetrahedral element with unknown nodal displacements, as illustrated in Figure 8.

Letting u_x, u_y, and u_z be the displacements in the x, y, and z directions, respectively, at any point in the element, the strain energy $\widetilde{U}$ can be expressed as follows:

$$\widetilde{U} = \frac{1}{2} E \iiint_{v_e} \left(\frac{\partial u_x}{\partial x} + \frac{\partial u_y}{\partial y} + \frac{\partial u_z}{\partial z} \right)^2 dx\, dy\, dz. \quad (1)$$

Here, E is the bulk modulus and v_e is the volume of the tetrahedral element. Letting $\dot{u}$ be the time derivative of the displacement, the kinetic energy $\widetilde{T}$ is expressed as follows:

$$\widetilde{T} = \frac{1}{2} \iiint_{v_e} \rho \{\dot{u}\}^T \{\dot{u}\}\, dx\, dy\, dz. \quad (2)$$

Here, ρ represents the density of the element, and T represents the transposition. In addition, the viscous energy $\widetilde{D}$ of the viscous compressible fluid is expressed as follows:

$$\widetilde{D} = \iiint_{v_e} \frac{1}{2} \left\{ \overline{T}^T \right\} \Gamma dx\, dy\, dz. \quad (3)$$

Here, $\{\overline{T}\}$ is the stress vector due to viscosity, and Γ is the strain vector.

Furthermore, using the potential energy $\widetilde{V}$ and applying Lagrange's equation, we obtain an equation of motion for the air elements that accounts for air viscosity:

$$\frac{d}{dt} \frac{\partial \widetilde{T}}{\partial \{\dot{u}_e\}} - \frac{\partial \widetilde{T}}{\partial \{u_e\}} + \frac{\partial \widetilde{U}}{\partial \{u_e\}} - \frac{\partial \widetilde{V}}{\partial \{u_e\}} + \frac{\partial \widetilde{D}}{\partial \{\dot{u}_e\}} = 0. \quad (4)$$

Here, $\{u_e\}$ and $\{\dot{u}_e\}$ are the nodal displacement vector and nodal velocity vector, respectively. Substituting (1)–(3) into (4), the following equation is obtained:

$$-\omega^2 [M_e] \{u_e\} + [K_e] \{u_e\} + [C_e] \{\dot{u}_e\} = \{f_e\}. \quad (5)$$

Here, $[M_e]$, $[K_e]$, and $[C_e]$ are the element mass matrix, the element stiffness matrix, and the element damping matrix, respectively. Furthermore, assuming a periodic

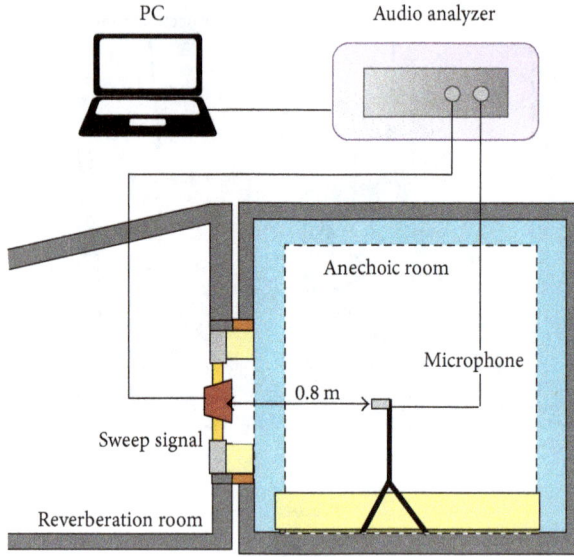

FIGURE 6: Sound pressure frequency measurement system.

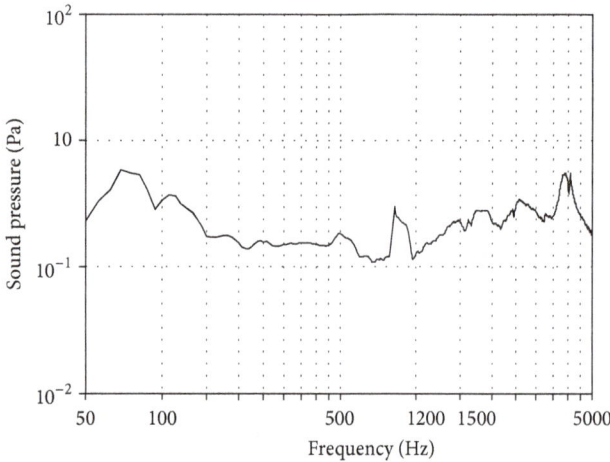

FIGURE 7: Comparison of measurement result and system characteristics of sound pressure frequency properties.

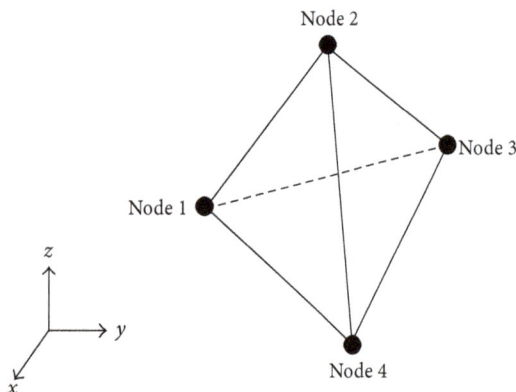

FIGURE 8: Tetrahedral element.

response with an angular frequency of ω and using $\{\dot{u}_e\} = j\omega\{u_e\}$, (5) can be expressed by the following equation:

$$-\omega^2 [M_e] \{u_e\} + [K_e] \{u_e\} + j\omega [C_e] \{u_e\} = \{f_e\}. \qquad (6)$$

Here, $-\omega^2 [M_e]\{u_e\}$ is the inertia term, $[K_e]\{u_e\}$ is the stiffness term, $j\omega[C_e]\{u_e\}$ is the viscosity term, and $\{f_e\}$ is the external force.

For the structural elements, we use the general formulation:

$$-\omega^2 [M_{se}] \{u_{se}\} + [K_{se}] \{u_{se}\} = \{f_{se}\}, \qquad (7)$$

where $\{u_{se}\}$, $\{f_{se}\}$, $[K_{se}]$, and $[M_{se}]$ are the structural elements' nodal displacement vector, nodal force vector, complex stiffness matrix, and mass matrix, respectively. For the structural elements, in addition to tetrahedral elements, we also used hexahedral isoparametric elements account for incompatible modes.

Superimposing these equations of motion on the elements in the places of interest and letting the displacements be the common unknown variables, the following equation of motion for the entire system, including structure and acoustics, is obtained:

$$-\omega^2 [M] \{u\} + [K] \{u\} + j\omega [C] \{u\} = \{f\}. \qquad (8)$$

By solving the equation of motion in (8) for displacement, all nodal displacements can be determined. Furthermore, the strain and pressure of each element can be calculated from the nodal displacements [13].

Therefore, we used solutions of (8) to carry out the coupled vibration and acoustic analysis described in this paper.

To verify the performance of the loudspeaker, we relied primarily on frequency response analysis. To calculate the pressure of the elements during the actual vibration and acoustic analysis, we divided up the frequency range to be analyzed and then found the nodal displacement $\{u\}$ corresponding to the sine wave excitation force $\{f\}$ at each frequency. By repeating these calculations, the nodal displacement and element pressure were calculated for the entire frequency range to be analyzed, yielding frequency response characteristic graphs for displacement and pressure.

4.2. Analytical Model. We created an analytical model with conditions substantially equivalent to the conditions in the reverberation room and anechoic room shown in Figure 5. As illustrated in Figure 9(b), the analytical model is composed of three parts: the sound-absorbing elements in the outer pink areas, the air elements in the inner green areas, and the cone loudspeaker setup in the airspace in the small reverberation room. Figure 9(a) shows the element model of the loudspeaker from the middle of Figure 9(b).

Within the loudspeaker elements, there is a narrow width of only 2×10^{-5} m between the voice coil and the magnetic circuit. In order to determine the behavior of the air around the voice coil in a way that accounts for the viscosity of the air, the fine air mesh shown in Figure 9(c) was created. For

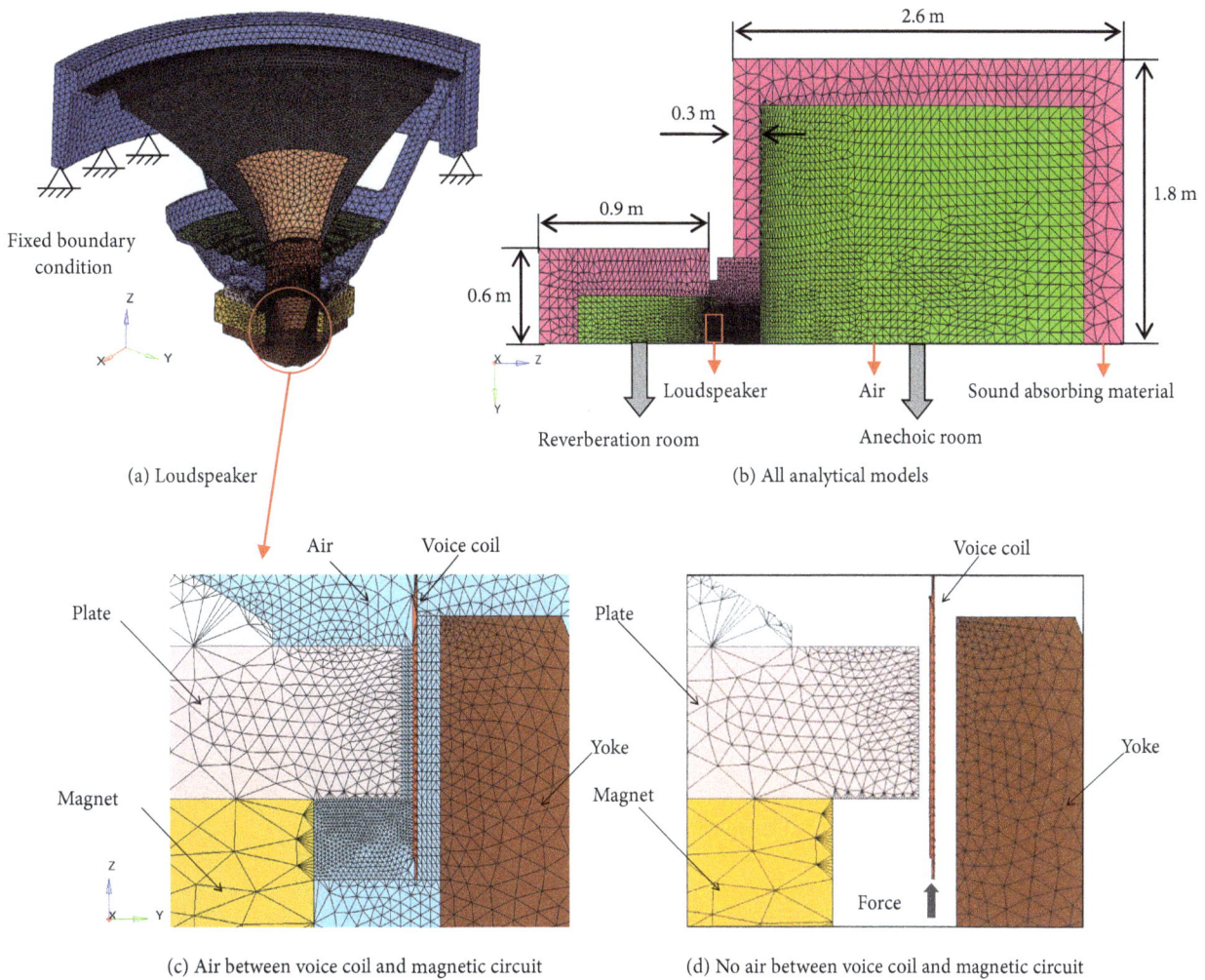

(a) Loudspeaker

(b) All analytical models

(c) Air between voice coil and magnetic circuit

(d) No air between voice coil and magnetic circuit

FIGURE 9: Analytical model.

comparison, Figure 9(d) shows the state when there is no air between the voice coil and magnetic circuit.

As a result, the structural part of the model contained 159,042 elements and 42,965 nodes, while the air part contained 704,942 elements and 143,585 nodes.

The sine wave excitation force F was applied to the entire circumference of the voice coil in the direction of the center axis of the loudspeaker, as shown in Figure 9(d).

With respect to the boundary conditions, the three degrees of freedom on the x-, y-, and z-axes were fixed at the underside of the frame as shown in Figure 9(a) and at the exterior air nodes as shown in Figure 9(b). In addition, to account for symmetry conditions, the displacement of nodes on the y-z and x-z planes was fixed in the x-axis and y-axis directions, respectively.

Since the loudspeaker is made of composite materials, the material properties used in our analysis were measured by the vibrating reed method [15] for the cone and whizzer cone, by the tensile test for the edge, and by FEM static analysis based on load displacement measurements for the spider. Other physical property values were obtained by consulting chronological scientific tables. Table 1 shows the

actual physical property data used in our analysis. Other physical property values included an effective air density of $1.2 \, \mathrm{kg/m^3}$, a bulk modulus of $1.4 \times 10^5 \, \mathrm{N/m^2}$, a coefficient of viscosity of $1.82 \times 10^{-5} \, \mathrm{N \cdot s/m^2}$, and a speed of sound of $340.0 \, \mathrm{m/s}$.

4.3. Verification of the Accuracy of Our Analysis of Cone Vibration Displacement. Figure 10 shows a comparison of the cone vibration displacement amplitudes from the measurement results and analysis results. The solid line represents the measurement results, and the dashed line represents the results calculated by considering air viscosity.

The results in Figure 10 show that the values from our analysis are in good agreement with the actual measured values. The fact that our calculation results yield roughly the same characteristics as the actual measured cone vibration displacement values suggests that highly accurate vibration analysis results can be obtained accounting for air viscosity.

4.4. Verification of the Accuracy of Our Analysis of Sound Pressure Frequency Properties. Figure 11 compares the results

TABLE 1: Material and geometric parameters.

Parts	Mass density (kg/m^3)	Young's modulus (N/m^2)	Poisson's ratio
Cone	3.1×10^2	1.16×10^9	0.3
Edge	8.0×10^2	1.39×10^8	0.3
Spider	6.60×10^2	8.90×10^7	0.3
Whizzer cone	5.70×10^2	2.00×10^9	0.3
Voice coil	1.87×10^3	1.29×10^{11}	0.34
Magnet	4.80×10^3	1.18×10^{11}	0.3
Plate	7.69×10^3	2.00×10^{11}	0.3
Yoke	7.94×10^3	2.00×10^{11}	0.3
Frame	9.10×10^2	1.66×10^9	0.3

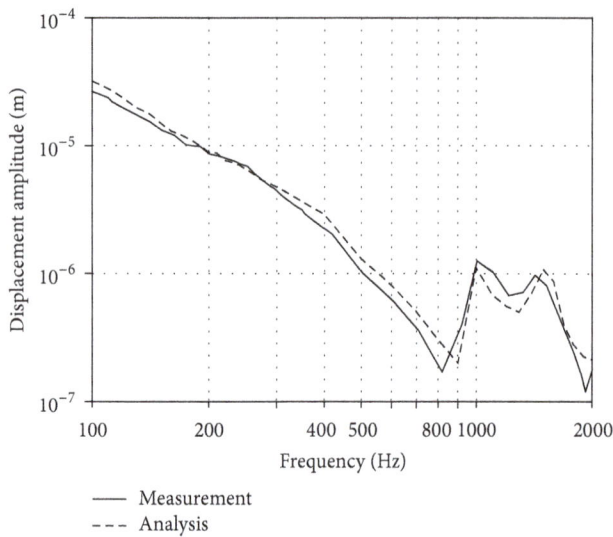

FIGURE 11: Comparison of measurement result and characteristics measurement system pressure frequency characteristics.

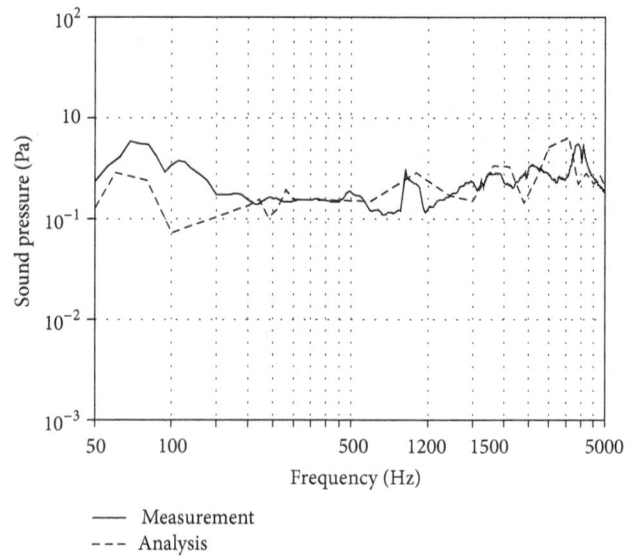

FIGURE 10: Comparison of the measurement result and analysis with viscosity of cone vibration displacement amplitude.

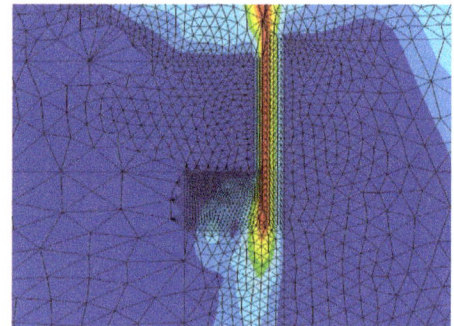

(a) Vibration mode around the voice coil

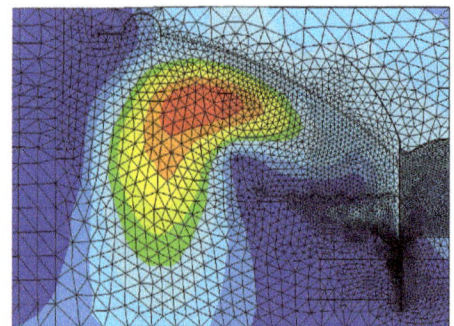

(b) Vibration mode around the cone

FIGURE 12: Effect by vibration mode of air with viscosity.

of our calculation of the sound pressure frequency response in the direction of the axis of the loudspeaker to the actual measurement results from the sweep signal excitation. The solid line represents the measurement results, and the dashed line represents the results calculated by considering air viscosity.

The changes and characteristics of the sound pressure frequency response from our analysis results are consistent with the traditional view in the literature that more accurate characteristics can be obtained by analyses that account for the viscous damping of the air [16].

The reason for this is that the narrow space between the voice coil and the magnetic circuit is so narrow that the air viscosity in the space affects the up-and-down motion of the voice coil. Our analysis is able to account for rotation as well as translational motion of air elements along the x-, y-, and z-axes. It is therefore able to represent the behavior of the air in the narrow passage, yielding calculation results that are closer to the actual measured values.

4.5. Impact of the Narrow Space between the Voice Coil and the Magnetic Circuit.
When designing a cone loudspeaker, it is very important to consider the behavior of the air between the circumference of the voice coil and the magnetic circuit.

Figure 12(a) shows the behavior of the air around the voice coil when air viscosity is taken into account. Red indicates areas of large displacement and blue indicates areas of small

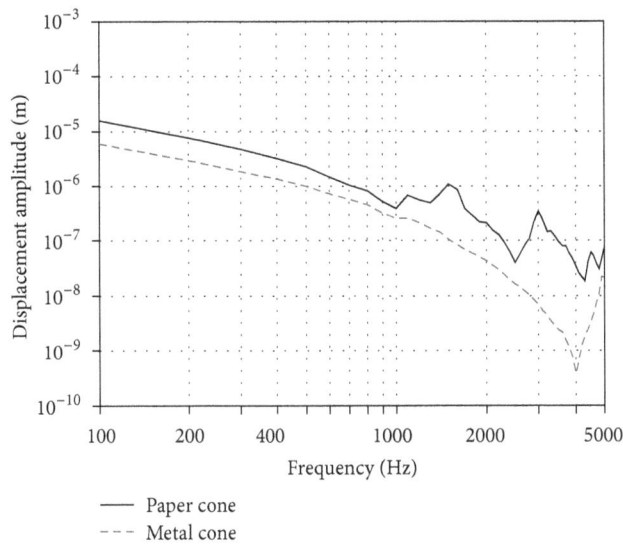

FIGURE 13: Comparison vibration characteristic of metal cone and paper cone.

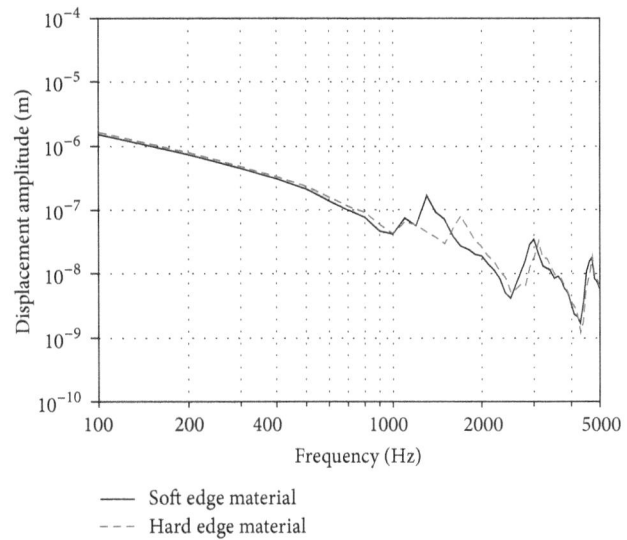

FIGURE 14: Comparison vibration characteristic of soft edge and hard edge material.

displacement. When air viscosity is taken into account, the differences in air pressure become apparent, and, owing to the natural tendency of air to flow from areas of high pressure to areas of low pressure, the direction of the air flow from the vibration of the voice coil becomes apparent. Figure 12(b) shows the behavior of the air around the cone. The figure clearly shows the direction of the air flow around the cone due to the behavior of the air around the voice coil.

The analysis results shown in Figure 12 demonstrate the potential for consciously designing the fine-grained characteristics of a loudspeaker by adjusting parameters such as the viscous resistance of the narrow acoustic pathway between the voice coil and magnetic circuit.

Using this analysis method, which accounts for air viscosity, it is possible to visualize the movement of air and capture the behavior of the air around the voice coil by solving the equation of motion with the nodal displacements as unknown quantities.

4.6. Impact of the Cone on the Acoustic Characteristics of the Loudspeaker. For purposes of comparison, we investigated cones made of both metal and paper-based materials. When metal is used, the material is highly uniform and exhibits greater flexural rigidity at the same thickness. On the other hand, when paper-based materials are used, the cone is easier to manufacture, but its vibration characteristics are likely to be more complicated because its flexural rigidity is lower than that of metal and the random arrangement of the pulp fibers renders the material nonuniform.

Figure 13 shows the cone displacement responses calculated by this method, accounting for air viscosity. The solid and dashed lines show the displacement responses for the metal and paper cones, respectively.

The figure shows that the metal cone exhibits piston motion up to 5 kHz, with virtually no peaks and dips in

displacement. In the case of the paper cone, more dips and peaks are observed, because the cone is softer and more prone to split vibration, in which the motion of the voice coil is not transmitted to the entire cone.

4.7. Impact of the Edge on the Acoustic Characteristics of the Loudspeaker. We compared the performance of an edge made of a soft material with Young's modulus of $6.95 \times 10^7 \, N/m^2$ and an edge made of a hard material with Young's modulus of $2.78 \times 10^8 \, N/m^2$. Figure 14 shows the vibration characteristics calculated by this method, taking into account air viscosity. The solid and dashed lines show the displacement response for the soft and hard materials, respectively.

The results in the figure show an initial peak at 1.3 kHz in the case of the soft edge material and an initial peak at 1.7 kHz in the case of the hard edge material. In other words, we used our analysis method to perform a coupled analysis of vibration and acoustics in which only the edge material was changed not the shape of the cone. These results demonstrate the potential for consciously designing the fine-grained characteristics of a loudspeaker, such as the first-peak position.

5. Discussion

In the previous sections of this paper, we took into account the viscosity of the air, verified the accuracy of the coupled analysis of vibration and acoustics of the loudspeaker, and saw the impact of the material of the cone and edge on the acoustic characteristics of the loudspeaker. Extending this line of research, a significant remaining challenge is to design the appropriate cone and edge characteristics to obtain the desired acoustic characteristics in the loudspeaker.

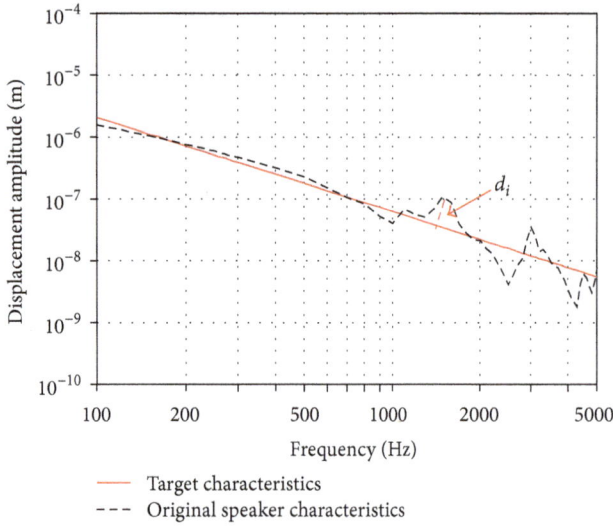

FIGURE 15: Cone vibration displacement amplitude standard analysis result and objective function.

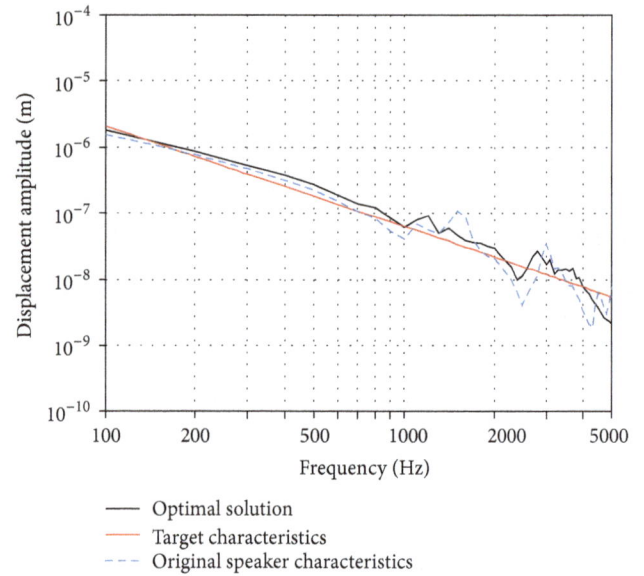

FIGURE 16: Analysis result of optimal solution.

To address this challenge, in this section, we consider how to optimize the cone and edge, expressing the problem mathematically as follows:

$$\text{Find} \quad x = \left[\rho_c, E_c, \rho_e, E_e\right]^T$$

$$\min \quad W = f(x) = \sqrt{\sum_{i=1}^{n} d_i^2} \quad (i = 1, 2, \ldots, n)$$

$$\text{S.t.} \quad 100 \left[\text{kg/m}^3\right] \leq \rho_c \leq 1200 \left[\text{kg/m}^3\right] \tag{9}$$

$$7.3 \times 10^8 \left[\text{N/m}^2\right] \leq E_c \leq 70 \times 10^8 \left[\text{N/m}^2\right]$$

$$300 \left[\text{kg/m}^3\right] \leq \rho_e \leq 1300 \left[\text{kg/m}^3\right]$$

$$3.2 \times 10^7 \left[\text{N/m}^2\right] \leq E_e \leq 20 \times 10^7 \left[\text{N/m}^2\right].$$

Here, ρ_c is the density of the cone, E_c is Young's modulus of the cone, ρ_e is the density of the edge, E_e is Young's modulus of the edge, and $W = f(x)$ is the objective function of optimization. In Figure 15, the solid line represents the target characteristic value, the dashed line (which was previously shown in Figure 10) represents the cone vibration displacement characteristics from our analysis, and d_i $(i = 1, 2, \ldots, n)$ represents the distance from the displacement characteristic to the target characteristic value at each frequency from 1 to 5 kHz. The number of sampling points was set to $n = 50$.

Although paper-based materials, metal materials, and other materials have been used to construct the sound-radiating cones in loudspeakers, paper-based materials are used most often. For that reason, the constraints imposed on the ranges of the cone's density and Young's modulus in (9) were set in the ranges for paper-based materials. Similarly,

the range of values established for the edge in (9) reflects the fact that the edge is typically constructed using cotton fabric and rubber.

We used the response surface methodology to optimize the physical properties of the cone and edge by carrying out the following calculation steps. First, we generated sample data in order to create the response surface for the optimization. Next, using the sample data, we carried out vibration analysis of the loudspeaker by changing the physical properties of each of its parts and, based on the results of the analysis, extracted the objective function values for the optimization calculation. Finally, we established a one-to-one correspondence between sample data and the characteristic values, created an interpolation approximation formula by the response surface method, and then used the approximation formula to find the optimum solution by performing the optimization calculation [17, 18].

Table 2 shows the sample data that was generated using the design variables and their upper and lower limit values, as well as the objective function value d, which is the result obtained from the coupled analysis of vibration and acoustics in accordance with the sample data.

We used these results to create the response surface and perform the optimization, the results of which are shown in Table 3.

Figure 16 shows the results of the vibration analysis using the optimal solution that was obtained. The black solid line represents the analysis result for the optimal solution and the red solid line represents the target characteristic values. The dashed line represents analysis results for the original loudspeaker shown in Figure 10. The results obtained for the optimal solution closely match the target characteristic values.

The optimal solution exhibits flatter characteristics compared to the original cone vibration displacement characteristics, which exhibited peaks at 1.5 and 3 kHz and a dip at

TABLE 2: Sample data and analysis results.

Number	Cone mass density (kg/m³)	Cone Young's modulus (10^8 N/m²)	Edge mass density (kg/m³)	Edge Young's modulus (10^7 N/m²)	Analysis results d
1	100	7.3	300	3.2	0.35
2	100	7.3	300	12	0.33
3	100	7.3	300	20	0.31
4	650	7.3	800	3.2	0.51
5	650	7.3	800	12	0.60
6	650	7.3	800	20	0.56
7	1200	7.3	1300	3.2	0.67
8	1200	7.3	1300	12	0.61
9	1200	7.3	1300	20	0.68
10	100	39	800	3.2	1.06
11	100	39	800	12	1.03
12	100	39	800	20	1.02
13	650	39	1300	3.2	1.29
14	650	39	1300	12	1.25
15	650	39	1300	20	1.27
16	1200	39	300	3.2	1.46
17	1200	39	300	12	1.43
18	1200	39	300	20	1.44
19	100	70	1300	3.2	0.98
20	100	70	1300	12	1.01
21	100	70	1300	20	1.00
22	650	70	300	3.2	1.55
23	650	70	300	12	1.56
24	650	70	300	20	1.55
25	1200	70	800	3.2	1.74
26	1200	70	800	12	1.77
27	1200	70	800	20	1.79

TABLE 3: Design variables of optimization.

	Initial value	Optimal value
ρ_c (kg/m³)	650	100
E_c (N/m²)	39×10^8	28×10^8
ρ_e (kg/m³)	800	800
E_e (N/m²)	12×10^7	20×10^7

2.5 kHz. The characteristics also appear flatter over the entire frequency range. The result of the objective function value d for the optimal solution was $d = 0.26$, which represents an improvement of approximately 35% over the original loudspeaker, for which the result was $d = 0.40$.

In the optimal solution, the density of the cone was 100 kg/m³, and Young's modulus was 28×10^8 N/m². These properties define a cone that is lighter in weight and vibrates more easily but whose stiffness enables suppression of split vibration in the nonaxisymmetric mode. This suggests that cones should be constructed from light but rigid materials.

The results for the edge were a density of 800 kg/m³ and Young's modulus of 20×10^7 N/m². Since the edge supports the cone, this suggests that edges should be constructed of materials with high rigidity.

In addition, the vibration mode of the cone was examined. The vibration modes of the cones are shown in Figure 17. Red and blue indicate areas of large and small displacement, respectively. Figure 17(a) shows the vibration mode of a standard loudspeaker cone at 1.5 kHz, and Figure 17(b) shows the vibration displacement mode of an optimal cone at 1.5 kHz. In the case of the standard loudspeaker, it can be seen that the displacement is smaller at the center of the cone and greater at the periphery (the part attached to the edge), whereas the optimal cone moves with the same amount of displacement.

By changing Young's modulus and density of the cone and edge, we were able to reduce the peaks that occur due to split vibration of the cone in the middle- and high-frequency regions above 1 kHz and achieve flatter frequency characteristics.

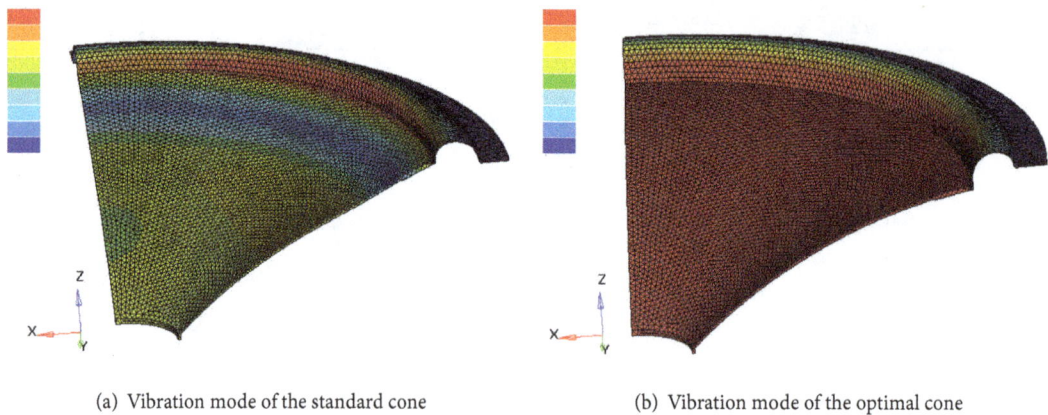

(a) Vibration mode of the standard cone   (b) Vibration mode of the optimal cone

FIGURE 17: Analysis result of optimal solution: cone A at 1500 Hz.

6. Conclusion

This study focused on the cone, the edge, and the behavior of air around the voice coil, which are important elements in the design of cone loudspeakers, and evaluated their effects on the acoustic characteristics of a loudspeaker. We performed a coupled analysis of the vibration and acoustics of a cone loudspeaker that accounted for the impact of air viscosity, which has received little attention in traditional acoustic analyses, and then sought optimal materials for the cone and edge. The following results were obtained:

(1) We took measurements of the vibration characteristics and sound pressure frequency response of a cone loudspeaker in order to determine its acoustic characteristics.

(2) We performed a coupled analysis of the vibration and acoustics of the loudspeaker that accounted for air viscosity and compared the vibration characteristics and sound pressure frequency response to the actual measured values. By accounting for air viscosity, the analysis was able to yield high accuracy.

(3) The coupled analysis of vibration and acoustics revealed the vibration characteristics of the cone, which are influenced by the characteristics of the cone and edge. Furthermore, with respect to the behavior of the air around the voice coil in the loudspeaker, the results of our analysis, which accounted for air viscosity, enabled us to clearly see the direction of air movement; this is not possible with traditional analyses. This led to an improvement in the accuracy of the analysis.

(4) We used the response surface methodology to optimize the design of Young's modulus and density of the cone and edge and obtained the optimum physical properties of the cone and edge in order to design a loudspeaker with the desired acoustic characteristics. The optimal solution that was obtained reduced the peaks and dips caused by split vibration of the cone and achieved flatter vibration characteristics.

With respect to future challenges, we hope to investigate the analysis of loudspeakers of complex shapes, beyond typical cone loudspeakers. In addition, we would like to alter the shape of the loudspeaker components in ways that will lead to an optimum design to achieve flatter sound pressure frequency properties.

References

[1] W. N. Brown, "Theory of conical sound radiators," *Journal of the Acoustical Society of America*, vol. 13, no. 1, pp. 20–22, 1941.

[2] P. G. Bordoni, "The conical sound source," *The Journal of the Acoustical Society of America*, vol. 28, no. 2, pp. 123–126, 1945.

[3] T. Nimura, E. Matsui, K. Shibayama, and K. Kido, "Study on the cone type dynamic loudspeakers," *The Journal of the Acoustical Society of Japan*, vol. 7, no. 2, pp. 16–28, 1952.

[4] Y. Kagawa, "Natural frequency and vibration modes of curved cone extensional vibration," *Tohoku University Dentsu Discourse Record*, vol. 31, no. 3, 1962.

[5] F. J. Frankfort, *Vibration and sound radiation of loudspeaker cones [thesis]*, Philips Research Reports Supplements no 2, 1975.

[6] T. Ueno, K. Takahashi, K. Ichida, and S. Ishii, "The vibration analysis of a cone loudspeaker by the finite element method," *The Journal of the Acoustical Society of Japan*, vol. 34, no. 8, pp. 470–477, 1978.

[7] K. Suzuki and I. Nomoto, "Computerized analysis and observation of the vibration modes of a loudspeaker cone," *Journal of the Audio Engineering Society*, vol. 30, no. 3, pp. 98–106, 1982.

[8] A. J. M. Kaizer, "Modeling of the nonlinear response of an electrodynamic loudspeaker by a volterra series expansion," *AES: Journal of the Audio Engineering Society*, vol. 35, no. 6, pp. 421–433, 1987.

[9] A. J. M. Kaizer, "Calculation of the sound radiation of a nonrigid loudspeaker diaphragm using the finite-element method," *Journal of the Audio Engineering Society*, vol. 36, no. 7-8, pp. 539–551, 1988.

[10] E. B. Skrodzka and A. P. Sęk, "Comparison of modal parameters of loudspeakers in different working conditions," *Applied Acoustics*, vol. 60, no. 3, pp. 267–277, 2000.

[11] R. M. Aarts, "Optimally sensitive and efficient compact loudspeakers," *Journal of the Acoustical Society of America*, vol. 119, no. 2, pp. 890–896, 2006.

[12] N. Kyouno, T. Usagawa, T. Yamabuchi, and Y. Kagawa, "Acoustic response analysis of a cone-type loudspeaker by the finite element method," *The Journal of the Acoustical Society of Japan*, vol. 61, no. 6, pp. 312–319, 2005.

[13] M. Sasajima, T. Yamaguchi, and A. Hara, "Acoustic analysis using finite element method considering effects of damping caused by air viscosity in audio equipment," *Applied Mechanics and Materials*, vol. 36, pp. 282–286, 2010.

[14] T. Yamamoto, *LoudSpeakers System*, Radio Technology, Tokyo, Japan, 1977.

[15] K. Hashimoto, M. Sakane, M. Ohnami, and T. Yoshida, "Development of vibrating reed machine for measuring Young's modulus of thin films," *Journal of the Society of Materials Science, Japan*, vol. 44, no. 507, pp. 1456–1463, 1995.

[16] T. Saeki, *LoudSpeakers & Enclosure*, SeiBunDo ShinKoSha, Tokyo, Japan, 1999.

[17] X. Zhao, "Some new problems and measure with structural optimization software development," in *Proceedings of the Conference on Computational Engineering and Science (ICCS '07)*, vol. 12, no. 1, pp. 129–132, Beijing, China, 2007.

[18] X. Zhao, Y. Hu, and I. Hagiwara, "Optimal design for crash characteristics of cylindrical thin-walled structure using origami engineering," *The Japan Society of Mechanical Engineers*, vol. 76, no. 761, pp. 10–17, 2010.

14

Acoustical Measurement and Biot Model for Coral Reef Detection and Quantification

Henry M. Manik

Department of Marine Science and Technology, Faculty of Fisheries and Marine Sciences, Bogor Agricultural University, Kampus IPB Darmaga, Bogor 16680, Indonesia

Correspondence should be addressed to Henry M. Manik; henrymanik@ipb.ac.id

Academic Editor: Marc Asselineau

Coral reefs are coastal resources and very useful for marine ecosystems. Nowadays, the existence of coral reefs is seriously threatened due to the activities of blast fishing, coral mining, marine sedimentation, pollution, and global climate change. To determine the existence of coral reefs, it is necessary to study them comprehensively. One method to study a coral reef by using a propagation of sound waves is proposed. In this research, the measurement of reflection coefficient, transmission coefficient, acoustic backscattering, hardness, and roughness of coral reefs has been conducted using acoustic instruments and numerical modeling using Biot theory. The results showed that the quantification of the acoustic backscatter can classify the type of coral reef.

1. Introduction

As an archipelago country, Indonesia has many types of coral reefs. Coral reefs consist of geological structures of calcium carbonate built over time by tiny living organisms. These structures are found in the marine waters which contain nutrients. Coral reefs had been used for marine tourism, shoreline protection, and fishing ground to the local communities. Coral reefs, which are called rainforests of the sea, are the most biologically diverse ecosystems. They support variety of marine live forms, provide pharmaceutical materials, and generate income from tourism [1].

Coral reefs are valuable marine resources; however, they are in the vulnerable condition [2]. In spite of their advantages and in spite of the fact that humans continue to benefit from coral reefs, the marine ecosystem has been subjected to serious anthropogenic threats. Nonsustainable activities and developments have inevitably caused degradation to the coral reef ecosystem. Blast fishing activities, mining of coral reef for building materials, and other threats like sedimentation and pollution in coastal areas are diminishing the ecosystem.

The greatest threats to coral reefs are rising seawater temperature and ocean acidification. With the recent global ocean warming, coral bleaching and mortality have become more frequent. Human induced disturbances, storm, and natural predators also affected the damage of coral reef [2].

However, the lack of accurate and comprehensive coral reef database has always been one of the limiting factors in their conservation and management efforts. The marine scientists and the coastal resource managers need to know the distribution and the status of these coral reefs.

State of the Art and Scientific Contribution. Detection and mapping of coral reef ecosystem should be at a scale that is adequate to detail information on the distribution of the major coral reef substrates. Unfortunately, the current coral reef research methodologies are still unable to fulfill this need. This is mainly due to the high cost involved and inherent shortcomings of the remote sensing technologies. The conventional coral reef research methods use a diver by applying the line transect, quadrate plot, belt transect, and manta tow survey [3]. All of these methods are used to derive both quantitative and qualitative data of coral reefs on a small scale surveyed area. In order to produce contiguous, broad scale resource map, marine scientists have explored the underwater videography and LIDAR survey using airborne and satellite remote sensing [4].

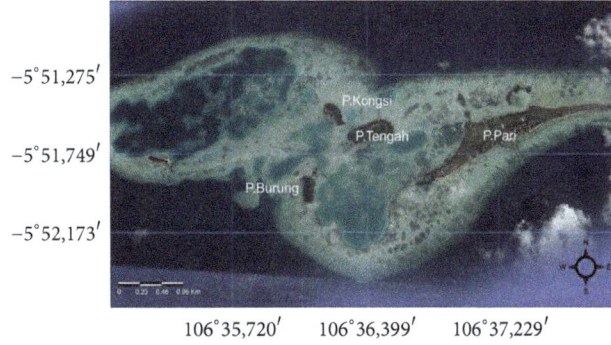

FIGURE 1: Research location in Seribu Island waters.

These methods yield different degrees of resolution and accuracy depending on the sensor type, cost, time, ground truthing, and postprocessing. LIDAR airborne and satellite remote sensing derived images. Several researchers mapped the coral reef using satellite remote sensing [5]. This remote sensing technology was originally designed for terrestrial application. A number of limitations become obvious, when remote sensing is applied to map underwater marine resources. The electromagnetic wave used in satellite remote sensing was more absorbed in ocean environment.

To overcome this problem, the application of underwater acoustic technologies is used to detect and quantify coral reef. Acoustic technologies offer a promising method for underwater remote sensing techniques [6–8]. The sound waves can propagate deeper compared with electromagnetic wave. This method allows for large area coverage in greater details within a shorter time at a relatively low cost. Acoustic detection using side scan sonar and multibeam echosounder has been successful for water column and seabed imaging [9, 10]. Bottom backscattering strength was computed for acoustic seabed classification [11, 12].

The objective of this research is to develop a methodology for effective ecosystem monitoring and mapping of tropical marine coral reefs using underwater acoustic technology in Seribu Island waters, Indonesia.

2. Research Method

The research was conducted at Ocean Acoustics and Instrumentation Laboratory, Department of Marine Science and Technology, Faculty of Fisheries and Marine Sciences, Bogor Agricultural University, and Seribu Island waters, Jakarta, Indonesia (Figure 1).

2.1. Flow Diagram of the Research. The flow diagram of this research in the laboratory is shown in Figure 2. Research output is to develop and apply the methodology of high resolution underwater acoustic technology, calibrating the instrument and measuring acoustic backscattering from coral reef.

2.2. Hydroacoustic Data Acquisition, Processing, and Analysis. Based on underwater acoustic theory, the ability of this method is applied to discriminate coral reef types, to map coral substrates, and to measure acoustic reflectivity of coral reef at Seribu Island waters. The result encourage where the acoustic instrument was able to pick up distinct echo when ensonifying over coral reef type. All ocean surveys were conducted on the research vessel. The ship is outfitted with a suite of underwater acoustics instrument. Matlab software was applied and used in postprocessing and visualization to overlay pings and backscatter returns in both space and time.

Raw data was corrected for ambient noise and interference artifacts. A subset of pings were identified in the acoustic data and concurrent pings are selected in the acoustic data records. Coral reef detection algorithms were developed and used in Matlab to delineate bottom fish and seabed backscatter.

Consecutive echoes representing individual coral are similarly delineated in the hydroacoustic image. In the case of dense coral, tracks are selected on the outer limits of the coral. The average backscattering strength is calculated for the series of echoes. Figure 3 shows a simplified block diagram for coral reef backscattering measurement by quantitative echo sounder.

The backscattered pressure signal from the bottom received by the transducer (Figure 3), P_{RB} [12], is

$$P_{RB}^2 = P_o r^{-2} \exp(-4\alpha r) \Phi S_S, \tag{1}$$

where P_o is the source pressure level.

The bottom echo signal is amplified by the amplifier to give

$$E_{RB} = P_{RB} M G_R, \tag{2}$$

where E_{RB} is the echo amplitude at the preamplifier output, M is the receiving sensitivity of the transducer, and G_R is the preamplifier gain.

The echo amplitude, E_{RB}, is shown from (1) and (2) as

$$E_{RB}^2 = K_{TR}^2 r^{-2} \exp(-4\alpha r) \Phi S_S, \tag{3}$$

where $K_{TR} = P_o M G_R$ is the transmitting receiving coefficient.

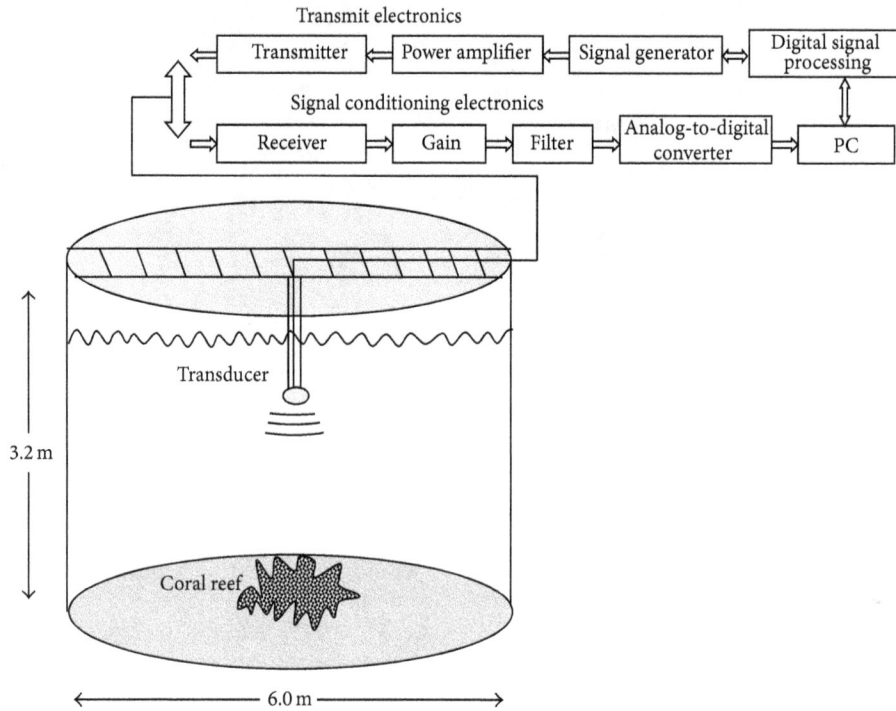

FIGURE 2: Flow diagram of underwater acoustic instrument for measuring coral reef.

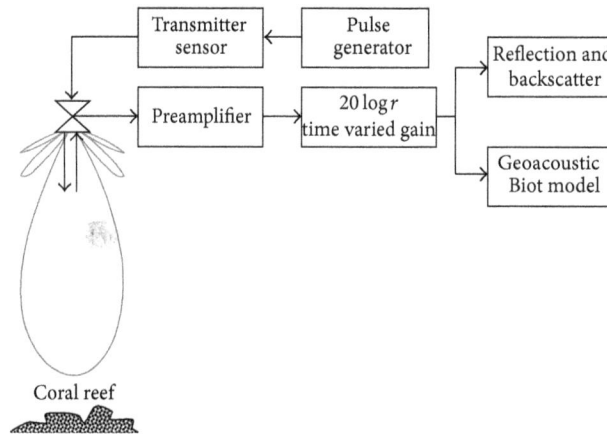

FIGURE 3: Methodology of acoustical measurement and Biot model.

The time varied gain (TVG) amplifier output of E_{RB} corrected for absorption and spreading losses, E_{TB}, is

$$E_{\mathrm{TB}} = G_{\mathrm{TM}} r \exp\left(2\alpha r\right) E_{\mathrm{RB}},$$
$$E_{\mathrm{TB}}^2 = \left(K_{\mathrm{TR}} G_{\mathrm{TM}}\right)^2 \Phi S_{\mathrm{S}}. \tag{4}$$

The raw SV value of the bottom echo, S_{VB}, is

$$S_{\mathrm{VB}} = \frac{E_{\mathrm{TB}}^2}{K_M^2}, \tag{5}$$

where K_M is the multiple echo coefficient and is given by

$$K_M^2 = \left(K_{\mathrm{TR}} G_{\mathrm{TM}}\right)^2 \Psi\left(\frac{c\tau}{2}\right), \tag{6}$$

where

$$\Psi = \int_0^{2\pi} \int_0^{\pi/2} b^2 \sin\theta \, d\theta \, d\phi. \tag{7}$$

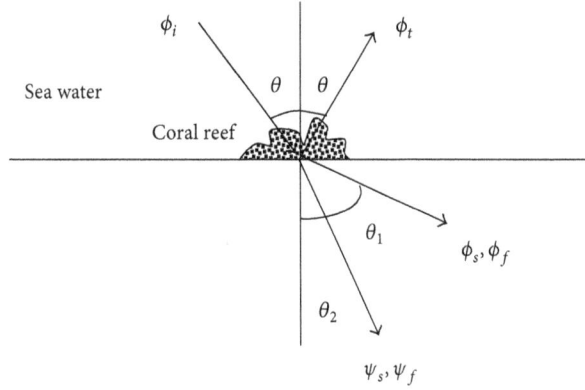

FIGURE 4: Incident, reflected, and transmitted wave in seawater and coral reef.

The above Ψ is the equivalent beam angle of the volume scattering.

Substitution of E_{TB} and K_M into (6) yields

$$S_{VB} = \frac{S_S \Phi}{\Psi (c\tau/2)}. \tag{8}$$

This equation is called ring surface scattering model.

2.3. Geoacoustic Biot Model. Coral reefs are porous, with the possibility that the fluid and granular phases will vibrate differently in response to acoustic excitation. Poroelastic theory or Biot theory treats both porosity and elasticity [13, 14].

The incident and reflected waves in the seawater and coral will have displacement potential (Figure 4):

$$\phi_i = A_i \exp\left[j\left(\omega t - k_w \cos\theta \cdot z - k_w \sin\theta \cdot x\right)\right],$$
$$\phi_r = A_r \exp\left[j\left(\omega t + k_w \cos\theta \cdot z - k_w \sin\theta \cdot x\right)\right], \tag{9}$$

where $k_w = \omega/c_w$. ω is the angular frequency, and c_w is the sound velocity in the seawater.

Biot developed a pair of coupled differential equations to describe acoustic wave propagation in an isotropic saturated porous medium with permeability κ, pore fluid viscosity η, and bulk density ρ_f:

$$\nabla^2 (H\varepsilon - C\varsigma) = \frac{\partial^2}{\partial t^2}\left(\rho\varepsilon - \rho_f\varsigma\right),$$
$$\nabla^2 (C\varepsilon - M\varsigma) = \frac{\partial^2}{\partial t^2}\left(\rho_f\varepsilon - m\varsigma\right) - \frac{F\eta}{\kappa}\frac{\partial\varsigma}{\partial t}, \tag{10}$$

where ς is the incremental volume of fluid which enters or leaves the frame and ε is the volumetric strain of the frame. Stoll developed the theory for Biot moduli in terms of measurable sediment properties:

$$H = \frac{(K_r - K_b)^2}{D - K_b} + K_b + \frac{4}{3}\mu,$$
$$C = \frac{K_r (K_r - K_b)^2}{D - K_b},$$
$$M = \frac{K_r^2}{D - K_b}, \tag{11}$$
$$D = K_r \left(1 + n\left(\frac{K_r}{K_f} - 1\right)\right),$$

where K_r is the grain bulk modulus, K_f is the modulus of the pore fluid, and η is porosity. The frame bulk modulus was calculated by

$$K_b = K_{br} + jK_{bi} = K_{br}\left(1 + j\frac{\delta_b}{\pi}\right) \tag{12}$$

and the frame shear modulus was calculated by

$$\mu = \mu_r + j\mu_i = \mu_r\left(1 + j\frac{\delta_s}{\pi}\right), \tag{13}$$

where δ_b and δ_s are the bulk and shear logarithmic decrement.

The boundary conditions are required at a seawater-coral reef interface.

(1) For continuity of fluid movement

$$\frac{\partial\phi_i}{\partial z} + \frac{\partial\phi_r}{\partial z} = \frac{\partial\phi_s}{\partial z} + \frac{\partial\psi_s}{\partial x} - \frac{\partial\phi_f}{\partial z} - \frac{\partial\psi_f}{\partial x}. \tag{14}$$

(2) For equilibrium of normal traction

$$H\left(\frac{\partial^2\phi_i}{\partial x^2} + \frac{\partial^2\phi_s}{\partial z^2}\right) - 2\mu\left(\frac{\partial^2\phi_s}{\partial x^2} - \frac{\partial^2\psi_s}{\partial x\partial z}\right)$$
$$- C\left(\frac{\partial^2\phi_f}{\partial x^2} + \frac{\partial^2\phi_t}{\partial z^2}\right) = \rho_f\left(\frac{\partial^2\phi_i}{\partial t^2} + \frac{\partial^2\phi_r}{\partial t^2}\right). \tag{15}$$

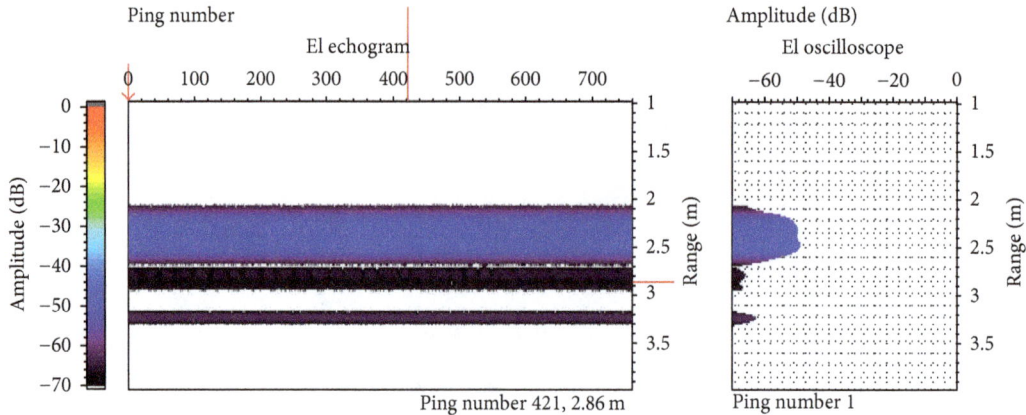

FIGURE 5: Acoustic calibration using sphere ball.

For equilibrium of fluid pressure

$$M\left(\frac{\partial^2 \phi_f}{\partial x^2} + \frac{\partial^2 \phi_f}{\partial z^2}\right) - C\left(\frac{\partial^2 \phi_s}{\partial x^2} + \frac{\partial^2 \phi_s}{\partial z^2}\right)$$

$$= -\rho_f\left(\frac{\partial^2 \phi_i}{\partial t^2} + \frac{\partial^2 \phi_r}{\partial t^2}\right), \tag{16}$$

where $M = K_t^2/(D - K_b)$.

(3) For equilibrium of tangential traction

$$2\mu\frac{\partial^2 \phi_s}{\partial x \partial z} - \mu\left(\frac{\partial^2 \psi_s}{\partial z^2} - \frac{\partial^2 \psi_s}{\partial x^2}\right) = 0. \tag{17}$$

Four linear complex equations can be obtained for acoustic reflection measurement:

$$\begin{pmatrix} c_{11} & c_{12} & c_{13} & c_{14} \\ c_{21} & c_{22} & c_{23} & c_{24} \\ c_{31} & c_{32} & c_{33} & c_{34} \\ c_{41} & c_{42} & c_{43} & c_{44} \end{pmatrix} \begin{pmatrix} A_1 \\ A_2 \\ A_3 \\ A_4 \end{pmatrix} = \begin{pmatrix} Y_1 \\ Y_2 \\ Y_3 \\ Y_4 \end{pmatrix}, \tag{18}$$

where the components of $\{c\}$ and $\{Y\}$ are given by the physical parameters of seawater and coral. A_r is the complex amplitude of the reflected wave. The reflection coefficient can be computed from this equation. Physical parameters for coral reef models were shown in Table 1.

3. Results and Analysis

In order to obtain accurate and precise quantitative measurements of coral reef resources, proper calibration of an acoustic system is necessary when using acoustic techniques such as echo detection and echo integration. Figure 5 shows the acoustic calibration using sphere ball method. The blue color shows the sphere detected on −42.5 dB and this value agreed

TABLE 1: Physical parameters for coral reef models.

Physical parameters	Soft coral	Hard coral
Grain diameter, ϕ	5.2	1.5
d (mm)	0.022	0.453
Grain density, ρ_r (kg/m^3)	2,675	2,675
Bulk modulus, K_r (Pa)	3.6×10^{10}	3.6×10^{10}
Density of pore fluid, ρ_f (kg/m^3)	1,000	1,000
Bulk modulus of pore fluid, K_f (Pa)	2.25×10^9	2.25×10^9
Viscosity, η (Pa-s)	1.00×10^{-3}	1.00×10^{-3}
Porosity, β	0.82	0.33
Permeability, κ (m^2)	4.45×10^{-12}	8.06×10^{-11}
Pore size, a (m)	1.30×10^{-5}	7.35×10^{-5}
Structure factor, α	1.30	1.30
Bulk modulus of frame, K_{br} (Pa)	1.45×10^7	5.8×10^7
Bulk logarithmic decrement, δ	0.15	0.15
Shear modulus, μ_r (Pa)	4.45×10^6	9.45×10^6
Shear logarithmic decrement, δ_s	0.16	0.16

with the theoretical value obtained by the manufacturer. The result of calibration was shown in Table 2.

Acoustic data were collected using four frequencies (10, 50, 100, and 200 kHz) and pulse transmission was simultaneous at all frequencies. All transducers were suspended from a small boat and data were collected while drifting slowly over 15–40 m deep water. During the 6 h of data collection, the boat drifted approximately 3 km. Figure 6 shows the acoustic image of soft coral and hard coral. The amplitude intensity for soft coral ranged from −35.0 to −30.0 dB and that for hard coral ranged from −28.0 to −10.0 dB.

The hardness and roughness of coral reefs were shown in Figure 7. The hardness of soft coral ranged from −33.0 dB to −15.0 dB (Figure 7(a)) and the hardness of hard coral was about −22.0 dB to −15.0 dB. The roughness of soft coral

TABLE 2: Configuration of the four acoustic frequencies.

Frequency (kHz)	Type of beam	Half-power beam width (°)	Source level at 1 m (dB re 1 μPa)	Acoustic power (W)	Pulse length (ms)	Noise level (dB) at 1 m depth	Absorption (dB/km)
10	Single	7.0 × 6.9	219.8	100	0.512	−142	0.45
50	Single	7.2 × 7.3	216.9	150	0.512	−141	4.5
100	Single	6.5 × 6.5	217.5	80	0.512	−138	12.4
200	Split	5.4 × 5.4	218.5	60	0.500	−122	57.5

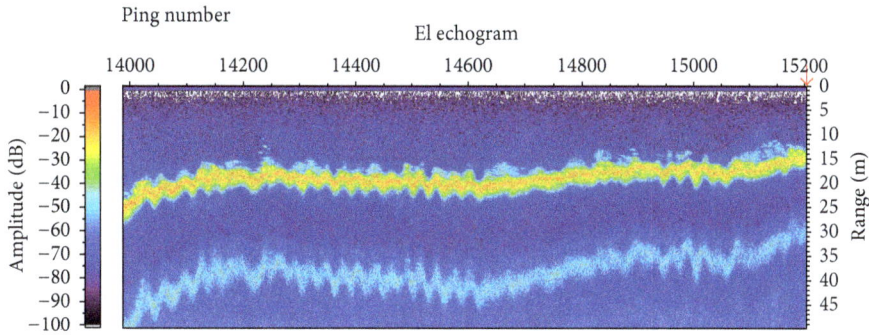

(a)

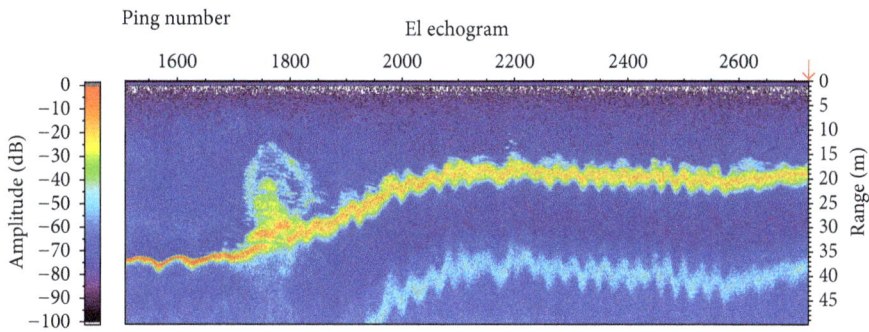

(b)

FIGURE 6: Acoustic image of soft coral reef (a) and hard coral reef (b).

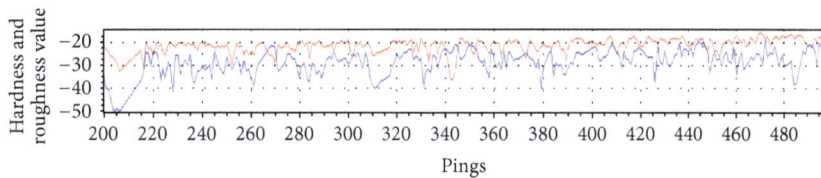

(a)

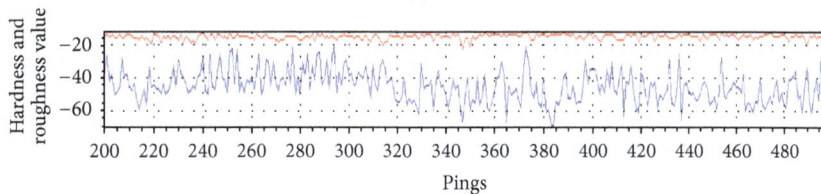

(b)

FIGURE 7: Hardness and roughness of coral reef.

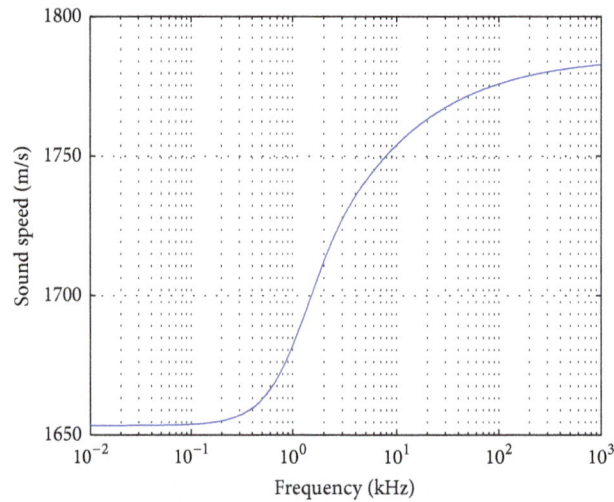

FIGURE 8: Sound speed as a function of frequency calculated using the Biot model.

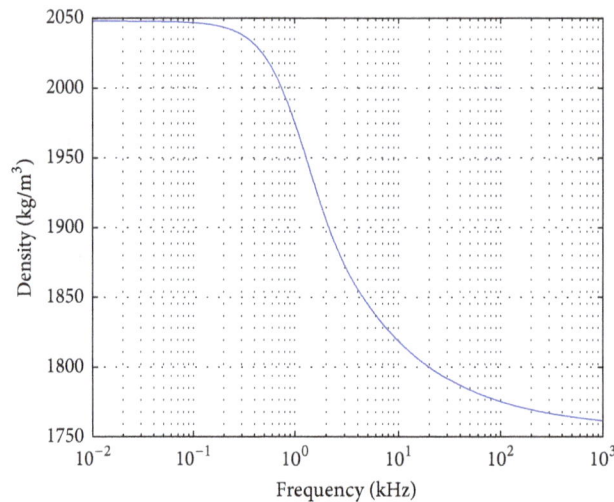

FIGURE 9: Density as a function of frequency calculated using the Biot model.

ranged from −50.0 to −20.0 dB and that of hard coral ranged from −65.0 to −22.0 dB. The harder coral had more energy due to higher acoustic impedance compared to soft coral. However, a very rough and hard surface can scatter much energy acoustically [15, 16].

The parameters in Table 1 were used to examine the behavior of the density as a function of frequency. Figure 8 showed a plot of the real parts of the sound speed as a function of frequency from 0.01 Hz to 1000 kHz. At low frequencies, the pore size parameter is smaller than the viscous skin depth and at high frequencies the skin depth is smaller than the pore size [17].

Figures 9 and 10 show the density and attenuation as a function of frequency for Biot model. The bulk and shear moduli of the frame obviously increase the sound speed

slightly. The effect of the larger difference in sound speed will be seen in the reflection, transmission, and backscattering strength.

The attenuation due to water viscosity in Biot model varies as frequency squared up to around 1 kHz. The attenuation due to frame moduli varies linearly with frequency. Numerical computation of Biot model confirms that the sound speed, density, and attenuation were frequency dependent.

Figure 11 shows the reflection coefficient as a function of grazing angle for four different frequencies (10, 50, 100, and 200 kHz) for the measurement and Biot model. The measurement of reflection coefficient was conducted by using the average intensity of the coral reef-water interface echo. The biggest difference is near the critical angle

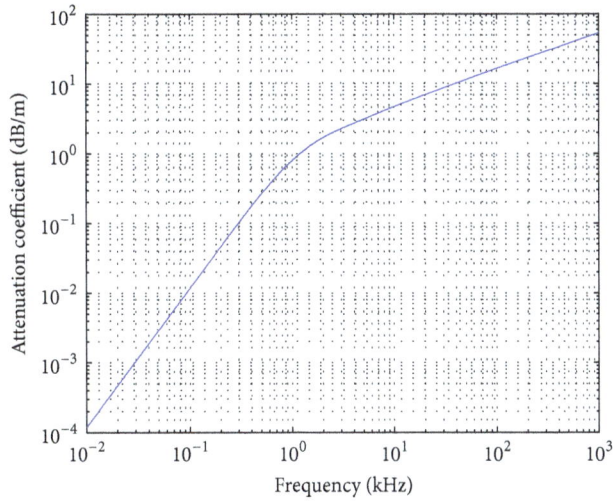

FIGURE 10: Attenuation coefficient as a function of frequency calculated using the Biot model.

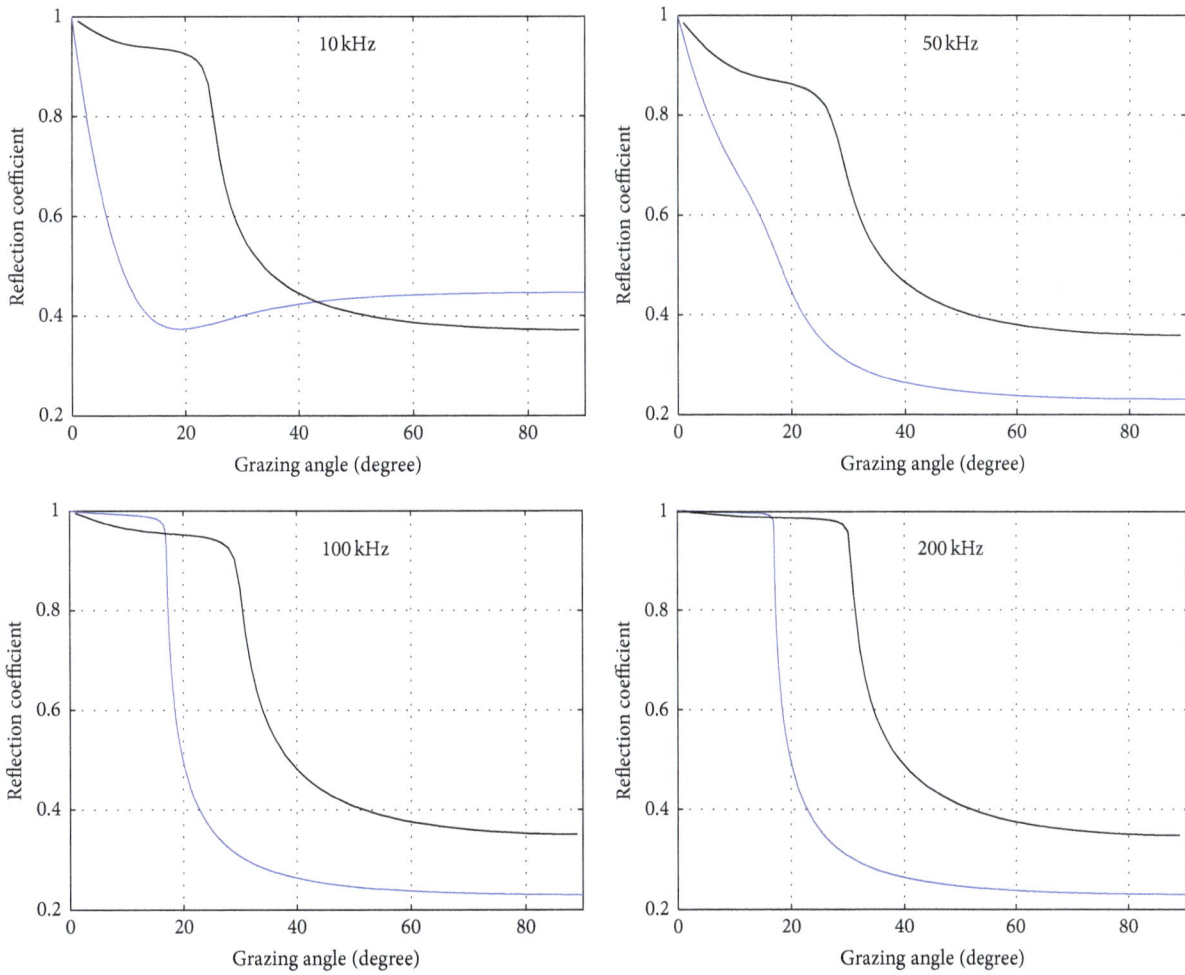

FIGURE 11: Reflection coefficient using Biot's model as a function of grazing angle at four different frequencies for soft coral (blue line) and hard coral (black line).

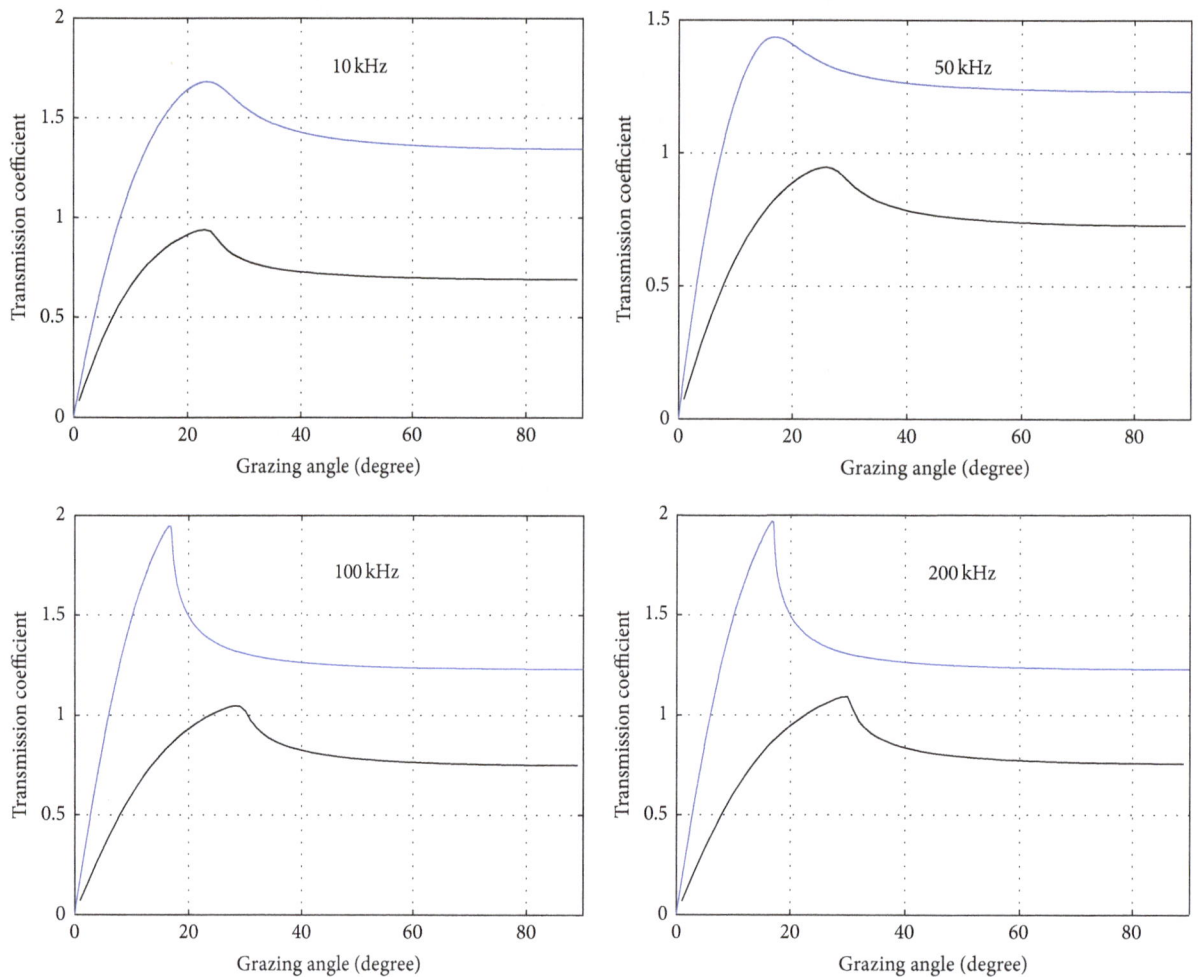

FIGURE 12: Transmission coefficient using Biot's model as a function of grazing angle at four different frequencies for soft coral (blue line) and hard coral (black line).

(near 20° to 30°) and this is due to the difference in sound speed between coral reef and seawater. The reflection coefficient is strongly correlated with sediment porosity and bulk density. Based on this research, the porosity of the coral reef was estimated using the reflection coefficient.

The general trend for the transmission coefficient is shown in Figure 12. For the Biot model, the procedure used to calculate the transmitting coefficient using the scalar potential and fast wave. This figure means more energy is transmitted to the coral reef than reflected.

Figure 13 showed the backscattering strength as a function of grazing angle. Backscattering is dependent on many confounding parameters such as the coral composition, surface roughness, slope, and spatial heterogeneity of coral. Backscattering versus grazing angle can be exploited to detect differences in coral properties and to solve geoacoustic model of coral properties. Frequency response may indicate coral reef type. The objective of our study was to quantify

backscattering for coral versus both frequency and grazing angle and to use parameters of Biot model to identify the type of coral reef. Backscattering strength of hard coral was higher than that of soft coral. These results were agreed on by previous researchers [8, 18].

4. Conclusion

Underwater acoustic remote sensing measurement and Biot model had been used to detect and quantify the coral reefs. The hardness of hard coral was higher than that of soft coral. The roughness of coral contributed to reflectivity and backscattering energy. Numerical computation of Biot model confirms that the sound speed, density, and attenuation are frequency dependent. Acoustic measurement agreed with Biot model in reflection and backscattering value of coral reef. This remote acoustic measurement and Biot model may be useful for estimating coral reef properties when core data are not available.

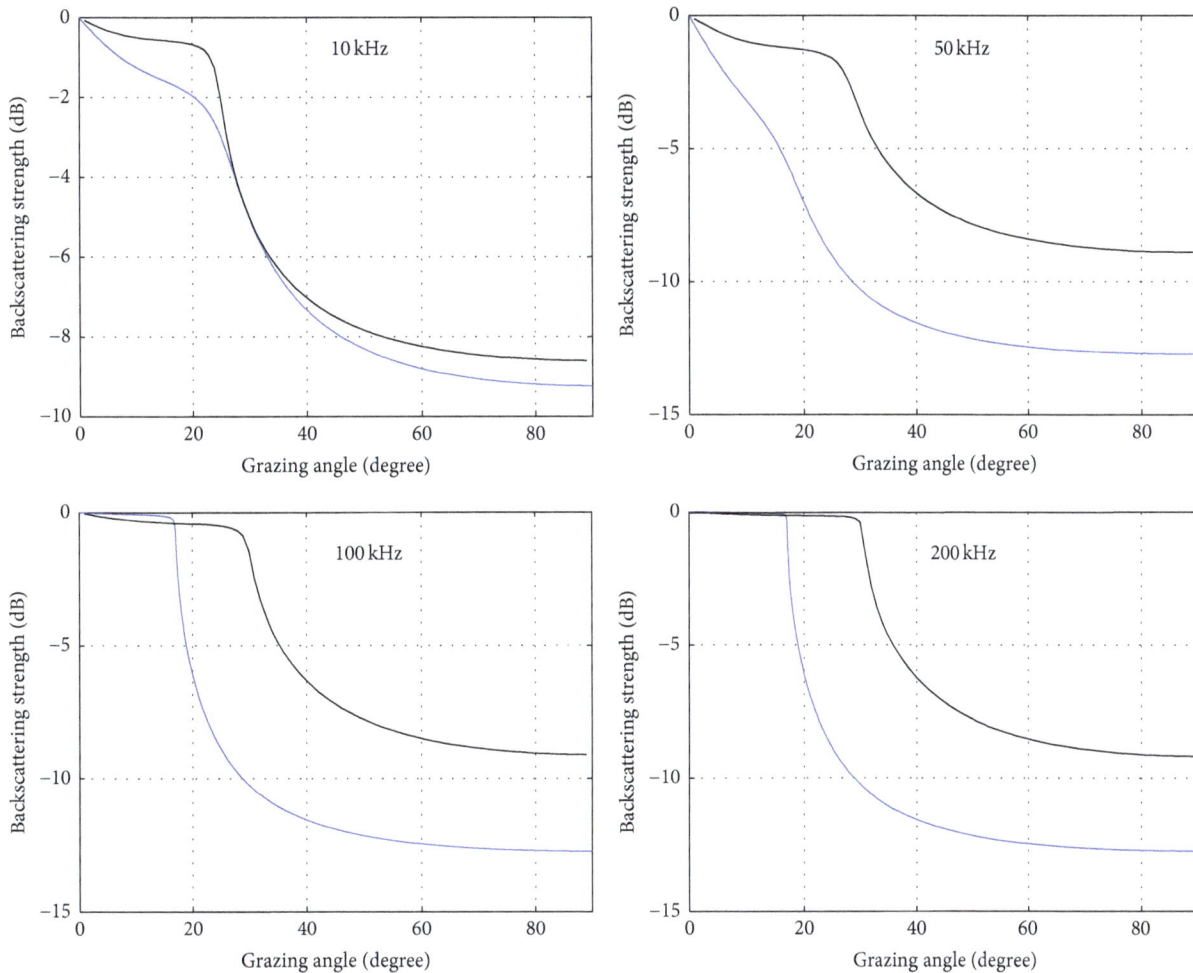

FIGURE 13: Backscattering strength as a function of grazing angle at four different frequencies for soft coral (blue line) and hard coral (black line).

Competing Interests

The author declares that there are no competing interests regarding the publication of this paper.

Acknowledgments

The author would like to thank the Indonesian Ministry of Research, Technology, and Higher Education and SEAMEO BIOTROP for sponsoring this research. The author wishes to acknowledge the many helpful discussions with Professor Darrell Jackson and Dr. Kevin L. Williams at Applied Physics Laboratory, University of Washington, USA, and Professor Masao Kimura at Poro-Acoustics Laboratory, Shimizu, Japan.

References

[1] N. A. Farmer, J. S. Ault, S. G. Smith, and E. C. Franklin, "Methods for assessment of short-term coral reef fish movements within an acoustic array," *Movement Ecology*, vol. 1, article 7, 2013.

[2] A. J. Irigoyen, D. E. Galván, L. A. Venerus, and A. M. Parma, "Variability in abundance of temperate reef fishes estimated by visual census," *PLoS ONE*, vol. 8, no. 4, Article ID e61072, 2013.

[3] D. Mallet and D. Pelletier, "Underwater video techniques for observing coastal marine biodiversity: a review of sixty years of publications (1952–2012)," *Fisheries Research*, vol. 154, pp. 44–62, 2014.

[4] C. J. Brown, S. J. Smith, P. Lawton, and J. T. Anderson, "Benthic habitat mapping: a review of progress towards improved understanding of the spatial ecology of the seafloor using acoustic techniques," *Estuarine, Coastal and Shelf Science*, vol. 92, no. 3, pp. 502–520, 2011.

[5] P. T. Harris and E. K. Baker, "GeoHab Atlas of seafloor geomorphic features and benthic habitats—synthesis and lessons learned," in *Seafloor Geomorphology as Benthic Habitat: GeoHab Atlas of Seafloor Geomorphic Features and Benthic Habitats*, P. T. Harris and E. K. Baker, Eds., pp. 871–888, Elsevier, Amsterdam, The Netherlands, 2011.

[6] D. Ierodiaconou, J. Monk, A. Rattray, L. Laurenson, and V. L. Versace, "Comparison of automated classification techniques for predicting benthic biological communities using hydroacoustics and video observations," *Continental Shelf Research*, vol. 31, no. 2, pp. S28–S38, 2011.

[7] V. M. Trenkel, P. H. Ressler, M. Jech, M. Giannoulaki, and C. Taylor, "Underwater acoustics for ecosystem-based management: state of the science and proposals for ecosystem indicators," *Marine Ecology Progress Series*, vol. 442, pp. 285–301, 2011.

[8] H. M. Manik, "Shallow-water acoustics investigations for underwater detection and seabed imaging," *International Journal of Applied Engineering Research*, vol. 10, no. 17, pp. 38302–38307, 2015.

[9] H. M. Manik, "Acoustic characterization of fish and seabed using underwater acoustics technology," *International Journal of Oceans and Oceanography*, vol. 9, no. 1, pp. 77–95, 2015.

[10] H. M. Manik, D. Yulius, and Udrekh, "Development and application of MB system software for bathymetry and seabed computation," *International Journal of Software Engineering and its Applications*, vol. 9, no. 6, pp. 143–160, 2015.

[11] H. M. Manik, S. Rohman, and D. Hartoyo, "Underwater multiple objects detection and tracking using multibeam and side scan sonar," *International Journal of Applied Information System*, vol. 2, no. 2, pp. 1–4, 2014.

[12] H. M. Manik, "Seabed identification and characterization using sonar," *Advances in Acoustics and Vibration*, vol. 2012, Article ID 532458, 5 pages, 2012.

[13] M. A. Zimmer, L. D. Bibee, and M. D. Richardson, "Measurement of the frequency dependence of the sound speed and attenuation of seafloor sands from 1 to 400 kHz," *IEEE Journal of Oceanic Engineering*, vol. 35, no. 3, pp. 538–557, 2010.

[14] R. D. Stoll and T.-K. Kan, "Reflection of acoustic waves at a water-sediment interface," *The Journal of the Acoustical Society of America*, vol. 70, no. 1, pp. 149–156, 1981.

[15] N. P. Chotiros, "Reflection and reverberation in normal incidence echo–sounding," *Journal of the Acoustical Society of America*, vol. 96, no. 5, pp. 2921–2929, 1994.

[16] N. P. Chotiros, "Inversion and sandy ocean sediments," in *Full Field Inversion Methods in Ocean and Seismo-Acoustics*, O. Diachok, A. Caiti, P. Gerstoft, and H. Schmidt, Eds., pp. 353–358, Kluwer, Norwell, Mass, USA, 1995.

[17] N. P. Chotiros, A. M. Mautner, A. Løvik, Å. Kristensen, and O. Bergem, "Acoustic penetration of a silty sand sediment in the 1-10-kHz band," *IEEE Journal of Oceanic Engineering*, vol. 22, no. 4, pp. 604–614, 1997.

[18] K. L. Williams, D. R. Jackson, E. I. Thorsos, D. Tang, and S. G. Schock, "Comparison of sound speed and attenuation measured in a sandy sediment to predictions based on the Biot theory of porous media," *IEEE Journal of Oceanic Engineering*, vol. 27, no. 3, pp. 413–428, 2002.

15

Vibroacoustic Analysis of a Refrigerator Freezer Cabinet Coupled with an Air Duct

Onur Çelikkan and Haluk Erol

Mechanical Engineering Faculty, Istanbul Technical University, Gumussuyu, 34439 Istanbul, Turkey

Correspondence should be addressed to Haluk Erol; erolha@itu.edu.tr

Academic Editor: Sven Johansson

In this study, the vibration and acoustic interactions between the structure and the cavity inside the freezer cabinet were investigated. Thus, a set of numerical and experimental analyses were performed. In the numerical analysis, the acoustic characteristics of the freezer cavity were solved, and the mixed finite element method was then implemented to analyse the coupled behaviour of the cavity with the air duct using the Acoustic Fluid-Structure Interaction (AFSI) technique. In the experimental analyses, an acoustic modal analysis of the freezer cavity and a structural modal analysis of the air duct were performed for the validation process. A good agreement was obtained among the results. Thus, the accuracy of the numerical model was confirmed. The validated models were used for optimizing the design. To solve the noise generation mechanism inside the freezer cabinet, the noise primarily generated by the freezer fan unit was measured under normal working conditions of the refrigerator, and the resonance frequencies were obtained. This information was compared with the normal modes of the air duct, and the overlapping frequencies were identified. To reduce the interaction between the source and the structure, a few design modifications were applied to the air duct. Thus, the structural-borne noise radiating from the air duct into the freezer cavity was reduced.

1. Introduction

One modern innovation is the no-frost (or frost free) type of refrigerator, which uses an autodefrost technique that regularly defrosts the evaporator in a refrigerator or freezer. This type of refrigerator is equipped with an additional ventilation fan mounted in the air duct to circulate the cooled air and aid in the defrosting process. In this type of configuration, the fan and the compressor become a source of noise, which is the primary contributor to the overall noise level of the refrigerator, thus increasing the vibration and resulting in sound level a few decibels compared to static cooling refrigerators. Furthermore, the increasing demand for larger fresh-food storage capacities results in refrigerators with larger volumes, which need faster ventilation fans to generate larger flow rates. This high-speed rotation generates more sound energy, and this situation increases the priority of the fan among the other noise sources in the refrigerator. This case demonstrates that applying new technologies in

refrigerators involves additional noise and vibration sources, which need to be investigated.

Baran et al. observed that the primary source of vibration typical for a no-frost refrigerator is the imbalance of the blades of the ventilation fan, which stimulates the plenum and effectively causes the entire structure to vibrate [1]. Seo et al. achieved a reduction in the refrigerator's sound pressure level by isolating the transmission of ventilation noise between the freezer compartment and the machinery room [2]. Takushima et al. searched for the sound sources using the sound intensity method, which indicated that the noise radiated through the openings of the front board [3]. Igarashi and Kitagawa performed CFD (Computational Fluid Dynamics) analyses by evaluating the flow fields of a propeller fan used in the freezing compartment of household refrigerators [4]. Kim et al. identified the source of excessive noise in a small fan-motor system for household refrigerators. They investigated an undesirable effect of cogging torque from the BLDC motor, which prevented the smooth rotation of the rotor and resulted in noise [5]. Gue et al. conducted experimental and numerical investigations on the aerodynamic noise of an axial fan to develop a low noise fan, which was used to cool a compressor and a condenser in the mechanical room of a household refrigerator [6]. Öztürk and Erol showed that the contribution of structure-borne noise from the vibrating panels to the overall noise levels is significant for end users because of its relationship to noise and comfort, especially at low frequencies [7].

The goal of this study is to analyse the acoustic characteristics of the freezer compartment coupled with the air duct. Hence, the study is divided into two sections. In the first section, fluid analyses have been performed with the fan blade in the air duct, and the pressure distribution is solved in the interior surfaces of the fan louver and the evaporator cover. In the second section, coupled acoustic modal analyses have been performed between the air duct and the freezer compartment using the acoustic-structure interaction techniques. The results obtained from the numerical solutions have been validated by the experimental results. In the validation process, an experimental modal analysis was performed for the air duct, and an experimental acoustic modal analysis was performed for the freezer compartment using an external sound source. Consequently, the vibration characteristics of the air duct have been resolved, and the contribution to the noise generation in the freezer compartment has been observed.

2. Theory

For the acoustic-structure system, the structure is described by the differential equation of motion for a continuum body assuming small deformations whereas the fluid is described by the acoustic wave equation. Coupling conditions at the boundary between the structural and fluid domains ensure the continuity in displacement and pressure between the domains. The governing equations and boundary conditions, as described in detail by Carlsson [8], can be written as follows:

Structure:
$$\begin{cases} \tilde{\nabla}^T \sigma_S + b_S = \rho_S \dfrac{\partial^2 u_S}{\partial t^2} & @\Omega_S \\ +\text{Boundary and initial conditions} \end{cases}$$

Fluid:
$$\begin{cases} \dfrac{\partial^2 p_F}{\partial^2 t} - c_0^2 \nabla^2 p_F = c_0^2 \dfrac{\partial q_F}{\partial t} & @\Omega_F \\ +\text{Boundary and initial conditions} \end{cases} \quad (1)$$

Coupling:
$$\begin{cases} u_S = u_F & @\partial\Omega_{FS} \\ \sigma_S = -p_F & @\partial\Omega_{FS}, \end{cases}$$

where $\sigma_S(x,y,z,t)$ denote stresses, $b_S(x,y,z,t)$ denote body forces, ρ_S denotes dynamic density, $u_s(x,y,z,t)$ and $u_F(x,y,z,t)$ show displacements for the structural domain and fluid, separately, x, y, z denote the Cartesian coordinates, t denotes time in seconds, $p_F(x,y,z,t)$ denotes dynamic pressure, $q_F(x,y,z,t)$ denotes added fluid mass per unit volume, Ω's show domains, c_o is the speed of sound, and ∇ denotes a gradient of a variable; that is,

$$\nabla = \begin{bmatrix} \dfrac{\partial}{\partial x} & \dfrac{\partial}{\partial y} & \dfrac{\partial}{\partial z} \end{bmatrix}^T \quad (2)$$

and the differential operator $\tilde{\nabla}$ can be written as

$$\tilde{\nabla} = \begin{bmatrix} \dfrac{\partial}{\partial x} & 0 & 0 \\ 0 & \dfrac{\partial}{\partial y} & 0 \\ 0 & 0 & \dfrac{\partial}{\partial z} \\ \dfrac{\partial}{\partial y} & \dfrac{\partial}{\partial x} & 0 \\ \dfrac{\partial}{\partial z} & 0 & \dfrac{\partial}{\partial x} \\ 0 & \dfrac{\partial}{\partial z} & \dfrac{\partial}{\partial y} \end{bmatrix}. \quad (3)$$

The finite element formulation of both the continuum body and the acoustic fluid is used for modelling the fan louver and freezer cavity. The structure of interest in most acoustic-structure problems is two-dimensional and therefore often described using the plate or shell theory.

The structure is described by the equation of motion for a continuum body. The finite element formulation is derived with the assumption of a small displacement. The governing system of equations can be written as follows [9]:

$$M_S \ddot{d}_S + K_S d_S = f_F + f_b, \quad (4)$$

where

$$M_S = \int_{\Omega_S} N_S^T \rho_S N_S \, dV,$$

$$K_S = \int_{\Omega_S} \left(\check{\nabla} N_S \right)^T D_S \check{\nabla} N_S \, dV,$$

$$f_F = \int_{\partial\Omega_S} N_S^T t_S \, dS, \quad (5)$$

$$f_b = \int_{\Omega_S} N_S^T b_S \, dV,$$

where N_S contains the finite element shape functions for the structural domain, $d_S(t)$ is the finite element approximation of the displacement, c_S is the nodal weight functions, t_S is the surface traction vector, and

$$D_S = \begin{bmatrix} \lambda+2\mu & \lambda & \lambda & 0 & 0 & 0 \\ \lambda & \lambda+2\mu & \lambda & 0 & 0 & 0 \\ \lambda & \lambda & \lambda+2\mu & 0 & 0 & 0 \\ 0 & 0 & 0 & \mu & 0 & 0 \\ 0 & 0 & 0 & 0 & \mu & 0 \\ 0 & 0 & 0 & 0 & 0 & \mu \end{bmatrix}, \quad (6)$$

where λ and μ are the Lame coefficients expressed in the modulus of elasticity, E, the shear modulus, G, and Poisson's ratio.

In Acoustic Fluid-Structure Interaction problems, the structural dynamic equation must be considered along with the Navier-Stokes equations of fluid momentum and the flow continuity equation. The governing equations for an acoustic fluid can be derived using the following assumptions for the compressible fluid: the fluid is inviscid; the fluid only undergoes small translations; and the fluid is irrotational. Thereby, the governing equations for an acoustic fluid are the equation of motion

$$\rho_0 \frac{\partial^2 u_F(t)}{\partial t^2} + \nabla p_F(t) = 0, \qquad (7)$$

the continuity equation

$$\frac{\partial \rho_F(t)}{\partial t} + \rho_0 \nabla \frac{\partial u_F(t)}{\partial t} = q_F(t), \qquad (8)$$

and the constitutive equation

$$p_F(t) = c_0^2 \rho_F(t). \qquad (9)$$

The system of equations for an acoustic domain becomes

$$M_F \ddot{p}_F + K_F p_F = f_q + f_s, \qquad (10)$$

where

$$M_F = \int_{\Omega_F} N_F^T N_F \, dV,$$

$$K_F = c_0^2 \int_{\Omega_F} (\nabla N_F)^T \nabla N_F \, dV,$$

$$\qquad (11)$$

$$f_S = c_0^2 \int_{\partial \Omega_F} N_F^T n_F^T \nabla p \, dS,$$

$$f_q = c_0^2 \int_{\Omega_F} N_F^T \frac{\partial q}{\partial t} \, dV,$$

where N_F contains the finite element shape functions for the fluid and n's denote normal vectors.

At the boundary between the structural and fluid domains, denoted as Ω_{SF}, the fluid particles and the structure move together in the normal direction of the boundary. Furthermore, the acoustic-structure problem can be described by an unsymmetrical system of equations as follows:

$$\begin{bmatrix} M_S & 0 \\ \rho_0 c_0^2 H_{SF}^T & M_F \end{bmatrix} \begin{bmatrix} \ddot{d}_S \\ \ddot{p}_F \end{bmatrix} + \begin{bmatrix} K_S & -H_{SF} \\ 0 & K_F \end{bmatrix} \begin{bmatrix} d_S \\ p_F \end{bmatrix} = \begin{bmatrix} f_b \\ f_q \end{bmatrix}, \qquad (12)$$

where H_{SF} denotes spatial coupling matrix.

3. Numerical Studies

The numerical simulations were performed using the FEM solver ANSYS Workbench R15.0. In the frame of this study, each component composing the freezer compartment was first investigated individually. Thus, acoustic analyses of the freezer cavity and the modal analysis of the fan louver and evaporator cover were performed so that the mode frequencies and mode shapes could be obtained. Then, to create a realistic model, the flow field inside the air duct was modelled and included in the analyses. Lastly, a coupled modal analysis was performed to solve the problem between the structure and the acoustic cavity. The change in the acoustic modes of the freezer cavity was observed.

The fluid-structure interaction simulations were performed using the multifield solver, which used an implicit sequential coupling to calculate the interactions between the fluid and structural analyses. The FSI techniques are used to compute the effects between the acoustic and structural domains using specialized acoustic elements.

In the details of the analysis process, the state variables were defined, and the mathematical model was built to describe the physical phenomena. The mathematical model may deviate from the actual model due to various assumptions, such as viscosity and compressibility for the fluid flow and stiffness and damping for the structure.

It is widely accepted that the element size in element-based acoustic computations is related to the wavelength. In modelling, the element size has been chosen very small to ensure the sufficient number of elements per wavelength that corresponding to upper limit of frequency. The properties such as density and bulk modulus have a significant role in specifying the wavelength also defined for the fluid media. Before performing an experimental analysis to validate the numerical studies, it is possible to perform a preliminary examination basically. To show the cause of the similarity of the freezer cavity to the rectangular box, the analytical solution is also available. The equivalent box model with the same outer size could be easily used to represent the cavity model [10].

The interior of a freezer compartment resembles a closed rectangular volume, which has a simple analytical solution for its natural frequencies and acoustical modes. The natural frequencies can be calculated as follows:

$$f_{ijk} = \frac{c}{2} \sqrt{\left(\frac{i}{L_x}\right)^2 + \left(\frac{j}{L_y}\right)^2 + \left(\frac{k}{L_z}\right)^2} \text{ (Hz)}. \qquad (13)$$

Furthermore, the mode shapes can be calculated as follows:

$$\Psi_{ijk} = \cos\frac{i\pi x}{L_x} \cos\frac{j\pi y}{L_y} \cos\frac{k\pi z}{L_z}, \qquad (14)$$

where $L_x = 0.661$ m and is the width of the freezer cabin; $L_y = 0.447$ m and is the height of the freezer cabin measured from the bottom to the top; and $L_z = 0.446$ m and is the depth of the freezer cabin measured from the fan louver to the freezer door. The speed of sound is calculated at room temperature (at 25°C), and the indexes for the normal modes of vibration $i = 0, 1, 2, \ldots$, $j = 0, 1, 2, \ldots$, and $k = 0, 1, 2, \ldots$. Figure 1 depicts the inner dimensions of the freezer compartment and the isometric 3D model.

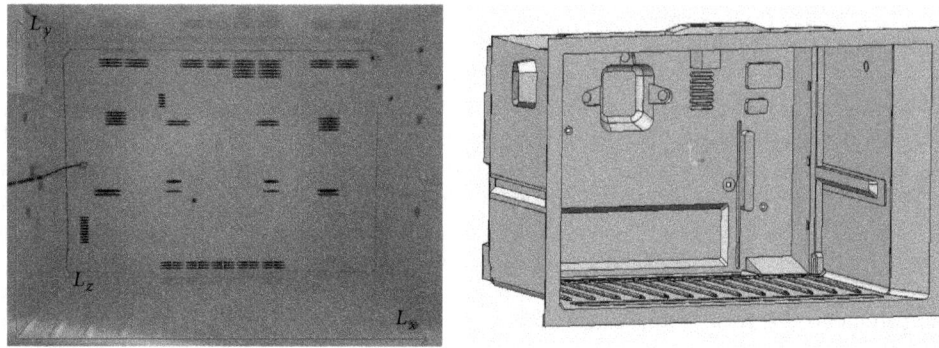

FIGURE 1: Inner dimensions of the freezer compartment and the isometric 3D model.

TABLE 1: Comparison of the analytical and numerical natural frequencies.

Mode (#)	Theory (Hz)	FE model (Hz)	Difference (%)
1	255.2	259.4	1.6
2	373.6	383.7	2.5
3	381.7	384.5	0.7
4	448.4	463.2	3.2
5	457.9	463.9	1.3
6	509.5	518.9	1.8
7	533.0	543.2	1.9
8	587.4	602.0	2.4
9	627.0	645.4	2.5
10	635.1	645.8	1.7

TABLE 2: Material properties of the fan louver and the evaporator cover.

Property	Fan louver	Evaporator cover
Density (kg/m^3)	932	4660
Young modulus (Pa)	900000000	700000000
Poisson ratio	0.42	0.30

TABLE 3: First ten natural frequencies of the fan louver and the evaporator cover.

Mode (#)	Fan louver (Hz)	Evaporator cover (Hz)
1	4.8	15.0
2	36.5	18.3
3	49.0	35.8
4	54.6	58.8
5	76.3	73.6
6	105.1	80.5
7	111.2	90.5
8	130.6	103.5
9	132.3	110.5
10	150.2	117.0

A convergence study has been performed on the number of elements to obtain a better accuracy between the numerical and experimental models. The convergence study can be described as follows: the acoustic modes of the cavity are computed for models with an increasing number of elements. Increasing the number of elements increases the accuracy of the model until a certain number of elements is reached; beyond this number, the accuracy does not considerably improve. A minimum number of elements should be used for a satisfactory accuracy.

To define the acoustic model in the simulation environment, the dimension of the cavity in the freezer cabin and the characteristics of the fluid (air at 25°C) were entered, and the natural frequencies and mode shapes were computed.

A comparison of the natural frequencies between the analytical solution and the numerical solution is presented in Table 1, and Figure 2 illustrates the first three numerical acoustic mode shapes of the freezer cavity.

The results obtained using the finite element model of the freezer cavity closely match those obtained analytically from the rectangular box model. In terms of the natural frequency, the largest difference occurs at 463.2 Hz, where the absolute difference is 14.8 Hz (3.2%). The primary reason for this discrepancy is the minor geometrical differences among the models. At other frequencies, the differences do not exceed the ratio of 2.5%, which indicates that the numerical model agrees extremely well with the analytical model.

In household refrigerators, the freezer air duct is composed of two components, fan louver and evaporator cover, as illustrated in Figure 3. The air, which is cooled by the evaporator behind the freezer compartment, is blown by the freezer fan unit and passes through the air duct to the freezer compartment. The evaporator cover has a few ribs that lead air to the openings, that is, discharge holes, located on the fan louver.

To identify the natural frequencies and mode shapes of the air duct, a three-dimensional finite element analysis and simulation were performed. In the numerical model, the evaporator cover and fan louver were modelled as coupled and flexible. Both of the components are made of polypropylene, and the material properties are listed in Table 2.

The first ten natural frequencies of the fan louver and the evaporator cover calculated from the finite element models are listed in Table 3.

The fluid field of the air cavity was extracted from the air duct, and the numerical simulation of the fluid model was

A: Model
Acoustic pressure 1
Expression: PRES
Unit: MPa
12.03.2014 09:52

44,365 Max
34,507
24,648
14,789
4,9305
−4,9282
−14,787
−24,646
−34,504
−44,363 Min

A: Model
Acoustic pressure 2
Expression: PRES
Unit: MPa
12.03.2014 09:53

55,058 Max
43,263
31,469
19,675
7,8806
−3,9137
−15,708
−27,502
−39,296
−51,091 Min

A: Model
Acoustic pressure 3
Expression: PRES
Unit: MPa
12.03.2014 09:53

52,961 Max
41,501
30,041
18,58
7,1201
−4,3401
−15,8
−27,26
−38,721
−50,181 Min

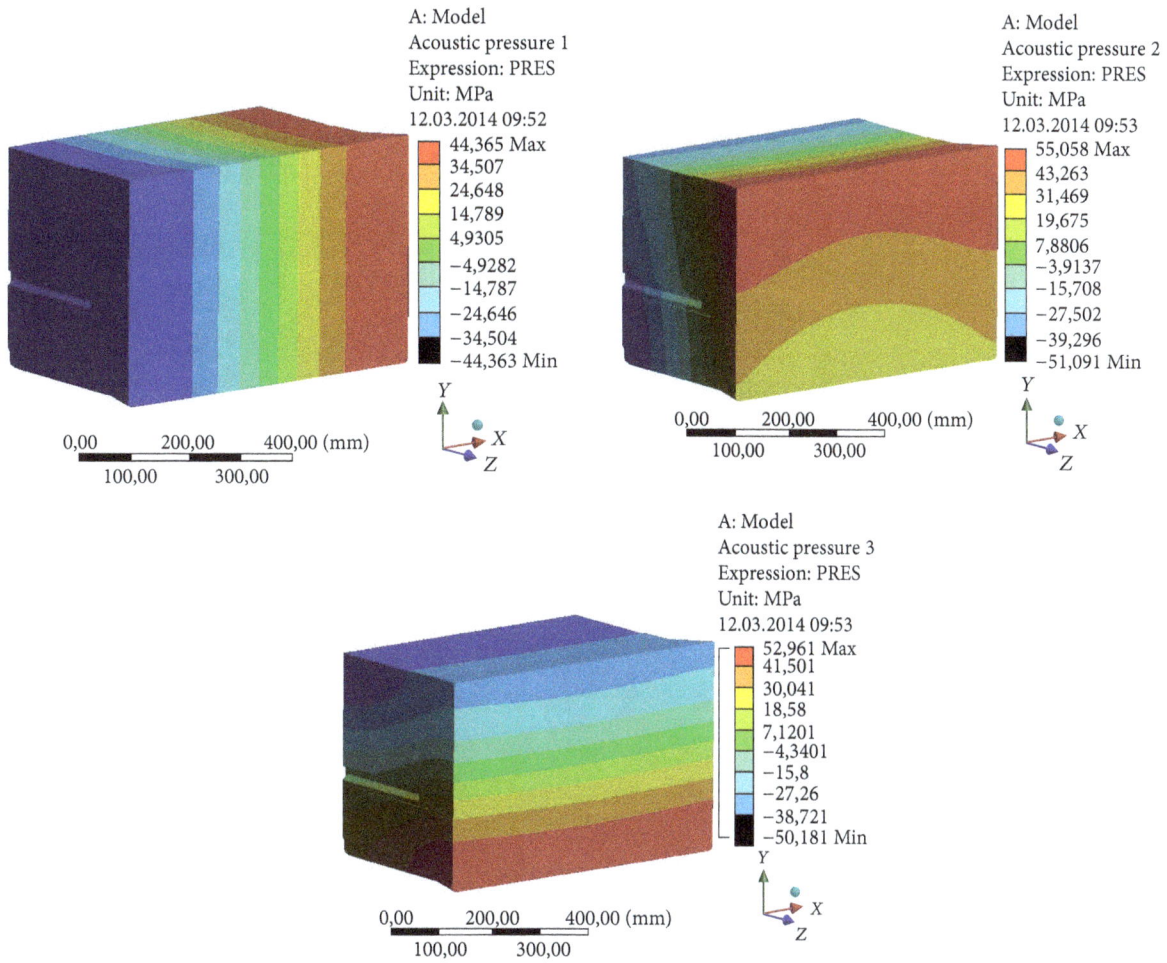

FIGURE 2: First three acoustic mode shapes of the freezer cavity.

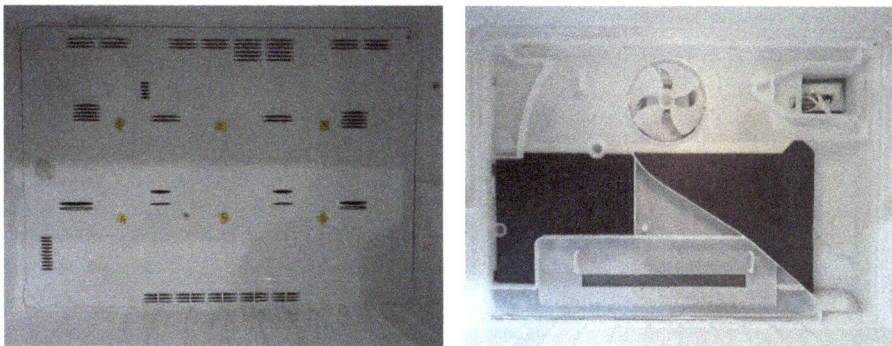

FIGURE 3: Fan louver and evaporator cover.

created by adding the axial freezer fan driven by a motor rotating at 2200 rpm. The location of the axial fan inside the evaporator cover is depicted in Figure 4. The axial fan has 4 blades with outlet diameters of 100 mm and is located near the centre of the air inlet hole.

A no-slip boundary condition is defined at the walls, pressure inlet, and pressure outlet. The pressure inlet boundary condition is used to define the fluid pressure at the air inlet hole, and negative values are set to simulate the back pressure

due to the pressure loss of the system. The pressure inlet boundary condition is applicable for both incompressible and compressible flow calculations. The pressure inlet boundary condition can be used when the inlet pressure is known and the flow rate and/or velocity are unknown.

The pressure outlet boundary conditions are defined for the discharge holes located on the fan louver. The operating pressure is set to normal working conditions, and a gauge pressure required in the pressure outlet boundary condition

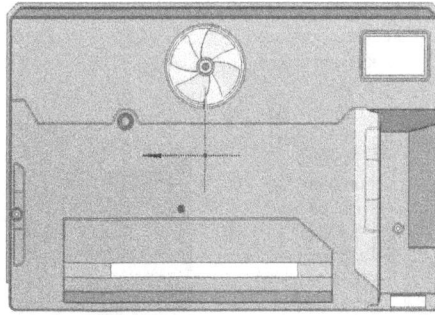

FIGURE 4: Location of the axial fan inside the evaporator cover.

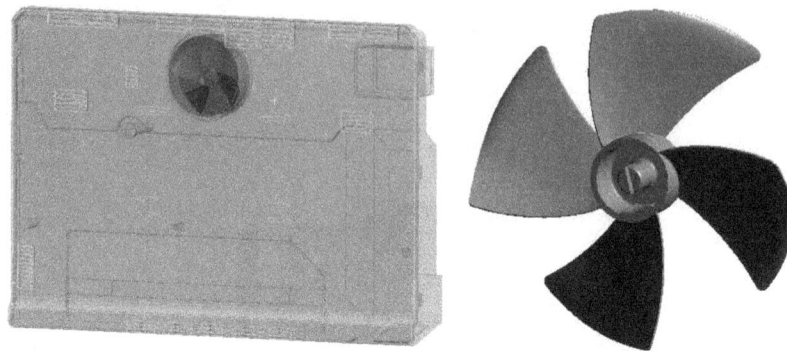

FIGURE 5: Isometric 3D models of the air duct cavity and the axial fan.

is specified as zero. Additionally, the backflow conditions due to the reversed flow of air across the boundary are defined. This operation minimizes the convergence errors during the iteration process. Figure 5 presents the isometric 3D models of the air duct cavity and the axial fan.

Using the finite volume method, the flow domain is discretized into approximately 2.5×10^6 control volumes. Furthermore, the inflation method is applied on all of the surfaces of the cavity to resolve the boundary layer. A flow over the propeller blades tends to cause asymmetries. To capture any possible asymmetries in the flow, the entire fan-duct geometry is simulated. A transition SST model is used to solve the flow field in the air duct. A spatial discretization is performed using the 2nd-order upwind scheme.

The transition SST model is a variant of the standard k-ω model. It combines the original Wilcox k-ω model for use near the walls and the standard k-ε model away from the walls using a blending function, and the eddy viscosity formulation is modified to consider the transport effects of the principle turbulent shear stress. This model is especially preferred for flows involving rotation, boundary layers under strong adverse pressure gradients and separation or recirculation.

In the generated model, the fluid flow conditions were supplied by the rotation of the fan blades. To simulate this motion, a "moving reference frame" technique was used. In this approach, the frame of reference is attached to the moving domain, and governing equations are modified to account for this moving frame. In fact, there is no moving part or mesh; that is, the local acceleration is added as source

terms to each grid cell. This technique provides a steady approximation of the interaction between the fan and the duct.

In this study, the simulations are performed in steady-state conditions using the transition SST turbulence model. In the transition SST model, four-transport equations are solved in addition to the RANS equations. This model provides a better performance at capturing near-wall behaviour compared to the two-equation k-ε model and provides more stable solutions. This property is especially important for solving the flow field around the blades.

The pressure-based coupled solver was used as a solution method. This method provides a faster convergence compared to other methods. The computation was conducted using the second-order discretization method to obtain a higher accuracy. The calculation of the solution was obtained after 250 iterations. Figure 6 illustrates the streamlines of the velocity field in the air cavity and magnitude of the velocity at the axial fan surface.

Figure 7 depicts the pressure distributions on the front surface and the rear surface of the cavity.

The fluid-structure interaction analysis is a multiphysics problem where the interaction between two different physics phenomena, performed in separate analyses, is considered. In this study, the response of a structure under the flow-induced loads is investigated. The pressure loads obtained from the CFD analysis of the cavity are imported to the corresponding structural analysis of the fan louver. This interaction is taken at the common boundaries that the fluid domain shares with

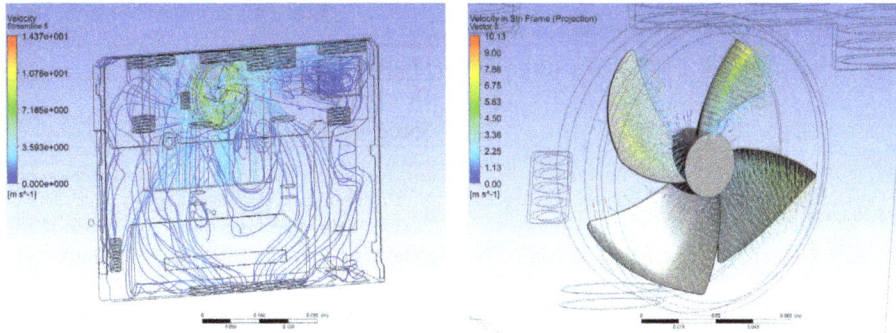

FIGURE 6: Streamlines of the velocity field in the air cavity and magnitude of the velocity at the axial fan surface.

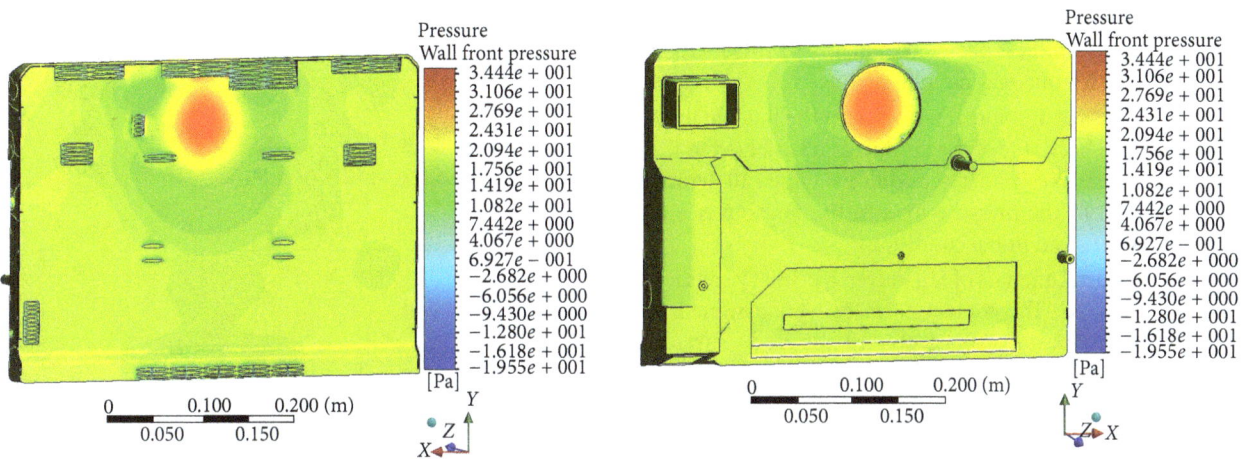

FIGURE 7: Pressure distributions on the front surface of the cavity and the rear surface of the cavity.

the structural domain. The results of the CFD analysis are applied to the structural analysis as an initial condition.

The partitioned approach is used for the simulation of the fluid-structure interaction. In this approach, the equations governing the flow and the displacement of the structure are solved separately with two distinct solvers. In this study, a CFD analysis is primarily performed, and the structural analysis is performed using this approach. This process indicates that the flow does not change while the structural solution is calculated. The results are interpolated to the structural analysis after the completion of the CFD analysis. In this approach, the information is exchanged at the interface between the two solvers, and this process is defined as "coupling." Figure 8 depicts the algorithm of the partitioned approach.

A one-way coupling method is performed to transfer the data from the fluid analysis to the structural analysis. In this method, the motion of a fluid flow influences a solid structure but the reaction of a solid upon a fluid is negligible. The resulting forces at the interface obtained from the fluid calculation of the cavity are interpolated to the structural mesh of the fan louver. Then, the static structural analysis is performed.

Furthermore, the fluid-structure interaction can be categorized by the degree of physical coupling between the fluid

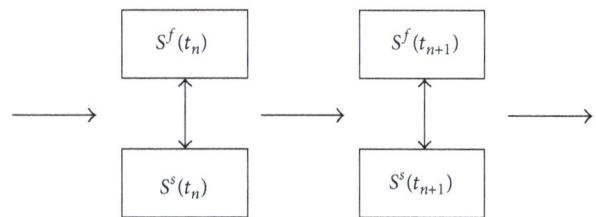

FIGURE 8: Algorithm of the partitioned approach.

and solid solution fields. The degree of coupling specifies how one field is sensitive to a change in the other field. Fields that are strongly coupled physically require a strong numerical coupling, and the solution is generally more difficult. In this study, the coupling degree is weak because of the physical shape of the geometries; thus, the sensitivity is less than that of strongly coupled systems.

The data transfer between the coupled participants is one of the critical parts of an FSI analysis. In our study, the static load transfer method was used in accordance with the nature of the coupling method. In the details of this data transfer method, the information is exchanged between two different types of mesh of different mediums at the interface of these mediums. This process is performed using a

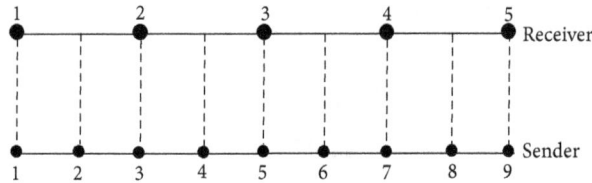

FIGURE 9: General grid interface.

systematic sequence. The first process of the data transfer is to match or pair the source and target mesh to generate weights. The source mesh feeds the data to the target mesh, and this matching is performed using the *General Grid Interface* mapping algorithm. The conservative nature of the *General Grid Interface* algorithm is depicted in Figure 9. This method is only available for one-way coupling.

The pressure distribution obtained from the fluid analysis on the front and back side of the air duct is transferred to the structural analysis as boundary conditions. Figure 10 depicts the importation of the pressure load to the fan louver and evaporator cover as an initial condition.

In the structural analysis of fan louver, a primary solution setup was prepared. The contact surfaces were assigned as a bond between the target surface and the body surface. As an initial condition, the fixed supports were defined at the surrounding frame of the evaporator cover to reflect the actual conditions in the freezer cabinet, and the imported pressure distribution results from the fluid analysis were applied as a load at the inner surface of the fan louver. Figure 11 depicts the contact surfaces on the fan louver and the evaporator cover.

In the analysis setting step, the "large deflection" option is set to "on." For simple linear static analyses, in which the deflection and strain are small, this option is typically set to "off" if the displacements are small enough that the resulting stiffness changes are insignificant. However, in our case, to obtain more accurate results, a prestress modal analysis will be generated in the next step, and this option must be available in this static structural analysis because the deformation in the modal analysis is high.

By setting the "large deflection" to on, the geometric non-linearities open and consider the stiffness changes resulting from changes in the element shape and the orientation due to the large deflection, large rotation, and large strain. Therefore, these results provide more accurate solutions. However, this effect requires an iterative solution. Additionally the load needs to be applied in small increments. Thus, the solution takes a longer time. In the analysis settings, a pressure load was applied in 10 steps with a time step of 0.1 seconds between each iteration starting at time 0.

In this section, the freezer cavity has been modelled and coupled with the solved prestressed structural analysis on the fan louver. The freezer cavity was defined as an acoustic body. Air properties of the freezer cavity are listed in Table 4. The reference pressure was $20 \times 10^{-6} \, N/m^2$.

TABLE 4: Air properties of the freezer cavity.

Property	Freezer cavity
Density (kg/m^3)	1.38
Speed of sound (m/s) @25°C	346.13

TABLE 5: First ten natural frequencies of the freezer cavity both coupled and uncoupled with the fan louver.

Mode (#)	Uncoupled freezer cavity (Hz)	Coupled freezer cavity (Hz)
1	259.4	262.6
2	383.7	386.7
3	384.5	388.3
4	463.2	468.7
5	463.9	469.9
6	518.9	526.1
7	543.2	551.8
8	602.0	613.0
9	645.4	657.6
10	645.8	658.7

For the freezer cavity, the frequency domain acoustics can be governed by the Helmholtz equation for the acoustic pressure,, as follows:

$$\nabla \cdot \left(-\frac{1}{\rho_o} \nabla p \right) - \frac{\omega^2}{\rho_o c^2} p = 0, \quad \omega = 2\pi f. \quad (15)$$

Additionally, a coupled acoustic analysis considers the fluid-structure interaction. The governing equation for acoustics has been discretized by considering the coupling of the acoustic pressure and the structural motion at the interface. By specifying the acoustic FSI label, the structural motion and the fluid pressure are coupled at the interface. The acoustic elements used in the freezer cavity have the capabilities of translating the pressure in the nonfluid medium and the translations in the x, y, and z directions at the interface.

The dissipative effects due to the fluid viscosity and the absorption resulting from the damping are neglected. The boundaries enclosing the acoustic cavity are assumed to be hard, and the pressure gradients on all boundaries without the FSI interface are hence set to zero. Before performing the analysis, all of the openings in the fan louver were closed to fulfil the requirements of the acoustic FSI at the interface.

Table 5 lists a comparison of the first ten natural frequencies of the freezer cavity both coupled and uncoupled with the fan louver.

When the obtained results are compared, it can be clearly observed that the impact of the fan louver on the freezer cavity increases the natural frequency of the freezer cavity. These differences result in an approximate increase in the higher frequencies (2%). However, there is almost never a

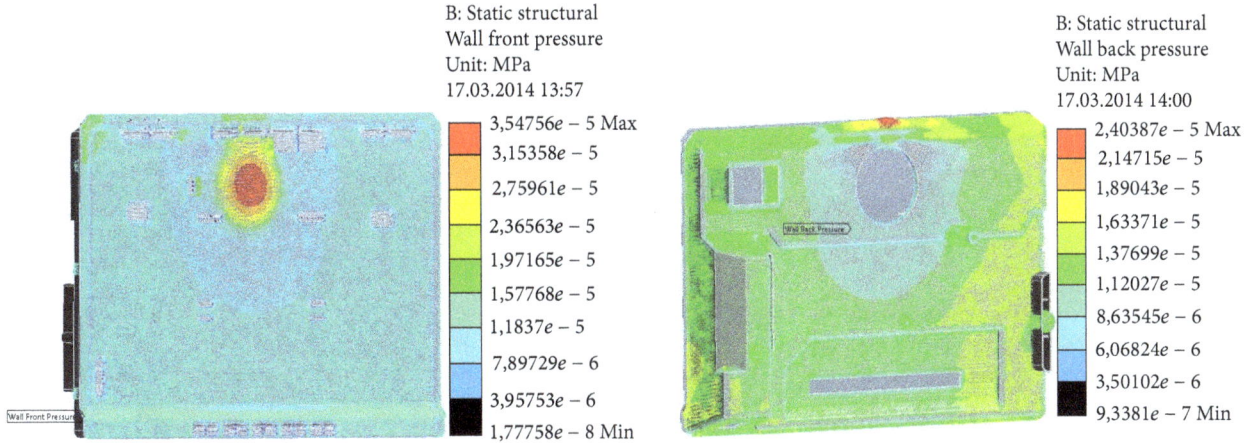

FIGURE 10: Pressure load imported to the fan louver and evaporator cover as an initial condition.

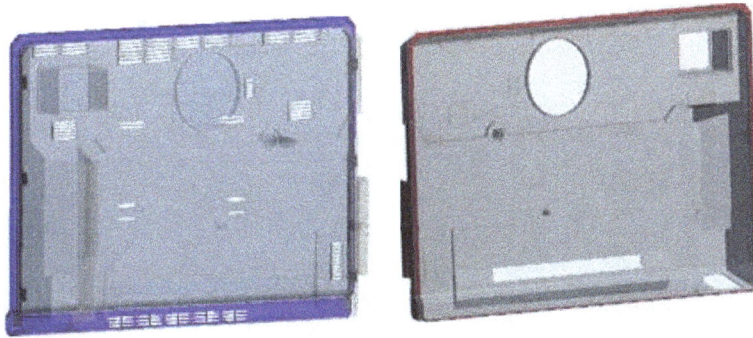

FIGURE 11: Contact surfaces on the fan louver and the evaporator cover.

change in the mode shapes of the freezer cavity in comparison to the uncoupled results of freezer cavity.

Figure 12 illustrates the first three mode shapes of the freezer cavity coupled with the fan louver calculated from the finite element model.

4. Experimental Studies

Experimental measurements were performed to validate the numerical model. Thus, the experimental acoustic modal analysis for the freezer cavity and the modal analysis for the fan louver and evaporator cover were conducted. The entire data acquisition process was achieved using a no-frost refrigerator. Figure 13 provides an image of the studied refrigerator.

The dynamic behaviour of cavity is expressed with three modal parameters; these are natural frequencies (ω_n), modal damping ratios (ζ_n), and mode shapes (Φ). The algorithm used to extract these parameters from the experimental model is based on a frequency domain curve fitting of the transfer function $H(\omega)$ between the pressure measured at point i inside the cavity and the pressure measured at point k in front of the sound source used to create the pressure field. The transfer function could be expressed in terms of the modal parameters by modal superposition as follows [10]:

$$H(\omega) = \frac{P_i}{P_k}(\omega) = \sum_{i=1}^{m} \frac{\Phi_{im}\Phi_{km}}{M_m\left(\omega_{m-}^2 \omega^2 + 2\zeta_m \omega \omega_m\right)}, \quad (16)$$

where M_m is the modal mass; ω_m is the natural frequency; ζ_m is the damping of mode m; and Φ_{im} is the modal participation of mode m at point i. The mode shapes can be expressed proportional to the modal participation as follows:

$$\left\{ \begin{array}{c} \text{modeshape}_m\left(x_1\right) \\ \cdots \\ \text{modeshape}_m\left(x_k\right) \\ \cdots \end{array} \right\} = \alpha_m \left[\begin{array}{c} \Phi_{m1}\Phi_{mk} \\ \cdots \\ \Phi_{mk}\Phi_{mk} \\ \cdots \end{array} \right], \quad (17)$$

where α_m is the coefficient of proportionality for the mode m. The system presented in Figure 14 is used to perform the acoustic modal analysis.

The signal from a white noise generator is limited between 0 and 1250 Hz, and an amplifier is used to adjust the level of

A: Modal
Acoustic pressure 2
Expression: PRES
Frequency: 0, Hz
Sweeping phase: 0,°
Unit: MPa
04.05.2014 17:07

43,983 Max
34,209
24,436
14,662
4,8884
−4,8853
−14,659
−24,433
−34,206
−43,98 Min

A: Modal
Acoustic pressure 3
Expression: PRES
Frequency: 0, Hz
Sweeping phase: 0,°
Unit: MPa
04.05.2014 17:09

54,594 Max
42,899
31,205
19,511
7,8162
−3,8781
−15,572
−27,267
−38,961
−50,655 Min

A: Modal
Acoustic pressure 4
Expression: PRES
Frequency: 0, Hz
Sweeping phase: 0,°
Unit: MPa
04.05.2014 17:09

52,514 Max
41,15
29,787
18,423
7,0599
−4,3035
−15,667
−27,03
−38,394
−49,757 Min

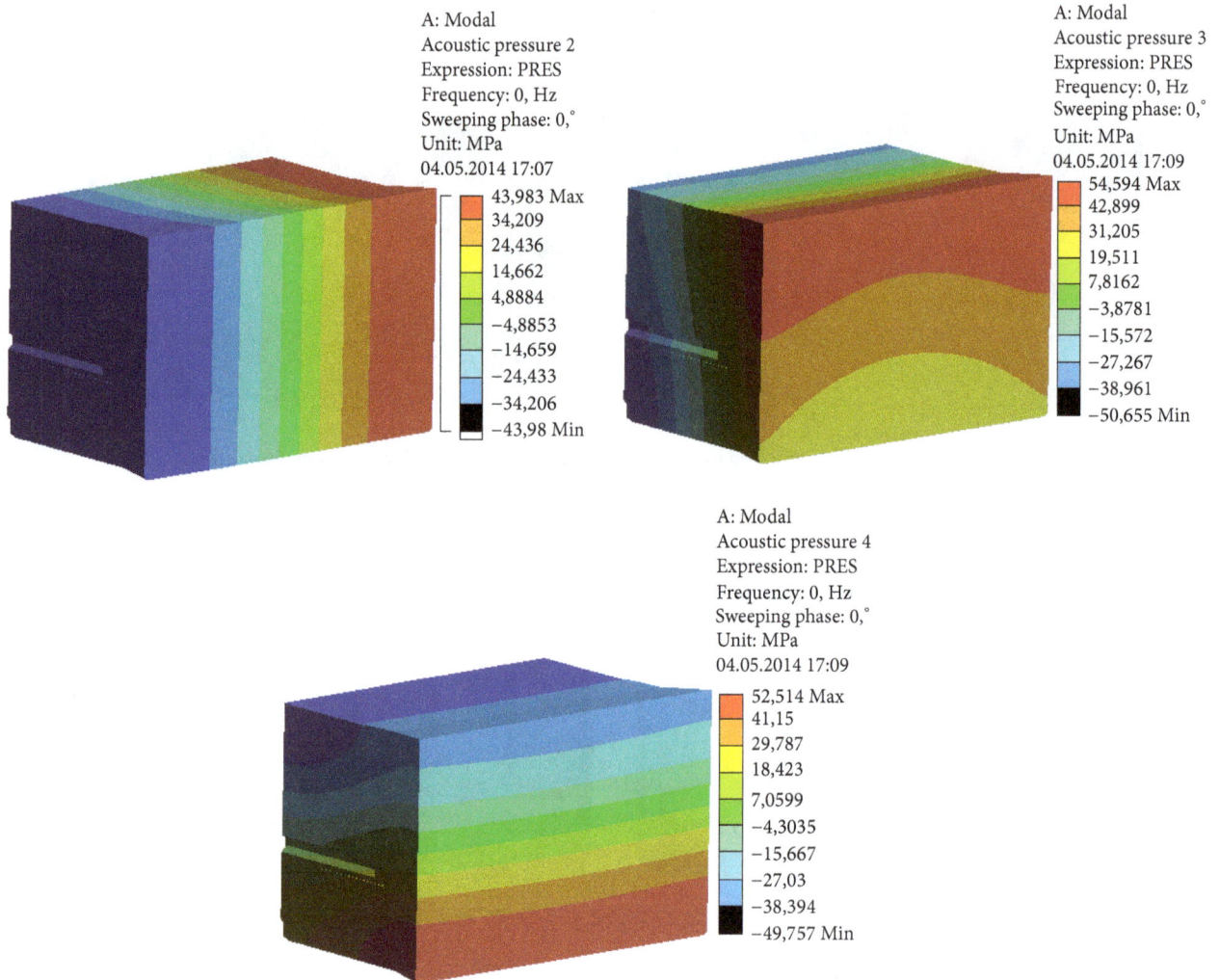

FIGURE 12: First three mode shapes of the freezer cavity coupled with the fan louver calculated from the finite element model.

the signal received by the loudspeaker to acoustically excite the freezer cavity. The loudspeaker was located at the location of the axial fan at the centre of the air inlet hole inside the air duct and isolated from the evaporator cover with soft foam. Figure 15 depicts the positions of the loudspeaker and the reference microphone.

Before the measurement, entire accessories, such as an ice tray and a glass shelf, were removed. The acoustic pressure measurement was performed at a total of 12 points over four rows with three different microphone positions inside the freezer cavity. Figure 16 shows the four rows of the microphone positions.

The signals were acquired using GRAS 46AE (1/2)″ free-field microphones with a 4-channel FFT analyser linked to a computer for postprocessing. Two microphones were used for one measurement: the first microphone, which was located close to the sound source, was used as a reference signal whereas the second microphone, which was located inside the freezer cavity, was used as a response signal. The acquisition processes were repeated 12 times for each measuring position. The FFT analyser computes the

autospectrum of each channel as well as the cross spectrum between the reference microphone connected to channel 1 and the response microphone connected to channel 2. This information is required to calculate both the transfer function, H_1, and the coherence, v. The frequency response function between the microphone responses at all of the grid points with respect to the reference microphone was measured (Pa/Pa)

$$H_1 = \frac{G_{xx}G_{yy}}{G_{xy}},$$

$$v = \frac{G_{xy}{}^2}{G_{xx}G_{yy}}.$$

(18)

In the measurement setup, data were collected for 10 seconds at each point by taking the linear average of the signal. The frequency resolution was set to 1,563 Hz between a frequency range of 0–1250 Hz. To satisfy the better periodicity requirement of the FFT analysis, the Hanning weighting function was used. This function minimizes the leakage error.

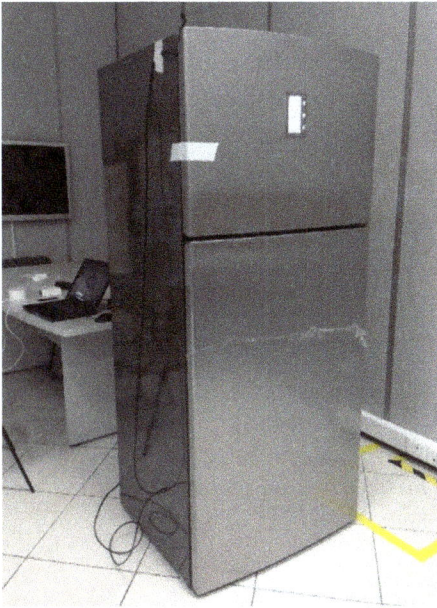

FIGURE 13: Refrigerator.

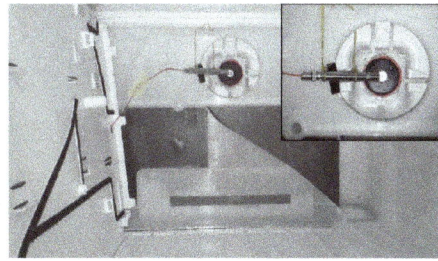

FIGURE 14: Acoustic modal analysis system configuration.

TABLE 6: First ten measured natural frequencies of the freezer cavity.

Natural frequencies (250–650 Hz)									
Experiment									
261.4	381.4	384.3	454.2	462.3	531.8	554.7	619.0	662.1	664.3

All of the measurements were taken at room temperature (25°C) inside the semianechoic room. Figures 17, 18, 19, and 20 illustrate the frequency response spectrum for all four measurement rows. Figure 21 shows one of the coherence functions.

From the analytical solution, we know the frequency interval that must be investigated. The analytical results indicate that the first ten modes of the freezer cavity are in a frequency range of 250–650 Hz. The first ten measured natural frequencies of the freezer cavity are presented in Table 6.

However, when the frequency response spectrum was investigated, certain peak frequencies were observed in a frequency range of 0–250 Hz. Generally, these lower peak values cannot belong to modes of the cavity because of the dimensions of the cavity. The primary reason for this

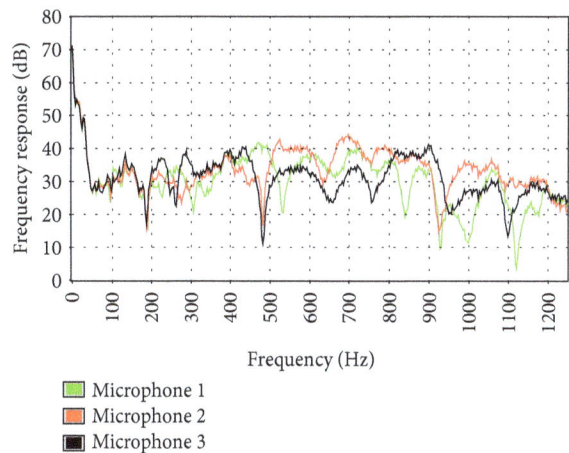

FIGURE 15: Position of the loudspeaker and the reference microphone.

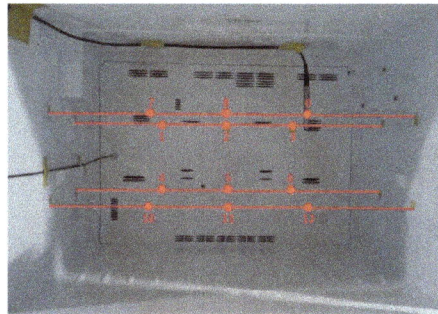

FIGURE 16: Four rows of the microphone positions.

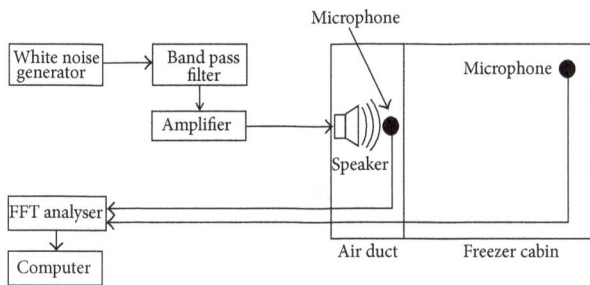

FIGURE 17: Frequency response spectrum (upper rear row).

TABLE 7: Measured peak frequencies obtained from the experiments in the range of 0–250 Hz.

Peak frequencies (0–250 Hz)									
31.2	56.2	65.6	87.5	103.1	118.7	134.3	150.0	178.1	203.1

phenomenon could be the structural-borne noise, which radiates from the vibration in the air duct. Thus, these lower peak frequencies are clearly observed at all of the measurement points. Hence, the dynamic behaviour of the air duct is also analysed within the scope of this study. Table 7 lists the measured peak frequencies obtained from the experiments in the range of 0–250 Hz.

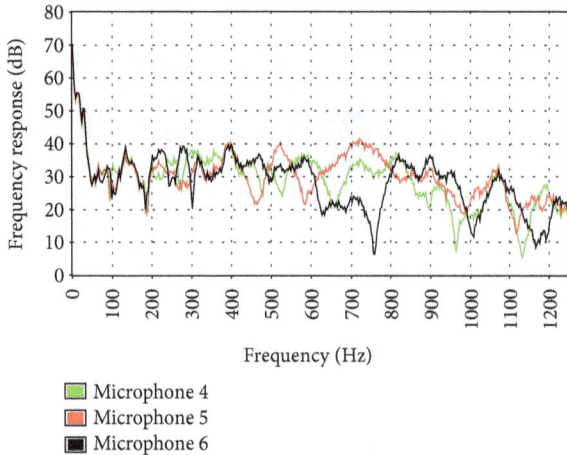

FIGURE 18: Frequency response spectrum (lower rear row).

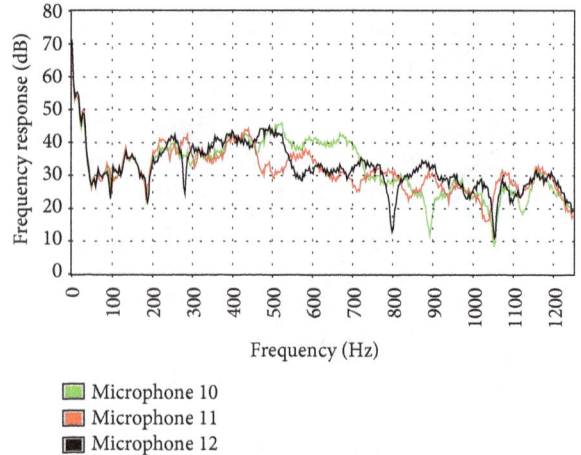

FIGURE 20: Frequency response spectrum (lower front row).

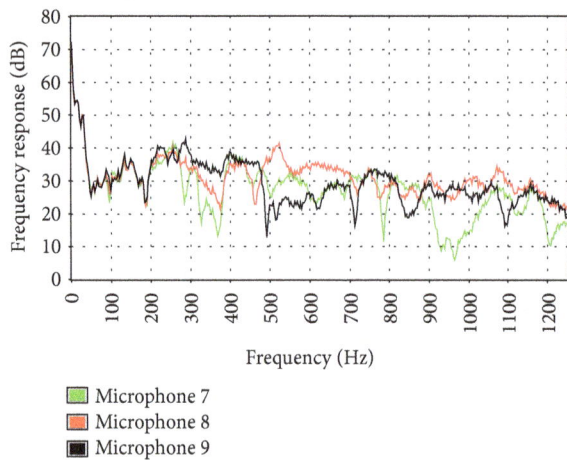

FIGURE 19: Frequency response spectrum (upper front row).

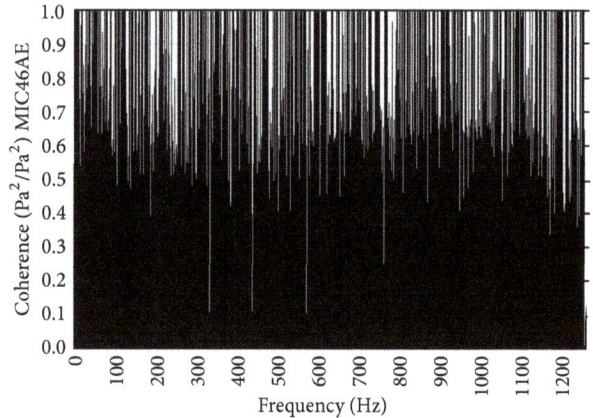

FIGURE 21: One of the coherence functions.

In the next step, the experimental modal analysis was applied to the air duct to define the dynamic characteristics of the structure under an external load. A series of frequency response functions were measured at various geometric locations using an instrumented impact hammer to supply an input force. The responses are measured in the z direction using a motion sensor, that is, an accelerometer. The system configuration presented in Figure 22 is used to perform the modal analysis.

The air duct is composed of two separate components: one is the evaporator cover and the other one is the fan louver. These two components were analysed individually. The evaporator cover and the fan louver were suspended with elastic cords from two points to simulate the free – free conditions as much as possible. The measurements were obtained from 84 points for the evaporator cover and 103 points for the fan louver. Figure 23 depicts the points on the fan louver and the evaporator cover where the measurements were obtained.

The experiment was performed with a 4-channel FFT analyser, where one channel was used to acquire the force signal and one channel was used for the acceleration signal

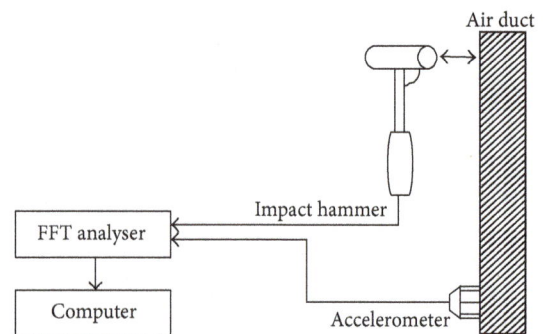

FIGURE 22: Modal analysis system configuration.

simultaneously. The frequency response functions were measured between 0 and 250 Hz using a frequency resolution of 0.78 Hz. The excitation was provided in the horizontal direction by an impact hammer from one point, and the response was measured at all of the defined points. A rubber

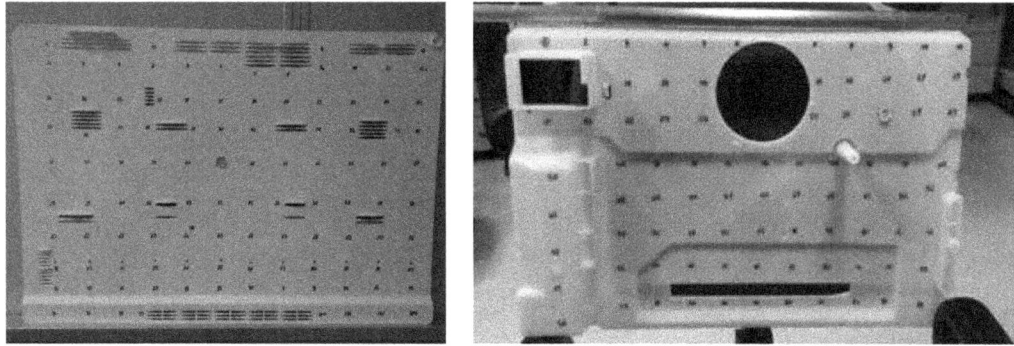

FIGURE 23: Measurement points on the fan louver and the evaporator cover.

TABLE 8: First ten experimentally measured natural frequencies and the corresponding damping ratios of the fan louver and the evaporator cover.

Natural frequencies (Hz)			
Fan louver		Evaporator cover	
Hz	%	Hz	%
5.0	2.1	14.1	2.6
37.0	2.4	18.8	1.7
46.5	1.5	32.8	3.1
55.5	2.5	56.3	2.2
77.0	0.8	70.3	2.7
104.0	1.3	79.7	1.3
112.0	2.4	89.8	3.0
129.0	4.1	101.0	3.2
133.0	2.9	107.0	1.9
149.0	3.1	114.0	3.1

TABLE 9: Comparison of the first ten natural frequencies obtained from the numerical and experimental analyses for the fan louver.

Natural frequencies (Hz)	
Numerical model	Experiment
4.8	5.0
36.5	37.0
49.0	46.5
54.6	55.5
76.3	77.0
105.1	104.0
111.2	112.0
130.6	129.0
132.3	133.0
150.2	149.0

5. Evaluation of the Results

To ensure the accuracy of the numerical model, the natural frequencies and mode shapes were compared with the measured experimental data for the validation process. In the experimental analyses, the modal animation software ME'scope was used for the visualization of the mode shapes. To solve the interaction between the air duct and the freezer cavity, the overlapping frequencies were identified, and a design optimization process was performed to change the natural frequencies of the structure; thus, the interaction between the domains was reduced. The comparison of the first ten natural frequencies obtained from the numerical and experimental analyses for the fan louver are provided in Table 9.

Figure 24 illustrates a comparison of the numerical and experimental first three mode shapes of the fan louver.

As shown in Table 9, the mode frequencies, which are acquired from the experimental analysis, agreed relatively well with the existing modes obtained from the numerical model of the fan louver. Although the polypropylene material is highly damped, the experimental mode shapes of the fan louver can be clearly observed. Particularly, in the third mode, there are a few local mode shapes that have been slightly observed. The comparison of the first ten natural frequencies

tip impact hammer was used to provide a long pulse and excite a narrow frequency range.

To prevent an incorrect estimation of the amplitude and frequency due to the effects of the nonperiodic signals, the measurement period was extended, and to better satisfy the periodicity requirement of the FFT process, the Hanning time weighting function was applied. This function attempts to heavily weight the beginning and end of the sample record to zero. Table 8 presents the experimentally measured ten natural frequencies and the corresponding damping ratios of the fan louver and the evaporator cover.

When the results are investigated, a common inference can be made: the modal density is higher than 100 Hz for both structures. There are five modes for the fan louver between 100–150 Hz and three modes for the evaporator cover between 100–115 Hz. Furthermore, the modal damping ratios obtained from the curve fitting algorithm using the modal animation software ME'scope were entered in the numerical analyses.

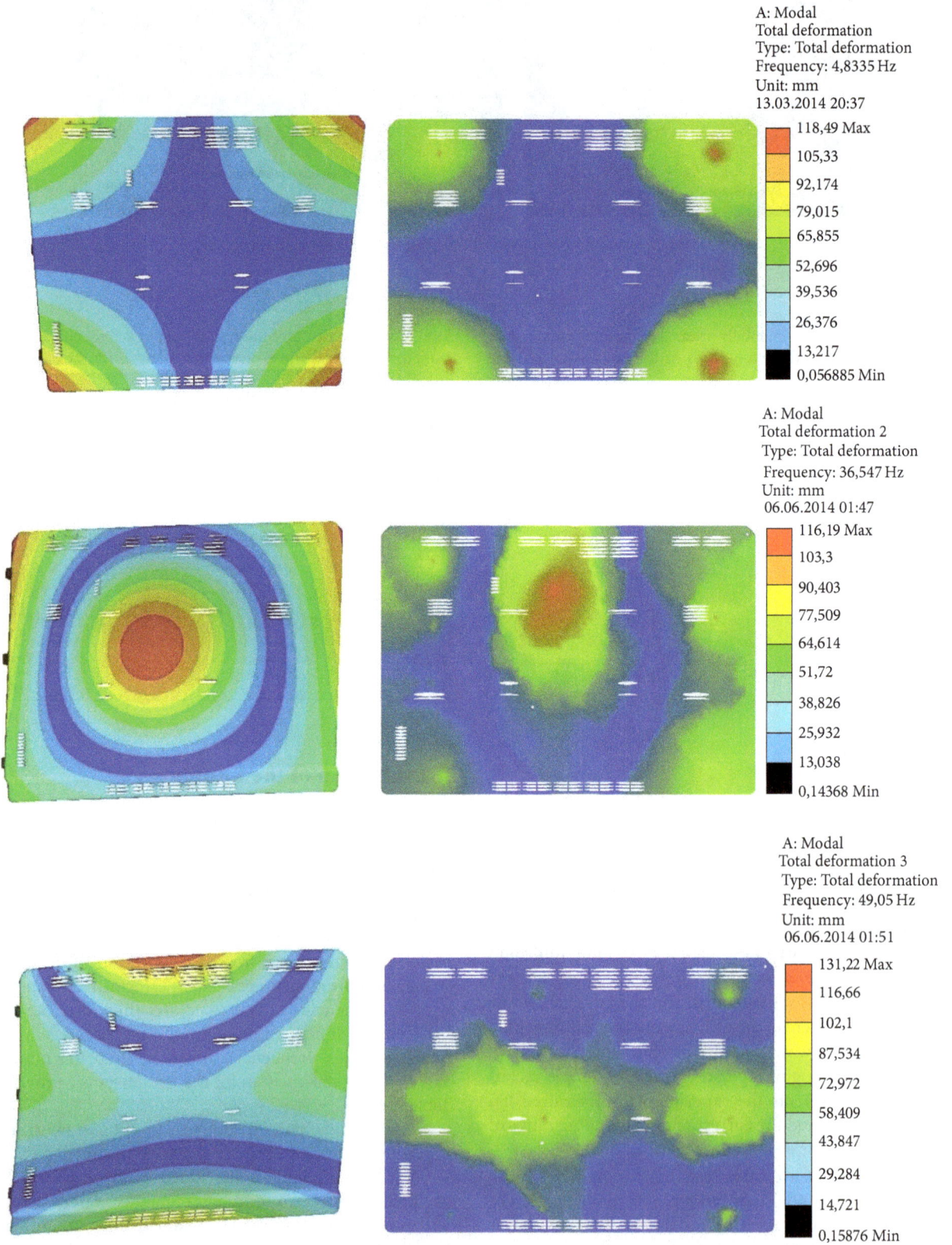

FIGURE 24: Numerical and experimental comparison of the first three mode shapes of the fan louver.

TABLE 10: Comparison of the first ten natural frequencies obtained from the numerical and experimental analyses for the evaporator cover.

Natural frequencies (Hz)	
Numerical model	Experiment
15.0	14.1
18.3	18.8
35.8	32.8
58.8	56.3
73.6	70.3
80.5	79.7
90.5	89.8
103.5	101.0
110.5	107.0
117.0	114.0

TABLE 11: Comparison of the numerical and experimental acoustic natural frequencies of the freezer cavity.

Natural frequencies (Hz)	
Numerical model	Experiment
262.6	261.4
386.7	381.4
388.3	384.3
468.7	454.2
469.9	462.3
526.1	531.8
551.8	554.7
613.0	619.0
657.6	662.1
658.7	664.3

TABLE 12: Comparison of the natural frequencies of the air duct and the peak frequencies of the freezer cavity.

Peak frequencies (Hz)	
Freezer cavity	Air duct
31.2	32.8 (evap. cover)
56.2	55.5 (fan louver)
65.6	—
87.5	89.8 (evap. cover)
103.1	104.0 (fan louver)
118.7	—
134.3	133.0 (fan louver)
150.0	149.0 (fan louver)
178.1	—
203.1	—

obtained from the numerical and experimental analyses for the evaporator cover is provided in Table 10.

Figure 25 presents a comparison of the numerical and experimental first three mode shapes of the evaporator cover.

In Table 10, the mode frequencies obtained from the experimental measurements are closely matched with the natural frequencies obtained from the numerical model of the evaporator cover. Despite the complexity of the geometry, the acquired data clearly reflect the motion of the structure in Figure 25. The experimental mode shapes agree with the numerical mode shapes extremely well. The numerical and experimental acoustic natural frequencies of the freezer cavity are presented in Table 11.

Between the range of 261.4 Hz and 462.3 Hz, the experimentally obtained modes are lower than the numerically obtained modes. Above this range, the experimental mode frequencies begin to diverge from the numerical mode frequencies.

Figure 26 depicts the numerical and experimental pressure distributions corresponding to the first three acoustic mode shapes of the freezer cavity.

In addition to the acquired frequency response data inside the freezer compartment, the sound pressure data were also collected under normal working conditions of the refrigerator. The measurement was performed when the freezer fun was running, and the sound pressure versus frequency spectrum data were collected by one microphone. The rotational speed of the freezer fan is 2200 rpm, and the blade passing frequency is 147 Hz for the impeller with four blades. From Figure 27, the blade passing frequency (147 Hz) and its second harmonic (294 Hz) can be easily observed.

Based on these results, if the interaction between the freezer cavity and the air duct is investigated in the range of 0–250 Hz, it can be observed that certain natural frequencies show good agreement with the peak frequencies obtained from the experimental acoustic modal analysis of the freezer cavity. Table 12 presents a comparison of the natural frequencies of the air duct and the peak frequencies of the freezer cavity.

In conclusion, the blade passing frequency of the freezer fan and the 10th normal mode of the fan louver are extremely close to each other. This situation indicates that the freezer fan could excite the fan louver in its nominal working frequency and cause the vibration of the fan louver. These vibrations turn into sound energy, which radiates from the fan louver and increases the amplitude of the sound pressure inside the cavity. In fact, this phenomenon occurs because one of the resonance frequencies observed in the cavity is 150 Hz. To reduce this interaction between the structure and the cavity, two methods could be used: one is decreasing the amplitude of the resonance frequency while the other is shifting the resonance frequency. The first method could be achieved by adding the damping material to the points that have a maximum deformation based on the mode shape of the fan louver at the resonance frequency. However, this method is not applicable due to the restrictions related to human health.

The technique that shifts the resonance frequency could be successful through certain design modifications. Thus, the rotation speed and the number of blades of the fan could be changed to alter the excitation frequency. However, all of these options also change the cooling performance of the refrigerator; hence, this is not a preferred method.

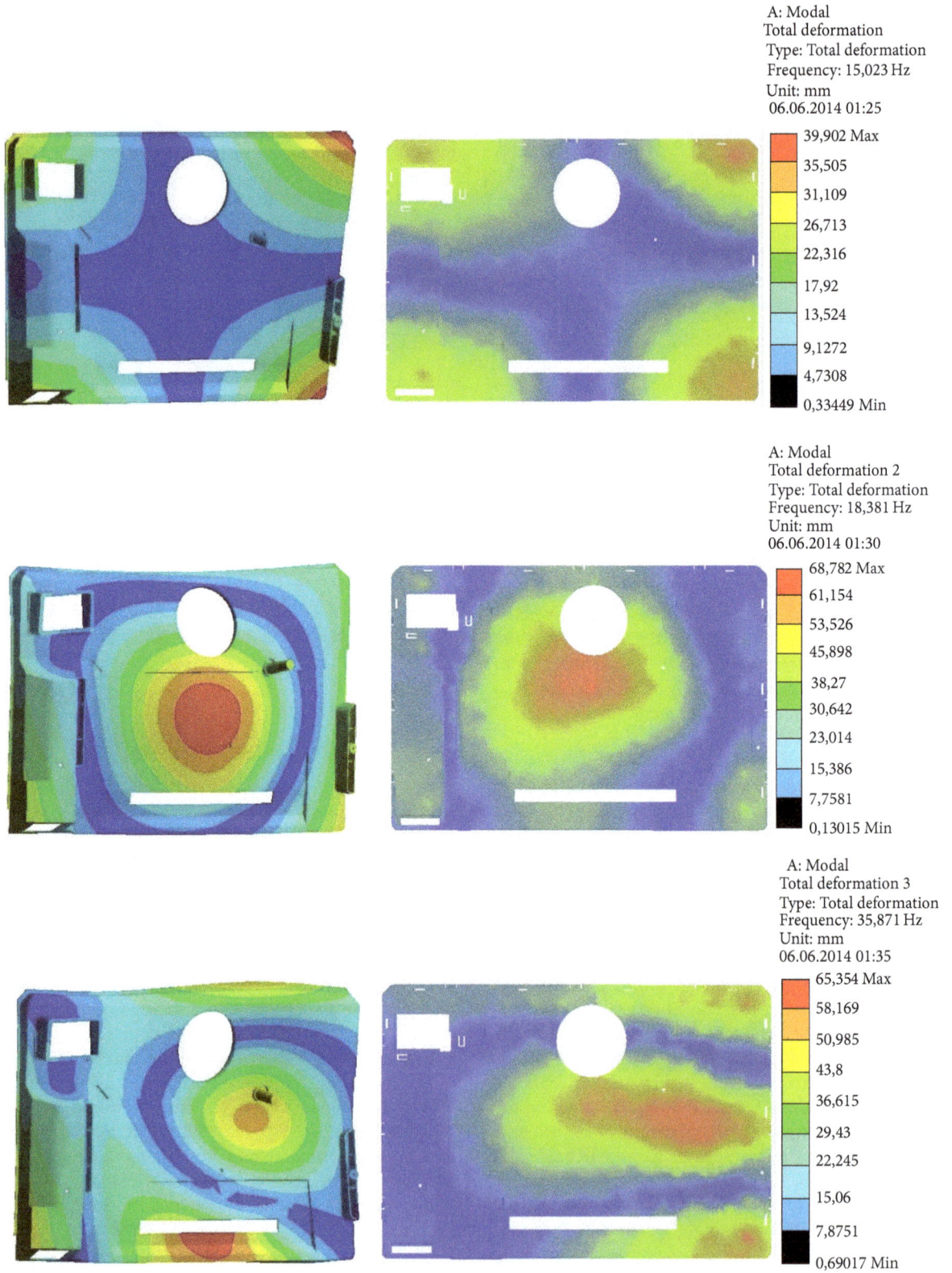

A: Modal
Total deformation
Type: Total deformation
Frequency: 15,023 Hz
Unit: mm
06.06.2014 01:25

39,902 Max
35,505
31,109
26,713
22,316
17,92
13,524
9,1272
4,7308
0,33449 Min

A: Modal
Total deformation 2
Type: Total deformation
Frequency: 18,381 Hz
Unit: mm
06.06.2014 01:30

68,782 Max
61,154
53,526
45,898
38,27
30,642
23,014
15,386
7,7581
0,13015 Min

A: Modal
Total deformation 3
Type: Total deformation
Frequency: 35,871 Hz
Unit: mm
06.06.2014 01:35

65,354 Max
58,169
50,985
43,8
36,615
29,43
22,245
15,06
7,8751
0,69017 Min

FIGURE 25: Comparison of the numerical and experimental first three mode shapes of the evaporator cover.

A: Modal
Acoustic pressure 2
Expression: PRES
Frequency: 0, Hz
Sweeping phase: 0,°
Unit: MPa
04.05.2014 17:07

43,983 Max
34,209
24,436
14,662
4,8884
−4,8853
−14,659
−24,433
−34,206
−43,98 Min

A: Modal
Acoustic pressure 3
Expression: PRES
Frequency: 0, Hz
Sweeping phase: 0,°
Unit: MPa
04.05.2014 17:09

54,594 Max
42,899
31,205
19,511
7,8162
−3,8781
−15,572
−27,267
−38,961
−50,655 Min

A: Modal
Acoustic pressure 4
Expression: PRES
Frequency: 0, Hz
Sweeping phase: 0,°
Unit: MPa
04.05.2014 17:09

52,514 Max
41,15
29,787
18,423
7,0599
−4,3035
−15,667
−27,03
−38,394
−49,757 Min

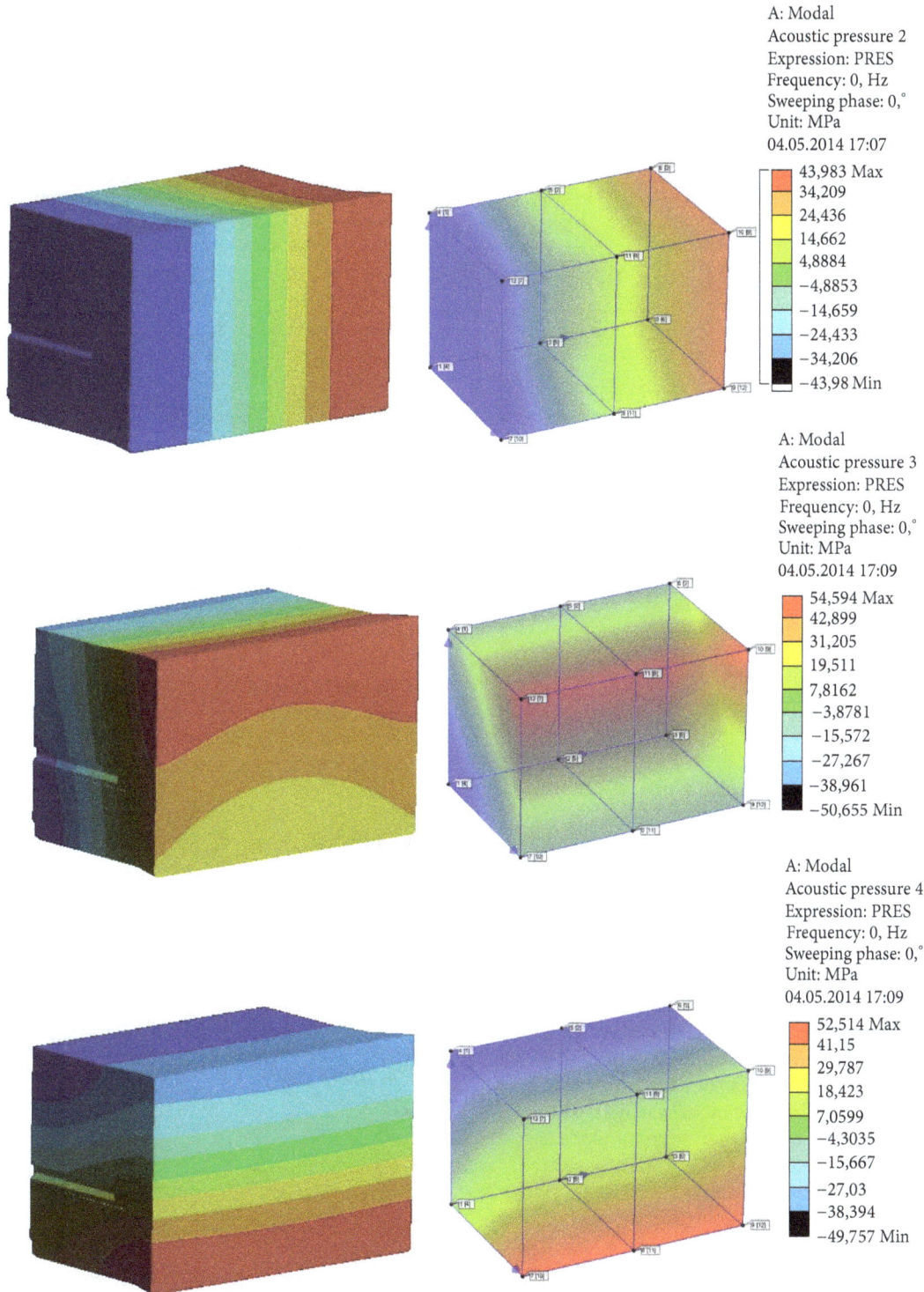

FIGURE 26: Numerical and experimental pressure distributions corresponding to the first three acoustic mode shapes of the freezer cavity.

In this study, the structural modification was applied to the fan louver because the other methods are not applicable to the refrigerator in reality. The thickness of the fan louver was increased to shift its natural frequencies. A change in frequencies was observed when the thickness was increased. The optimum thickness was determined as 2.1 mm (the current thickness is 2.0 mm). Table 13 presents the natural frequencies of the modified fan louver.

Using this modification, the 10th normal mode of the fan louver moved to 157.7 Hz in the frequency domain, and none

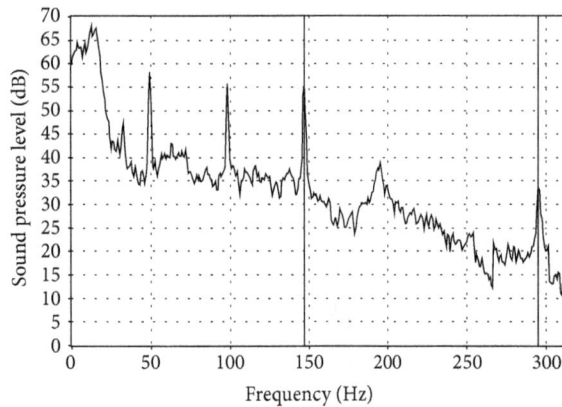

FIGURE 27: Frequency spectrum of the sound pressure level.

TABLE 13: Natural frequencies of the modified fan louver.

Natural frequencies (Hz)									
Numerical model									
5.0	38.3	51.4	57.3	80.1	110.3	116.8	137.1	138.9	157.7

of the submodes overlapped with the blade passing frequency. Thus, the interaction between the structure and the cavity was reduced.

6. Conclusions

In this study, the acoustic characteristics of the refrigerator freezer compartment are investigated. A reliable and accurate numeric model is developed to simulate the coupled behaviour of the structure and the acoustic cavity for the freezer compartment. A mixed finite element method is used to solve the acoustic-structure interaction of the acoustic cavity coupled with the elastic body. The numerical results indicate how the coupled structures change the natural frequencies and the mode shapes of the acoustic cavity.

The numeric results are validated with the experimental analyses. Thus, an experimental modal analysis of the fan louver and the evaporator cover is performed, and the natural frequencies are found. Lastly, an experimental acoustic modal analysis is implemented to the freezer cavity; thus, the normal modes that belong to both the acoustic cavity and coupled structures are obtained.

These results demonstrate that the normal modes of the fan louver and the evaporator cover change the acoustic characteristic of the freezer compartment, especially in low frequency ranges.

Lastly, the noise generation mechanism inside the freezer compartment is investigated under normal working conditions of the refrigerator. The resonance frequencies generated by the fan unit are obtained. The overlapping frequencies between the source and the structures are shifted using certain design modifications. Thus, the structural-borne noise, which radiates from the air duct into the freezer cavity, is reduced.

Competing Interests

The authors declare that they have no competing interests.

References

[1] M. Baran, R. Bolejko, P. Pruchnicki, B. Zoltogorski, and A. Dobrucki, "Refrigerator noise and vibration control and optimization," in *Proceedings of the 16th International Congress on Sound and Vibration*, Krakow, Poland, July 2009.

[2] S. Seo, T. Kwak, C. Kim, J. Park, and K. Jo, "Noise and vibration reduction of a household refrigerator," in *Proceedings of the HVAC Refrigeration Engineering*, pp. 1133–1140, 2000.

[3] A. Takushima, Y. Shinobu, S. Tanaka, M. Eguchi, and K. Matsuki, "Fan noise reduction of household refrigerator," *IEEE Transactions on Industry Applications*, vol. 28, no. 2, pp. 287–292, 1992.

[4] S. Igarashi and K. Kitagawa, "Numerical analysis for propeller fan in freezing compartment of household refrigerator," in *Proceedings of the 9th of International Symposium on Transport Phenomena and Dynamics of Rotating Machinery*, Honolulu, Hawaii, February 2002.

[5] Y.-H. Kim, B.-S. Yang, and C.-J. Kim, "Noise source identification of small fan-BLDC motor system for refrigerators," *International Journal of Rotating Machinery*, vol. 2006, Article ID 63214, 7 pages, 2006.

[6] F. Gue, C. Cheong, and T. Kim, "Development of low-noise axial cooling fans in a household refrigerator," *Journal of Mechanical Science and Technology*, vol. 25, no. 12, pp. 2995–3004, 2011.

[7] S. Öztürk and H. Erol, "Numerical and experimental studies on the structure-borne noise control on a residential kitchen hood," *International Journal of Acoustics and Vibrations*, vol. 18, no. 1, pp. 3–6, 2013.

[8] H. Carlsson, *Finite Element Analysis of Structure-Acoustic Systems; Formulations and Solution Strategies*, TVSM 1005, Structural Mechanics, LTH, Lund University, Lund, Sweden, 1992.

[9] N. Ottosen and H. Peterson, *Introduction to the Finite Element Method*, Prentice Hall, New York, NY, USA, 1992.

[10] J. C. Couche, *Active control of automobile cabin noise with conventional and advanced speakers [M.S. thesis]*, Virginia Polytechnic Institute and State University, 1999.

16

Dynamic Modal Correlation of an Automotive Rear Subframe, with Particular Reference to the Modelling of Welded Joints

Vincenzo Rotondella,[1] Andrea Merulla,[2] Andrea Baldini,[2] and Sara Mantovani[1]

[1] MilleChili Lab, Department of Engineering "Enzo Ferrari", Università degli Studi di Modena e Reggio Emilia, Via Vivarelli 10, 41124 Modena, Italy
[2] Ferrari S.p.A, Via Abetone Inferiore 4, 41053 Modena, Italy

Correspondence should be addressed to Sara Mantovani; sara.mantovani@unimore.it

Academic Editor: Marc Thomas

This paper presents a comparison between the experimental investigation and the Finite Element (FE) modal analysis of an automotive rear subframe. A modal correlation between the experimental data and the forecasts is performed. The present numerical model constitutes a predictive methodology able to forecast the experimental dynamic behaviour of the structure. The actual structure is excited with impact hammers and the modal response of the subframe is collected and evaluated by the PolyMAX algorithm. Both the FE model and the structural performance of the subframe are defined according to the Ferrari S.p.A. internal regulations. In addition, a novel modelling technique for welded joints is proposed that represents an extension of ACM2 approach, formulated for spot weld joints in dynamic analysis. Therefore, the Modal Assurance Criterion (MAC) is considered the optimal comparison index for the numerical-experimental correlation. In conclusion, a good numerical-experimental agreement from 50 Hz up to 500 Hz has been achieved by monitoring various dynamic parameters such as the natural frequencies, the mode shapes, and frequency response functions (FRFs) of the structure that represent a validation of this FE model for structural dynamic applications.

1. Introduction

The employment of Finite Element (FE) models to predict the dynamic properties of a vehicle has continuously become more important in modern automotive industries. Whenever there is a new design or modification of an existing one, the structural dynamic properties of the car should be examined to fulfil some criteria proposed by the industry itself before the product can be launched on the market.

The traditional methodology for evaluating the structural dynamic properties of a vehicle is to perform various dynamic tests on prototypes of the product and to demonstrate their capacity to withstand these tests. Until the experimental results show that the prototypes can comply with the relevant criteria, the component has to be redesigned and another design-test loop must be followed. In this design-test-redesign loop, the higher percentage of time and financial resources is spent in producing prototypes and performing tests.

With the growing capabilities of computing techniques, and the strength of the competition between companies, FE model predictions are used more and more frequently to substitute practical dynamic test data. Furthermore, the FE modelling technique may also be used to predict the dynamic response of structures when working beyond a limit situation that makes the simulations by experiment extremely difficult, if not impossible. All of these results depend on the accuracy of FE model predictions. The validation of FE models and their capability to predict the dynamic behaviour of the structures are crucial topics for industrial purposes and especially for aerospace and automotive applications.

In 1990s, Baker [1], Imregun and Visser [2], and Friswell and Mottershead [3] performed a complete review of different correlation methods and validation criteria for structural dynamic behaviour, in order to investigate the methodologies that might improve the prediction capability of FE models.

Brughmans et al. [4] focused on test-analysis correlation for vibroacoustic application, integrating methods and algorithms in a high performance computational environment. In their research, Jambovane et al. [5] employed experimental modal analysis to validate FE model of an engine oil pan. The correlation between experimental and numerical modal analyses on panels has been performed by Siano et al. [6] and Splendi et al. [7] in order to verify the FE model reliability for Noise Vibration Harshness (NVH) applications.

Schedlinski et al. [8] presented the validation of an FE model of a Body-in-White using the computational modal updating procedure to improve the quality and prediction of the model.

In addition to the previous works, [9, 10] confirmed the experimental modal analysis as a fundamental requirement for the validation of a FE model, built to perform dynamic analysis. The present paper is based on original experimental data, and it describes a detailed methodology that aims to investigate the capability and reliability of a structural FE model for NVH applications. The present research is performed at MilleChili laboratory from the University of Modena and Reggio Emilia and in collaboration with Ferrari S.p.A.

2. Materials and Methods

The analyzed structure is an aluminium rear subframe that can be disassembled from the chassis; it is made of eleven extruded beams, plates, gussets, and two casting components, as shown in Figure 1. There are eight points on each side that allow the connection of the subframe to the chassis. The suspension mounting points are located on the casting component. This is a prototype component of the chassis and the different parts are jointed together through metal inert gas (MIG) welding. MIG is one of the most widely used forms of welding in industry, and it is considered one of the easiest forms of welding to learn. The total weight of the rear subframe is 26.3 kg, including the testing devices (*e.g.*, suspension "biscuits" and gearbox mounting "clocks") and the suspension mounting bolts.

2.1. Experimental Modal Analysis (EMA). As pointed out by Schwarz and Richardson [11], natural modes of vibrations are inherent properties of a structure. Modes or resonances are defined by physical properties (mass, stiffness, and damping) and boundary conditions of the component. The modal properties of a structure are natural frequencies, modal damping, and mode shapes. A modal testing could be summarized in two main different phases:

(i) Structure test to obtain FRF measurements

(ii) FRF curve fitting to extract experimental modal parameters

A fast and convenient way to find the modes of a structure is impact testing. It was developed in the 1970's and it has become the most popular modal testing used for evaluating

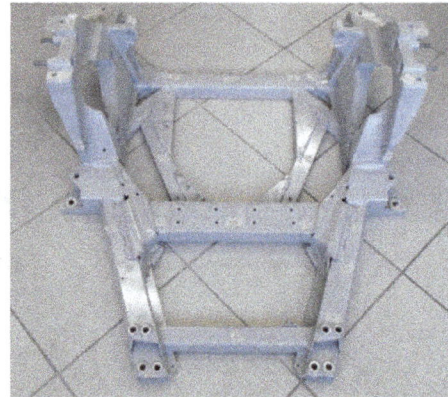

FIGURE 1: Aluminium rear subframe.

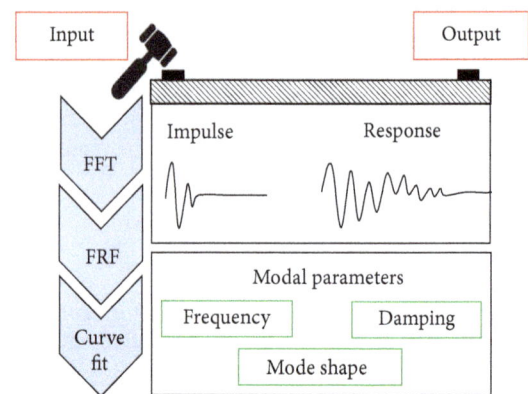

FIGURE 2: EMA through impact test.

the modal properties of a structure. The equipment required to perform an impact test is

(i) Impact hammer

(ii) Accelerometers

(iii) FFT analyzer

(iv) Postprocessing modal software

Figure 2 shows the whole process of EMA through impact test.

The modal parameter estimation is obtained from a set of frequency response function (FRF) measurements. The FRF, as explained in [11], "describes the input-output relationship between two points on a structure as a function of frequency." The basic formula of FRF is

$$H(\omega) = \frac{Y(\omega)}{X(\omega)}, \tag{1}$$

where

(i) $H(\omega)$ is the frequency response function,

(ii) $Y(\omega)$ is the output of the system in the frequency domain,

(iii) $X(\omega)$ is the input of the system in the frequency domain.

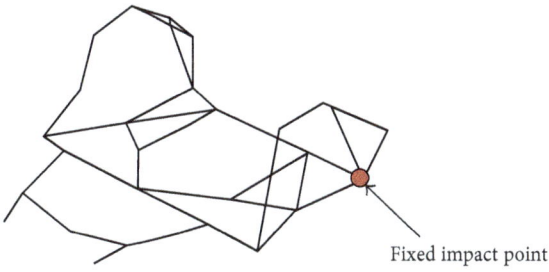

FIGURE 3: Test geometry of the rear subframe.

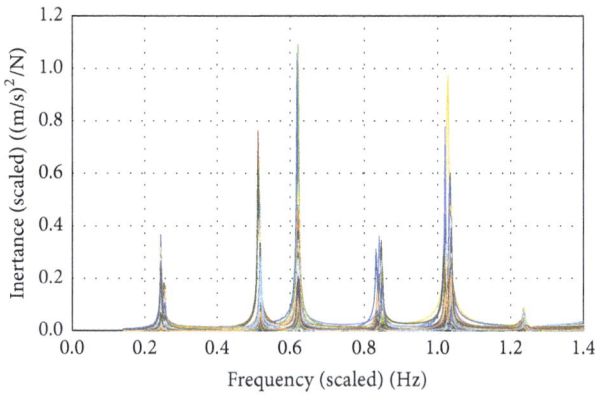

FIGURE 4: Plots of FRFs of the 35 reference points.

The modal test was performed at the NVH experimental department of Ferrari S.p.A. The structure was suspended by four soft elastic bungees in order to simulate the free-free condition; this condition means that the structure is not connected to the ground at any of its coordinates and it is, in effect, freely suspended in space. In this condition, the structure will exhibit rigid body modes, which are determined merely by its mass and inertia properties. It had been verified in a pretest phase that the suspension system did not interfere with the modes of vibration of the structure. A single impact point on the lower side of the left casting in the Z-direction was used during the test. The responses were measured in 35 positions, using three-axis accelerometers. The geometry (or wireframe) of the experimental setup is shown in Figure 3, where the red dot indicates the location of the impact point. The software used for impact testing is LMS Test.Lab.

Parameter estimation techniques for modal analysis are based on the extraction of natural frequency, the damping, and the mode shapes from the experimental data, which is in a processed form such as frequency response functions (FRFs). Figure 4 shows the FRFs collected from the structure during the impact test. The considered FRFs are the so-called inertance or receptance, because they measure the acceleration response of the structure at an output point, per unit of excitation force at an input point. The numerical values presented in this paper have been scaled by an arbitrary factor for secrecy reasons.

The estimation method used in this phase was PolyMAX (Peeters et al. [12]), which is available in LMS Test.Lab. As

described in [12], the PolyMAX method employs measured FRFs as primary data and it is called the z-domain method (i.e., a frequency-domain model is derived from a discrete-time model). The poles and modal participation factors could be retrieved following the right matrix-fraction model:

$$[H(\omega)] = \sum_{r=0}^{p} z^r [\beta_r] \cdot \left(\sum_{r=0}^{p} z^r [\alpha_r] \right)^{-1}, \qquad (2)$$

where

(i) $[H(\omega)]$ is the matrix containing the FRFs between all m inputs and all l outputs,

(ii) $z = e^{-j\omega\Delta t}$ is the frequency-domain model,

(iii) $[\alpha_r]$ and $[\beta_r]$ are the denominator and the numerator matrix polynomial coefficients,

(iv) p is modal order.

Once the coefficients $[\alpha_r]$ and $[\beta_r]$ are determined solving (2), it is possible to calculate the poles λ_i, λ_i^*, which are related to the eigenfrequencies ω_i and to the damping ratio ξ_i as follows:

$$\lambda_i, \lambda_i^* = \xi_i \omega_i \pm \sqrt{1 - \xi_i^2} \omega_i. \qquad (3)$$

The mode shapes could be determined by considering pole-residue model:

$$[H(\omega)]$$
$$= \sum_{i=1}^{n} \frac{\{v_i\} \langle l_i^T \rangle}{j\omega - \lambda_i} + \frac{\{v_i^*\} \langle l_i^{T*} \rangle}{j\omega - \lambda_i^*} - \frac{[\text{LR}]}{\omega^2} + [\text{UR}], \qquad (4)$$

where

(i) $[H(\omega)]$ is the matrix containing the FRFs between all m inputs and all l outputs,

(ii) n is the number of modes,

(iii) $\{v_i\}$ and $\{v_i^*\}$ are the complex conjugate mode shapes,

(iv) $\langle l_i^T \rangle$ and $\langle l_i^{T*} \rangle$ are the complex conjugate modal participation factors,

(v) λ_i and λ_i^* are the complex conjugate poles,

(vi) [LR] and [UR] are the lower and upper residuals which model the influence of the out-of-band modes.

Estimated poles are calculated from (2) and (3) and the results of this operation are presented in a so-called stabilization diagram. The interpretation of the stabilization diagram yields to a set of complex poles and participation factors, which are inserted in (4). Solving (4) in a linear least-squares sense, it is possible to retrieve the mode shape for each eigenfrequency.

Figure 5 shows the stabilization diagram employed to determine the modal parameters using PolyMAX method inside LMS Test.Lab software. Table 1 lists the natural frequencies, from 50 Hz up to 500 Hz, scaled, which was identified through EMA.

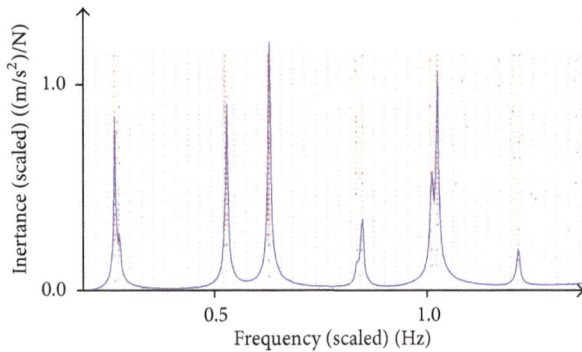

FIGURE 5: Stabilization diagram for PolyMAX method; the blue curve represents the sum of FRFs.

TABLE 1: Natural frequencies (scaled) from EMA.

Mode	Frequency (scaled) [Hz]	Damping ratio (%)
1	0.244	0.20
2	0.254	0.35
3	0.519	0.04
4	0.623	0.08
5	0.834	0.08
6	0.849	0.03
7	1.023	0.09
8	1.037	0.07
9	1.236	0.13

2.2. Finite Element Modal Analysis. The FE modelling techniques used for this study followed the internal criteria of the CAE Department of Ferrari S.p.A.

The extruded beams were modelled according to shell formulation using CQUAD4 and CTRIA3 elements. The castings were modelled with tridimensional elements; second-order CTETRA were used in order to compensate the spatial discretization. The description of the Finite Element model of welded joints is widely discussed in the next paragraph. Testing devices (e.g., suspension "biscuits" and gearbox mounting "clocks") were added to the FE model in order to accurately reproduce the testing setup. They were connected to the structure through CBAR elements. Nonstructural mass NSML1 was applied to align the numerical mass to the weight of the physically tested structure, taking into account the mass added from coating process.

Figure 6 shows the full three-dimensional view of the FE model of the rear subframe.

2.3. FE Modelling Technique for Welded Joints. The FE approach for modelling the welded joints is a relevant research topic for the present study. There are two different techniques for welding models: for stress analysis and for stiffness-based analysis. Obviously, the choice of the modelling technique depends on the aim of FE analysis performed. In literature [13], different methods for modelling the welded joints are proposed, mainly addressing stress

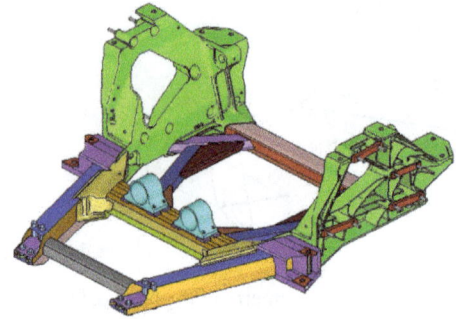

FIGURE 6: FE model of rear subframe.

analysis. Welded joints should be modelled in cases where stress is influenced by bending behaviour, when it is not easy to distinguish the nonlinear stress, caused from the notch to the welding foot, by stress concentrations effects that arise from geometric irregularities. In these cases, the stiffness of the welded joint section should be taken into account and the welds have to be modelled using several techniques. Although the stress field is detailed, the evaluation of the stiffness is not necessary accurate; for modal analysis this is a mandatory parameter; then this approach is not advised for this activity. Stiffness-based models of welding require an accurate representation of the stiffness (and the mass) of the joints. Furthermore, one fundamental requirement is a limited time spent in the FE model assessment. In the past, very simple models of welded joints have been used extensively in automotive industry; they consist of elastic or rigid one-dimensional elements or coincident nodes. The disadvantage of these models is the inadequacy to represent the behaviour of the welding and the underestimation of the stiffness. For modal and static stiffness analysis, a brick model by Pal and Cronin [14] was proposed. In this technique, a brick element was used to represent the spot weld and the connection between the shell plates and the solid was performed through rigid elements. A very accurate stiffness was guaranteed; however a mesh congruence between shell and brick elements was necessary and thus a high time of model setup was required. To overcome this problem, Backhans and Cedas [15] proposed the ACM1 method that allowed the connection between noncongruent shell meshes through rigid beam forming an umbrella shape. In recent years, the three-dimensional approach for welding is becoming more commonly employed in FE models for industrial analysis because they well represent the stiffness, the geometry, and also the mass of the connection. Heiserer et al. [16] proposed the model known as ACM2 for spot welds. As described in [17], the model consists of a brick element connecting the upper and lower plates via RBE3 elements. The RBE3 element distributes the applied loads throughout the model. Forces and moments applied to the brick nodes are distributed to the shell nodes in a way that depends on the RBE3 geometry and weight factors assigned to the shell nodes. The weights are the values assumed by the shape function corresponding to each shell node at the location of the brick node. The force acting on the brick node can be transferred to the

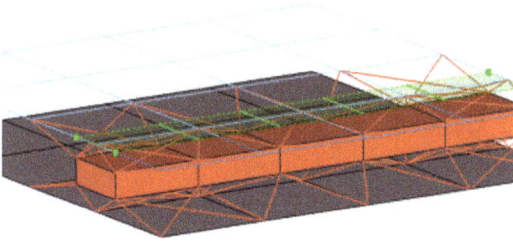

FIGURE 7: Details of ACM2 extended method: CHEXA for the body of the welded joint and RBE2/RBE3 for the head.

TABLE 2: Natural frequencies (scaled) from FEA.

Mode	Frequency (scaled) [Hz]
7	0.244
8	0.252
9	0.503
10	0.612
11	0.817
12	0.833
13	1.000
14	1.019
15	1.234

weighted centre of gravity of the shell nodes together with the moment produced by the force offset. The force is distributed to the shell nodes in proportion to the weighting factors. The moment is distributed as forces, whose magnitudes are proportional to their distance from the centre of gravity times their weighting factors.

The novel approach used in this paper is to extend the ACM2, proposed for spot welds, to MIG welding. A row of CHEXA elements is created for the body of welded joints and numerous RBE2/RBE3 elements for the head, as shown in Figure 7. The head elements project and connect the nodes of the body to the nodes of the adjoining shell elements. If there is a direct normal project then RBE2 elements are used; if there are only nonnormal projections, then RBE3 elements are created. The CHEXA elements are projected in a way that they touch the shell elements of the connecting parts.

This approach allows the development of a FE model which is useful both for stress analysis and for structural dynamic analysis.

Figure 8 shows the FE approach adopted for modelling the welded joints on the structure.

The FE model consists of 453852 nodes and 379708 elements. The structure was analyzed in free-free conditions, so six clear rigid body modes have been expected in the results. The results of the FE modal analysis, scaled, from 50 Hz up to 500 Hz are listed in Table 2. The solver used to obtain the modal numerical solution is Altair Optistruct 13.0 [17], included in Altair HyperWorks 13.0 suite.

A comparison between the novel ACM2 approach and the standard rigid one-dimensional elements has been performed. Figure 9 shows a detail of the same MIG welded joint, modelled through ACM2 method and rigid elements method.

Table 3 lists the natural frequencies, extracted from FEA, using two different methods. Both FE models present the same weight. The lack of mass due to the use of rigid elements for welded joints has been compensated through the application of nonstructural mass. ACM2 method presents generally natural frequencies higher than rigid elements method. The ACM2 model is stiffer than the second model because each ACM2 welded joint involves a greater number of nodes than the node-to-node rigid one-dimensional element.

The FE model using ACM2 method for MIG welded joints will be used for experimental-numerical correlation activity and FEA results will refer to it.

2.4. Modal Correlation. The correlation phase is focused on comparing, understanding, and evaluating the correlation between test and FE data. A modal based index, which is used for comparing experimental and numerical modal shapes, is Modal Assurance Criterion (MAC). MAC was originally developed for orthogonality check and, in the late 1970s, was proposed by Allemang and Brown [18] as a correlation coefficient for modal analysis. The function of the MAC [19] is to provide a measure of consistency between one modal and another reference modal vector. It is defined as a scalar constant which takes on values from zero, representing no consistent correspondence, to one, representing fully consistent correspondence. The MAC value between experimental and numerical FE modal vector is obtained by the following equation:

$$
\begin{aligned}
&\text{MAC}\left(\{\Phi\}_{\text{test}}, \{\Phi\}_{\text{FE}}\right) \\
&= \frac{\left|\{\Phi\}_{\text{test}}^{T} \cdot \{\Phi\}_{\text{FE}}\right|^{2}}{\left(\{\Phi\}_{\text{test}}^{T} \cdot \{\Phi\}_{\text{test}}\right) \cdot \left(\{\Phi\}_{\text{FE}}^{T} \cdot \{\Phi\}_{\text{FE}}\right)},
\end{aligned}
\tag{5}
$$

where

  (i) $\{\Phi\}_{\text{test}}$ is the modal vector of a EMA modal shape,

  (ii) $\{\Phi\}_{\text{test}}^{T}$ is the transpose of $\{\Phi\}_{\text{test}}$,

  (iii) $\{\Phi\}_{\text{FE}}$ is the modal vector of a FEA modal shape,

  (iv) $\{\Phi\}_{\text{FE}}^{T}$ is the transpose of $\{\Phi\}_{\text{FE}}$.

Generally, values of MAC should be above 0.7 for representing a good correlation. It is also worth noting that every degree of freedom (DOF) gives a contribution to the MAC index. So, relatively considerable in-phase displacements of a DOF give a positive contribution to the correlation; relatively considerable out-of-phase displacements give a negative contribution; and relatively small displacements give less important contributions.

3. Results and Discussion

The correlation results have been evaluated comparing the natural frequency values and the error percentages, the mode shapes, and the MAC matrix.

Plotting the FEA-EMA values of natural frequencies, the theoretical best fit is represented by a 45-degree line. The

TABLE 3: Comparison between natural frequencies (scaled) from FEA using different methods.

Mode	ACM2 method Scaled freq. [Hz]	Rigid elements method Scaled freq. [Hz]	Difference (%)
7	0.244	0.237	2.742
8	0.252	0.249	1.297
9	0.503	0.500	0.609
10	0.612	0.617	−0.896
11	0.817	0.787	3.630
12	0.833	0.820	1.610
13	1.000	0.978	2.193
14	1.019	1.019	0.029
15	1.234	1.226	0.655

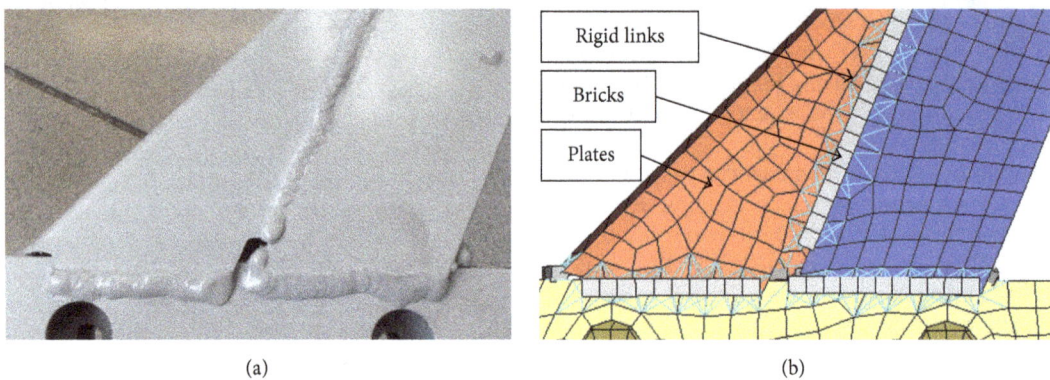

FIGURE 8: Details of MIG welding on the structure: (a) the actual welded joint, (b) the ACM2 approach.

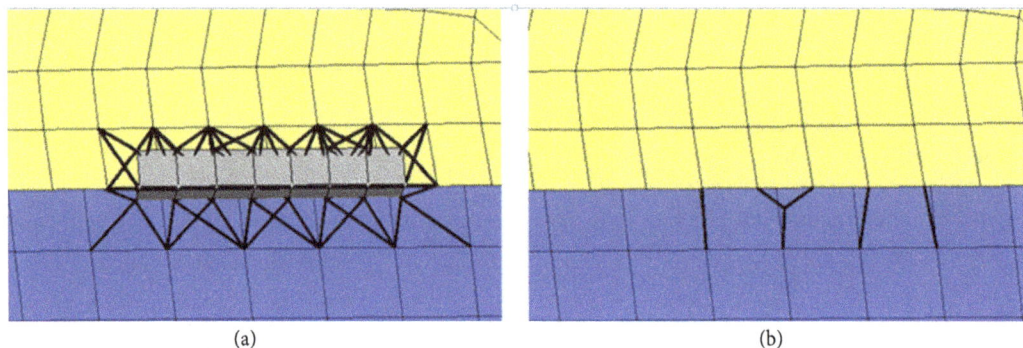

FIGURE 9: Details of MIG welding models: (a) the ACM2 approach, (b) the rigid elements approach.

natural frequency plot in Figure 10 exhibits an acceptable correlation, in which the frequency couples constitute a line very close to the reference one.

As pointed out in Figure 11, analyzing every single mode, the absolute percentage error between experimental and numerical eigenfrequency is less than 4 percent. The low percentage error represents a simple index to verify the correlation in automotive applications.

Figure 12 shows a numerical and experimental comparison among the deformed shape of the first modes couple. In particular, it corresponds to the first global torsional mode of the structure (EMA scaled frequency is 0.24 and FEA scaled normalized frequency is 0.244).

In addition, to a simple visual analysis of each mode shape, a comparison based on MAC values has been performed in order to deeply investigate the correlation achieved. Figure 13 shows a good correlation between EMA and FEA frequencies because the values on the main diagonal of the MAC matrix are greater than 0.7. This value represents a reliable MAC index for a satisfying correlation for complex structures, as described also in [6, 8]. Only the fifth mode pair (EMA Mode#5, FEA Mode#11) shows a MAC value on the main diagonal of the matrix that is lower than 0.7. The reason of this lack of correlation in terms of MAC could be found in the impact test execution. In Figure 5, the blue curve

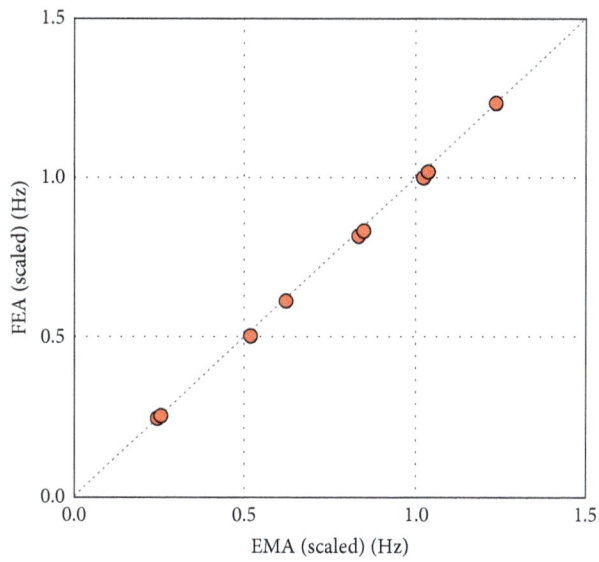

FIGURE 10: Natural frequency pairs plot.

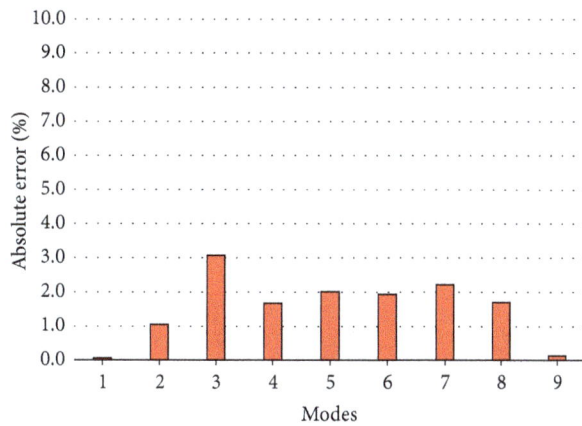

FIGURE 11: Absolute percentage error histogram.

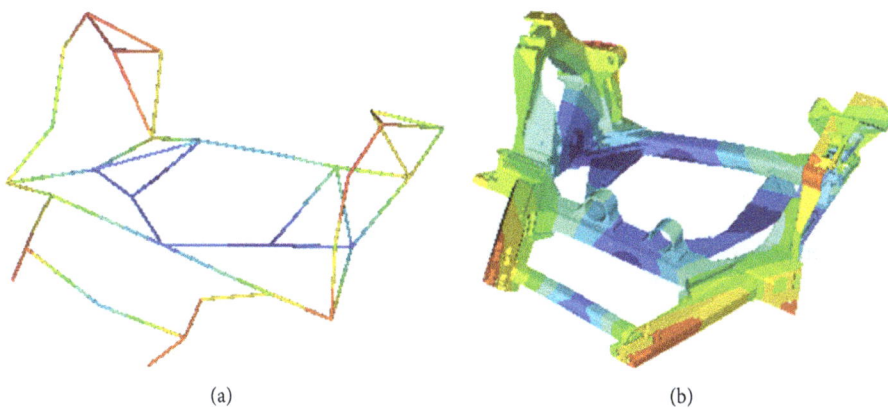

(a)   (b)

FIGURE 12: Mode shapes for the first pair of modes: it represents the first torsional mode: (a) experimental model on the left, (b) FE model.

TABLE 4: Summary of correlation between EMA and FEA results.

Mode pair	EMA scaled freq. [Hz]	FEA scaled freq. [Hz]	Error (%)	MAC
1	0.244	0.244	0.060	0.90
2	0.254	0.252	1.055	0.89
3	0.519	0.503	3.076	0.84
4	0.623	0.612	1.677	0.96
5	0.834	0.817	2.016	0.55
6	0.849	0.833	1.934	0.76
7	1.023	1.000	2.221	0.73
8	1.037	1.019	1.709	0.73
9	1.236	1.234	0.134	0.82

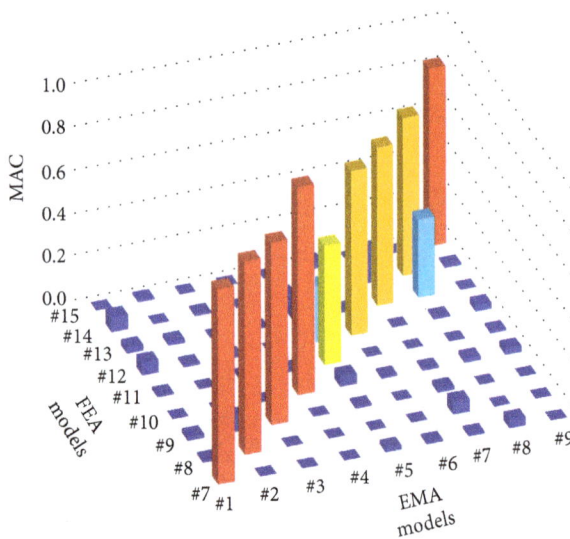

FIGURE 13: MAC matrix.

represents the sum of all the FRFs and it is worth noting that Mode#5 is not well excited during the hammer test.

Table 4 summarizes the obtained results, evidencing the natural frequency of each mode, the absolute percentage error, between EMA and FEA, and the MAC values for each mode pair.

4. Conclusions

In the present paper, a correlation activity has been performed between experimental and numerical modal analyses of an aluminium rear subframe in free-free condition. In particular, a novel modelling technique of welded joints has been applied; the ACM2 approach, formulated by [16] for spot welds, has been extended to MIG. A comparison between the novel ACM2 approach and the standard rigid one-dimensional method has been performed. The FEA using two different approaches lead to similar results with small differences due to stiffness contents in welding joints models. The novel ACM2 approach is used as reference for FEA results. Both an accurate FE representation of the test setup, due to the explicit modelling of testing devices, and a correct mass distribution on the structure have ensured a good

agreement between EMA and FEA. Comparing the results in terms of natural frequencies, a satisfying correlation has been found; the percentage error is lower than 4 percent and it evidences a very good agreement in terms of dynamic behaviour prediction of the FE model. A detailed investigation on the modal shapes has been carried out using MAC. MAC has been formulated by [18] for evaluating the consistency of different modal vectors. The values on the main diagonal of the MAC matrix evidence a good correlation in terms of modal shapes. All the mode pairs present MAC values greater than 0.7, which represents a reference value for an acceptable correlation, evaluated by MAC. This paper represents a suitable and stable approach for an accurate validation of FE model for structural dynamic applications; evaluating the correlation between EMA and FEA for industrial purposes, it confirms the capability to predict the dynamic behaviour of the structure in the frequency range of interest.

Further improvements should be made on the experimental side in order to reduce the data noise, exciting in a proper way all the modes of the structure in the frequency range of interest. Although the correlation between EMA and FEA has been found to be good, additional developments could regard the improvement of the FE model through sensitivity and updating procedure, so as to define a more accurate numerical model for dynamic purposes. Also, better correlation could be achieved using different correlation indexes: an example is the Coordinate Modal Assurance Criterion (CoMAC) that attempts to identify which measurements contribute negatively to a low value of MAC.

Acknowledgments

The authors would like to thank the Ferrari NVH Team Leader Mr. Tarabra Marco and NVH Senior Engineer Mr. Roncaglia Valerio for the technical support and the guidance throughout the course of this research. The authors wish to express their sincere gratitude to Cavazzoni Luca from MilleChili Lab for the invaluable assistance in the final review of the paper.

References

[1] M. Baker, "Review of test/analysis correlation methods and criteria for validation of finite element models for dynamic analysis," in *Proceedings of the 10th International Modal Analysis Conference*, San Diego, Calif, USA, February 1992.

[2] M. Imregun and W. J. Visser, "A review of model updating techniques," *Shock and Vibration Digest*, vol. 23, no. 1, pp. 141–162, 1990.

[3] M. I. Friswell and J. E. Mottershead, *Finite Element Model Updating in Structural Dynamics*, vol. 38 of *Solid Mechanics and Its Applications*, Kluwer Academic, Dordrecht, Netherlands, 1995.

[4] M. Brughmans, J. Leuridan, T. Van Langenhove, and F. Turgay, "Validation of automotive component FE models by means of test-analysis correlation and model updating techniques," SAE Technical Paper 1999-01-1797, 1999.

[5] S. Jambovane, D. Kalsule, and S. Athavale, "Validation of FE models using experimental modal analysis," SAE Technical Paper 2001-26-0042, 2001.

[6] D. Siano, M. Viscardi, P. Napolitano, and M. A. Panza, "Experimental/FE numerical correlation of a composite sandwich panel of a high-speed train," in *Proceedings of the 10th International Conference on Applied and Theoretical Mechanics (MECHANICS '14)*, pp. 978–960, Salerno, Italy, 2014.

[7] L. Splendi, L. D'Agostino, A. Baldini, L. Castignani, F. Pellicano, and M. Pinelli, "Simplified modeling technique for damping materials on light structures: experimental analysis and numerical tuning," in *Proceedings of the ASME International Mechanical Engineering Congress and Exposition (IMECE '13)*, San Diego, Calif, USA, November 2013.

[8] C. Schedlinski, F. Wagner, K. Bohnert et al., "Test-based computational model updating of a car body in white," *Sound and Vibration*, vol. 39, no. 9, pp. 19–23, 2005.

[9] S. Mariano, M. da Silva, A. de Costa Moreira et al., "Modal correlation of an aerospace structure," SAE Technical Paper 2006-01-2786, 2006.

[10] C. Azoury, A. Kallassy, B. Combes, I. Moukarzel, and R. Boudet, "Experimental and analytical modal analysis of a Crankshaft," *IOSR Journal of Engineering*, vol. 2, no. 4, pp. 674–684, 2012.

[11] B. J. Schwarz and M. H. Richardson, "Experimental Modal Analysis," in *Proceedings of the CSI Reliability Week*, Orlando, Fla, USA, 1999.

[12] B. Peeters, P. Guillaume, H. Van der Auweraer, B. Caubergue, P. Verboven, and J. Leuridan, "Automotive and aerospace applications of the PolyMAX modal parameter estimation method," in *Proceedings of the 22nd International Modal Analysis Conference*, pp. 17–21, Dearborn, Mich, USA, January 2004.

[13] M. Aygül, *Fatigue analysis of welded structures using the finite element method [Licentiate of Engineering thesis]*, Department of Civil and Environmental Engineering, Chalmers University of Technology, Gothenburg, Sweden, 2012.

[14] K. Pal and D. L. Cronin, "Static and dynamic characteristics of spot welded sheet metal beams," *Journal of engineering for industry*, vol. 117, no. 3, pp. 316–322, 1995.

[15] J. Backhans and A. Cedas, *A finite element model of spot welds between non-congruent shell meshes calculation of stresses for fatigue life prediction [M.S. thesis]*, Volvo Car Corporation, Gothenburg, Sweden, Analysis report, 93841-2000, 2000.

[16] D. Heiserer, M. Charging, and J. Sielaft, "High performance, process oriented, weld spot approach," in *Proceedings of the 1st MSC Worldwide Automotive User Conference*, Munich, Germany, September 1999.

[17] M. Palmonella, M. I. Friswell, J. E. Mottershead, and A. W. Lees, "Finite element models of spot welds in structural dynamics: review and updating," *Computers and Structures*, vol. 83, no. 8-9, pp. 648–661, 2005.

[18] R. J. Allemang and D. L. Brown, "Correlation coefficient for modal vector analysis," in *Proceedings of the 1st International Modal Analysis Conference & Exhibit*, pp. 110–116, Orlando, Fla, USA, November 1982.

[19] R. J. Allemang, "The modal assurance criterion (MAC): twenty years of use and abuse," in *Proceedings of the IMAC 20, The International Modal Analysis Conference*, pp. 397–405, Los Angeles, Calif, USA, 2002.

Permissions

All chapters in this book were first published in AAV, by Hindawi Publishing Corporation; hereby published with permission under the Creative Commons Attribution License or equivalent. Every chapter published in this book has been scrutinized by our experts. Their significance has been extensively debated. The topics covered herein carry significant findings which will fuel the growth of the discipline. They may even be implemented as practical applications or may be referred to as a beginning point for another development.

The contributors of this book come from diverse backgrounds, making this book a truly international effort. This book will bring forth new frontiers with its revolutionizing research information and detailed analysis of the nascent developments around the world.

We would like to thank all the contributing authors for lending their expertise to make the book truly unique. They have played a crucial role in the development of this book. Without their invaluable contributions this book wouldn't have been possible. They have made vital efforts to compile up to date information on the varied aspects of this subject to make this book a valuable addition to the collection of many professionals and students.

This book was conceptualized with the vision of imparting up-to-date information and advanced data in this field. To ensure the same, a matchless editorial board was set up. Every individual on the board went through rigorous rounds of assessment to prove their worth. After which they invested a large part of their time researching and compiling the most relevant data for our readers.

The editorial board has been involved in producing this book since its inception. They have spent rigorous hours researching and exploring the diverse topics which have resulted in the successful publishing of this book. They have passed on their knowledge of decades through this book. To expedite this challenging task, the publisher supported the team at every step. A small team of assistant editors was also appointed to further simplify the editing procedure and attain best results for the readers.

Apart from the editorial board, the designing team has also invested a significant amount of their time in understanding the subject and creating the most relevant covers. They scrutinized every image to scout for the most suitable representation of the subject and create an appropriate cover for the book.

The publishing team has been an ardent support to the editorial, designing and production team. Their endless efforts to recruit the best for this project, has resulted in the accomplishment of this book. They are a veteran in the field of academics and their pool of knowledge is as vast as their experience in printing. Their expertise and guidance has proved useful at every step. Their uncompromising quality standards have made this book an exceptional effort. Their encouragement from time to time has been an inspiration for everyone.

The publisher and the editorial board hope that this book will prove to be a valuable piece of knowledge for researchers, students, practitioners and scholars across the globe.

List of Contributors

Rujia Wang and Shaoyi Bei
School of Automotive and Transportation, Jiangsu University of Technology, Changzhou, Jiangsu, China

Hirotarou Tsuchiya, Hiroyuki Moriyama and Satoru Iwamori
Course of Science and Technology, Graduate School of Tokai University, 4-1-1 Kitakaname, Hiratsuka, Kanagawa, Japan

Nasmi Herlina Sari
Department of Mechanical Engineering, Faculty of Engineering, Mataram University, Nusa Tenggara Barat, Indonesia

I. N. G. Wardana, Yudy Surya Irawan and Eko Siswanto
Department of Mechanical Engineering, Faculty of Engineering, Brawijaya University, East Java, Indonesia

Sven Münsterjohann and Stefan Becker
Friedrich-Alexander-Universität Erlangen-Nürnberg, Cauerstr. 4, 91058 Erlangen, Germany

Jens Grabinger
SIMetris GmbH, AmWeichselgarten 7, 91058 Erlangen, Germany

Manfred Kaltenbacher
Vienna University of Technology, Getreidemarkt 9, 1060Wien, Austria

Angga Dwinovantyo
Graduate School of Marine Technology, PMDSU Batch II, Bogor Agricultural University, IPB Darmaga Campus, Bogor 16680, Indonesia

Tri Prartono
Department of Marine Science and Technology, Faculty of Fisheries and Marine Sciences, Bogor Agricultural University, IPB Darmaga Campus, Bogor 16680, Indonesia

Susilohadi Susilohadi
Marine Geological Institute, Ministry of Energy and Mineral Resources of the Republic of Indonesia, Jl. Dr. Djunjunan No. 236, Bandung 40174, Indonesia

Shreedhar Kolekar
Mechanical Engineering Department, Jain University, Bengaluru, Karnataka State, India

Mechanical Engineering Department, Satara College of Engineering & Management Limb, Satara, Maharashtra State 415015, India

Krishna Venkatesh
Centre for Incubation, Innovation, Research & Consultancy, Bengaluru, Karnataka State, India

Jeong-Seok Oh
Division of Automotive & Mechanical Engineering, Kongju National University, Cheonan-si, Chungnam 31080, Republic of Korea

Seung-Bok Choi
Department of Mechanical Engineering, Inha University, No. 253, Yonghyun-dong, Nam-gu, Incheon 402-751, Republic of Korea

Nasmi Herlina Sari
Department of Mechanical Engineering, Faculty of Engineering, Mataram University, Nusa Tenggara Barat, Indonesia

Praveena Raviprolu, Nagaraja Jade and Venkatesham Balide
Department of Mechanical & Aerospace Engineering, Indian Institute of Technology Hyderabad, Kandi, Telangana 502285, India

Meng Koon Lee, Mohammad Hosseini Fouladi and Satesh Narayana Namasivayam
School of Engineering, Taylor's University, No. 1 Jalan Taylor's, 47500 Subang Jaya, Selangor, Malaysia

Jihad Al-Oudatallah and Fariz Abboud
Department of Electronics and Communications, Damascus University, Damascus, Syria

Mazen Khoury and Hassan Ibrahim
Higher Institute for Applied Science and Technology, Damascus, Syria

Usama Kadri
Department of Mathematics, Massachusetts Institute of Technology, Cambridge, MA 02139, USA

Ahmet Demir
Engineering Faculty, Department of Mechatronics, Karabuk University, 78100 Karabuk, Turkey

Manabu Sasajima, Tatsushi Sasanuma and Akira Hara
Foster Electric Company, Limited, 1-1-109 Tsutsujigaoka, Akishima, Tokyo 196-8550, Japan

Yue Hu
Foster Electric Company, Limited, 1-1-109 Tsutsujigaoka, Akishima, Tokyo 196-8550, Japan
College of Mechanical Engineering, Saitama Institute of Technology, 1690 Fusaiji, Fukaya, Saitama 369-0293, Japan

Xilu Zhao
College of Mechanical Engineering, Saitama Institute of Technology, 1690 Fusaiji, Fukaya, Saitama 369-0293, Japan

Takao Yamaguchi
Department of Mechanical Science and Technology, Faculty of Science and Technology, Gunma University, 1-5-1 Tenjin-cho, Kiryu, Gunma 376-8515, Japan

Henry M. Manik
Department of Marine Science and Technology, Faculty of Fisheries and Marine Sciences, Bogor Agricultural University, Kampus IPB Darmaga, Bogor 16680, Indonesia

Onur Çelikkan and Haluk Erol
Mechanical Engineering Faculty, Istanbul TechnicalUniversity, Gumussuyu, 34439 Istanbul, Turkey

Index

www.ingramcontent.com/pod-product-compliance
Lightning Source LLC
Chambersburg PA
CBHW050458200326
41458CB00014B/5226

9 781632 408242